Regulatory Mechanisms in Insect Feeding

Regulatory Mechanisms in Insect Feeding

edited by

R.F. Chapman
Gerrit de Boer

CHAPMAN & HALL

THOMSON PUBLISHING

New York • Albany • Bonn • Boston • Cincinnati • Detroit • London • Madrid • Melbourne
Mexico City • Pacific Grove • Paris • San Francisco • Singapore • Tokyo • Toronto • Washington

Cover Design: Andrea Meyer, emDASH inc.

Copyright © 1995
Chapman & Hall

Printed in the United States of America

For more information, contact:

Chapman & Hall
115 Fifth Avenue
New York, NY 10003

Thomas Nelson Australia
102 Dodds Street
South Melbourne, 3205
Victoria, Australia

Nelson Canada
1120 Birchmount Road
Scarborough, Ontario
Canada, M1K 5G4

International Thomson Editores
Campos Eliseos 385, Piso 7
Col. Polanco
11560 Mexico D.F. Mexico

Chapman & Hall
2-6 Boundary Row
London SE1 8HN
England

Chapman & Hall GmbH
Postfach 100 263
D-69442 Weinheim
Germany

International Thomson Publishing Asia
221 Henderson Road #05-10
Henderson Building
Singapore 0315

International Thomson Publishing-Japan
Hirakawacho-cho Kyowa Building, 3F
1-2-1 Hirakawacho-cho
Chiyoda-ku, 102 Tokyo
Japan

1 2 3 4 5 6 7 8 9 10 XXX 01 00 99 97 96 95

Library of Congress Cataloging-in-Publication Data

Regulatory mechanisms in insect feeding / edited by Reginald F. Chapman and
    Gerrit de Boer.
        p.   cm.
    Includes bibliographical references and index.
    ISBN 0-412-03141-8
    1. Insect—Food.   2. Insects—Physiology.   I. Chapman, R. F. (Reginald
    Frederick)   II. de Boer, Gerrit
    QL496.R365   1995
    595.7'0132—dc20   94-38783
                                                                CIP

British Library Cataloguing in Publication Data available

Please send your order for this or any Chapman & Hall book to **Chapman & Hall, 29 West 35th Street,
New York, NY 10001, Attn: Customer Service Department.** You may also call our Order Department at
1-212-244-3336 or fax your purchase order to 1-800-248-4724.

For a complete listing of Chapman & Hall's titles, send your requests to **Chapman & Hall, Dept. BC,
115 Fifth Avenue, New York, NY 10003.**

Contents

Preface

The only book to deal comprehensively with insect feeding was published by C. T. Brues in 1946. His *Insect Dietary* was an account of insect feeding habits. Since that time there has been a revolution in biology, and almost all aspects of our understanding of insect feeding have expanded to an extent and into areas that would have been unthinkable in Brues' day. Yet, our book does not replace *Insect Dietary* but, instead, complements it, because our aim is to bring together information on the mechanisms by which food quality and quantity are regulated. We deliberately focus attention on the feeding process; to include food-finding would have required a much larger book and would have moved the focus away from more proximate mechanisms.

This book is dedicated to the late Vincent G. Dethier. As a pioneer in studying the physiological basis of animal behavior, he focused on regulation of feeding in flies and caterpillars. His work on the blowfly, together with that by his many students and co-workers, still provides the most completely described mechanism of insect feeding. The citation of his work in almost every chapter in this book illustrates the importance of his findings and ideas to our current understanding of regulation of insect feeding. The authors in this book provide many innovative and stimulating ideas typifying Dethier's approach to the study of feeding behavior.

We thank all the authors for their efforts and perseverance in contributing to this book. We specifically thank Greg Payne, Science Editor at Chapman and Hall, for his guidance and advice.

Gerrit de Boer, *Lawrence, Kansas*
Reg Chapman, *Tucson, Arizona*

24 August 1994

Acknowledgments

Drs. A. R. Fontaine and J. Hendrichs kindly allowed us to use the photographs in Figs. 1.3 and 8.4b, respectively.

The following publishers and societies have allowed us to use illustrations from previously published work:

Fig. 1.3. By courtesy of the Entomological Society of British Columbia.

Figs. 1.4a, 1.8, 4.9, 8.1, 10.4. By courtesy of Springer-Verlag GmbH & Co. KG.

Fig. 1.4b. By courtesy of the National Research Council of Canada.

Fig. 1.5a. By courtesy of Science.

Figs. 1.5b, 1.12, 1.13. By courtesy of the Ecological Society of America.

Fig. 1.7. Reprinted from the Journal of Insect Physiology, vol. 35, Wheater, C.P. and Evans, M.E.G., The mandibular forces and pressures of some predacious beetles, pp. 815–820. Copyright (1989), with kind permission from Elsevier Science Ltd, The Boulevard, Langford Lane, Kidlington OX5 1GB, UK.

Fig. 1.10. By courtesy of the Entomological Society of America.

Fig. 1.11. By courtesy of the New York Entomological Society.

Fig. 1.16. Reprinted from the Journal of Insect Physiology, vol. 30, Cheeseman, M.T. and Pritchard, G. Proventricular trituration in adult carabid beetles, pp. 203–209. Copyright (1984), with kind permission from Elsevier Science Ltd, The Boulevard, Langford Lane, Kidlington OX5 1GB, UK.

Fig. 4.3, 4.4. Reprinted from the International Journal of Insect Morphology & Embryology, vol. 18, Chapman, R. F. and Fraser, J. The chemosensory system of the monophagous grasshopper, *Bootettix argentatus*, pp. 111–118. Copyright (1989), with kind permission from Elsevier Science Ltd, The Boulevard, Langford Lane, Kidlington OX5 1GB, UK.

Figs. 4.7, 10.7. Reprinted by permission of Kluwer Academic Publishers.

Fig. 4.8a. Reprinted from the Journal of Insect Physiology, vol. 36, Blaney, W.M. and Simmonds, M.S.J. A behavioural and electrophysiological study of the role of the tarsal chemoreceptors in feeding by adults of *Spodoptera, Heliothis virescens* and *Helicoverpa zea,* pp. 743–756. Copyright (1990), with kind permission from Elsevier Science Ltd, The Boulevard, Langford Lane, Kidlington OX5 1GB, UK.

Fig. 4.8b. By courtesy of the Company of Biologists Ltd.

Figs. 4.11. By courtesy of Springer-Verlag New York, Inc.

Figs. 5.1, 9.4. By courtesy of Bailliere, Tindall & Cassell.

Figs. 8.2, 10.9. By courtesy of Academic Press Ltd.

Fig. 8.4a. By courtesy of Cambridge University Press.

Figs. 9.1, 9.2, 9.3, 9.5. By courtesy of the Royal Society of London.

Figs. 9.6, 9.7, 9.10, 10.11. By courtesy of Blacwell Scientific Publications Ltd.

Fig. 10.1. By Courtesy of Elsevier Science.

Fig. 10.5. By courtesy of Akadémiai Kiadó.

Contributors

D. A. Avé, Ecogen Inc., 2005 Cabot Boulevard West, Langhorne, Pennsylvania 19047

L. Barton Browne, Division of Entomology, Commonwealth Scientific and Industrial Research Organisation, Long Pocket Laboratories, Private Bag No. 3, P. O. Indooroopilly, Queensland 4068, Australia

E. A. Bernays, Department of Entomology, University of Arizona, Tucson, Arizona 85721

P. G. Chambers, Department of Zoology, University of Oxford, South Parks Road, Oxford, OX1 3PS, United Kingdom

R. F. Chapman, Division of Neurobiology, University of Arizona, Tucson, Arizona 85721

S. Chyb, Department of Entomology, Pennsylvania State University, University Park, Pennsylvania 16802

T. L. Daniel, Department of Zoology, NJ-15, University of Washington, Seattle, Washington 98195

E. E. Davis, SRI International, 333 Ravenswood Avenue, Menlo Park, California 94025

G. de Boer, Department of Entomology, University of Kansas, Lawrence, Kansas 66045

J. L. Frazier, Department of Entomology, Pennsylvania State University, University Park, Pennsylvania 16802

W. G. Friend, Department of Zoology, University of Toronto, *Present address:* Department of Biological Sciences, Simon Fraser University, Burnaby, British Columbia V5A 1S6, Canada

J. G. Kingsolver, Department of Zoology, NJ-15, University of Washington, Seattle, Washington 98195

D. Raubenheimer, Department of Zoology, University of Oxford, South Parks Road, Oxford, OX1 3PS, United Kingdom

J. M. C. Ribeiro, Department of Entomology, University of Arizona, Tucson, Arizona 85721

S. J. Simpson, Department of Zoology, University of Oxford, South Parks Road, Oxford, OX1 3PS, United Kingdom

J. G. Stoffolano, Jr., Department of Entomology, University of Massachusetts, Amherst, Massachusetts 01003

W. F. Tjallingii, Department of Entomology, Agricultural University, P.O.B. 8031, 6707 EH Wageningen, The Netherlands

Introduction

G. de Boer

Feeding is one of the most basic behaviors for survival of an individual animal and for the species. Animals must ingest chemicals necessary for energy production, for maintenance of biochemical processes, and for growth and development. At the same time, they should reject foods that contain toxic or useless compounds. Insects as a group feed on a great variety of materials, such as animal or plant tissues or dead organic matter. Despite this unique ability to utilize almost any organic substrate, most insect species restrict themselves to a particular category of food. Some insects even specialize on a certain plant or animal species. This variety in feeding habits that we see today is the result of anatomical and physiological adaptations to food sources that are not all equally beneficial to the insect. Food selection is therefore of paramount importance to the survival of the individual. The physiological need for food by insects varies constantly due to changing factors such as nutritional state, food deprivation, developmental state, and food experience. Therefore, the study of insect feeding has attracted many students of animal behavior using different approaches, from neurophysiology to evolutionary biology.

Diversity of Feeding Habits

Insects comprise about 75% of all animal species on earth (Strong et al., 1984). About one-half of this group uses living plants as their food, whereas the other half feeds upon dead plant material or upon animals, both dead and alive. Four major classes of feeding habits are recognized by Brues (1946): plant feeders, predators, scavengers, and parasites. Within each of these classes, various types of feeding can be found: biting and chewing on leaf or animal tissues and sucking from plant or animal cells or tissues.

Phytophagous insects are found predominantly in the following orders: Orthoptera, Lepidoptera, Homoptera, Thysanoptera, Phasmida, Isoptera, Coleoptera (families Cerambycidae, Chrysomelidae and Curculionidae), Hymenoptera, and some Diptera (Chapman, 1982). The terms *monophagy, oligophagy,* and *polyphagy* are generally used to indicate the degree of diet breadth of a particular insect species, but definitions may vary with authors (see Bernays and Chapman, 1994). Monophagous insects restrict their feeding to a particular plant species or genus. Oligophagous insects feed on a number of plant species usually in different genera but within one plant family. Polyphagous insects consume plants belonging to different plant families. Many phytophagous insect species, such as caterpillars, grasshoppers, and beetles, feed by biting and chewing plant material. Others, such as Hemiptera, feed strictly by sucking plant sap.

Insect predators consume mainly other insects and are found predominantly in the orders Odonata, Mantodea, Heteroptera (Reduviidae and others), larval Neuroptera, Mecoptera, Diptera (Asilidae and Empididae), Coleoptera (Adephaga and Coccinellidae), and Hymenoptera (Sphecidae and Pompilidae) (Chapman, 1982). Biting and sucking are the two types of feeding found among predator species.

Ectoparasites feed by sucking or biting on their host while residing on the outside of their prey. Examples are found in the Siphonaptera, Anoplura, Mallophaga, some Dermaptera, Heteroptera, various Diptera (such as mosquitoes), Simuliidae, Ceratopogonidae, Tabanidae, and the Pupipara.

Saprophagous insects are scavengers that feed on dead or decaying organic material. These insects are found in many orders—for example, in the Blattodea, Isoptera, Coleoptera, and Diptera. Insects with this feeding habit will not be further discussed because there is a general lack of information on control mechanisms.

Most insects obtain sufficient water with their food. However, some drink water, especially if they are dehydrated. Water intake and its control are discussed here only in the larger context of regulation of feeding. More information on regulation of drinking can be found in Bernays (1985).

Anatomical and Physiological Adaptations to Different Feeding Habits

The mechanics of food handling by chewing insects are discussed by Chapman in Chapter 1. In solid-food feeders, transfer of food into the gut requires removing fragments of the food and cutting and grinding them into particles suitable for movement through the gut. This is important mechanically, but also to ensure that digestion can continue effectively. As in the jaws of vertebrates, the mandibles of these insects are adapted to deal with the types of food they eat, but the mandibles become worn with use so their efficiency is impaired. Movement of food through the mouth into the foregut seems to be the result of activity of the various mouthparts, although food chemicals may trigger swallowing movements. Cuticular structures in the foregut may continue the process of food fragmentation so that efficiency of digestion is maximized.

Fluid-feeding insects require mouthparts that are clearly different in form and function from those used by insects feeding on solid foods. Fluid feeders show modifications of the mouthparts which commonly involve tubes of extremely fine bore. The size of the tubes imposes constraints on the nature of the food on which the insects can feed, with viscosity being especially important. In Chapter 2, Kingsolver and Daniel discuss the relationship between the mechanics of fluid feeding and functional morphology using simple mathematical models.

Many insects feed on foods differing widely in mechanical and chemical properties. This poses a variety of problems for insects to efficiently utilize their foods. The use of saliva has solved some of these problems. Ribeiro discusses the role of insect saliva, as well as its chemistry and physiology, in Chapter 3. Saliva may generally serve two distinct functions in insect feeding: solubilizing or adding a liquid vehicle to solid foods, and making food available for ingestion by enzymatic action. Some special adaptations for the use of saliva are found in blood-feeding and other fluid-feeding insects.

The Cyclic Nature of Feeding Activity

Feeding by many insects occurs in discrete meals separated by periods of nonfeeding. This is of physiological importance in maximizing the efficiency of digestion and conserving energy. Also, feeding in bouts may serve other purposes such as avoidance of predators or hazardous environmental conditions such as extreme heat. Meal duration varies with insect species. For example, nymphs of the locust, *Locusta migratoria,* have meals lasting from 10 to 20 min, whereas meal duration by the cabbage white caterpillar, *Pieris brassicae,* is about 5 min. Initiation of feeding may require phagostimulatory input at a time when the insect is in a state of readiness to feed. This state of readiness depends on various factors such as the degree of satiation and the presence of acceptable food. The signal causing feeding to start may be an exogenous or an endogenous stimulus not directly related to feeding itself. Appropriate sensilla sensing food stimuli provide the insect with the necessary information on food acceptability. Feeding will be maintained for some period until physiological feedback from continued food intake triggers meal termination. This cyclic pattern of initiation, maintenance, and termination of feeding is characteristic of many chewing insects. The distribution of activities related to a meal occurs over a relatively short period and is therefore called *short-term feeding*. In contrast to most chewing insects, some fluid-feeding insects such as aphids do not take meals. Feeding on phloem sap by these insects can last for several days or even for a complete larval or adult stage (see Chapter 7).

Regulation of a Meal

The chemical composition of food is one of the most important factors determining food suitability. Food intake is mainly controlled by the detection of chemical stimuli. In general, nutrients stimulate feeding whereas non-nutrients do not, except for particular plant secondary compounds which may act as "token stimuli" signaling food acceptance (see Chapter 4). The balance between nutrient quality and feeding deterrent compounds is of overriding importance. The different chemicals affecting feeding behavior have been the subject of many reviews, dealing mostly with phytophagous insects (Hsiao, 1985; Montllor, 1991; Morgan and Warthen, 1990; Städler, 1992; Warthen, 1990). Therefore, this topic will not be examined here in a separate chapter, but instead the appropriate chemicals are included in the discussion of feeding stimuli for the various insects.

Taste and smell largely affect feeding behavior by providing the central nervous system with information on quality and quantity of food chemicals. Initiation of feeding and probably also maintenance of feeding are controlled to a large degree by these chemosensory messages. The peripheral and central sensory mechanisms involved in controlling short-term feeding are discussed by Chapman in Chapter 4. Special emphasis is given to the distribution and function of taste sensilla and the coding and processing of gustatory information. The chemosensory regulation of feeding is reviewed for insects according to their feeding habits.

Because the mechanisms involved in the regulation of a meal (initiation, maintenance, and termination of feeding) vary among insects with different feeding habits, the discussion of this topic is better served by focusing separately on each individual group of feeders: chewing insects, blood-feeding insects, plant-fluid feeders, and carbohydrate-feeding insects.

Simpson in Chapter 5 discusses the regulation of a meal in chewing insects by focusing on the acridids and caterpillars. The propensity to begin a meal by these insects when food is available decreases with shorter intermeal intervals. This means that taking a meal inhibits further feeding. Such an inhibition may be the result of nutrient feedback and/or stretching of the crop and hindgut due to filling with food. The amount of food eaten (meal size) and ingestion rate appear to be regulated physiologically, and these two factors determine meal duration. In locusts, various excitatory factors, such as food stimuli and light intensity, can trigger feeding. Meal size and ingestion rate depend on the degree of this excitation before and during the meal. A model simulating locust feeding is presented illustrating the various parameters regulating a meal.

Meal regulation by blood-feeding insects is discussed by Davis and Friend in Chapter 6. They focus on the physiological state of the insect, as well as on the environmental signals and the physical events that mediate and control blood feeding. Most of the information presented deals with feeding by mosquitoes. Because many hematophagous insects also feed on plant juices, the control mechanisms of nectar feeding are also discussed. A heuristic model in the form of a decision diagram for a generic mosquito is presented to illustrate the various behavioral stages (including host location) through which the insect must pass together with the decisions required during feeding.

Some phytophagous insects, such as aphids, feed exclusively on plant sap. They insert their mouthparts into the tissues in search of phloem sap. Although ingestion is basically passive due to the hydrostatic pressure of this sap in the sieve elements, some control mechanism must be active because the duration of sap feeding varies significantly. Physicochemical properties of the plant seem to affect the duration of feeding. Tjallingii in Chapter 7 discusses the control mechanisms of food intake in aphids using data from electrical penetration graphs in combination with radiotracers and honeydew recordings.

The physiological mechanisms underlying control of carbohydrate feeding by adult Diptera, Lepidoptera, and Hymenoptera are examined by Stoffolano in Chapter 8. The best understood mechanism of carbohydrate feeding is found in the Diptera with the blowfly, *Phormia regina* (Dethier, 1976). An updated model illustrating the regulation of carbohydrate feeding in this insect is used as a basis for the discussion of control mechanisms for short-term feeding in other dipterous insects and carbohydrate feeders.

Long-Term Regulation of Feeding

The distribution of meals over time can be affected by many different factors, such as deficiency of specific nutrients, previous food experience, and specific needs associated with growth and development of the insect.

The patterns of insect feeding and nutritional compensation are examined by Simpson, Raubenheimer, and Chambers in Chapter 9. Nutritional homeostasis is approached as a problem of multidimensional geometry. At any given time a certain tissue needs a particular quantity and mixture of nutrients. This multidimensional geometric view of nutritional homeostasis provides the framework against which the question of how insects reach their nutritional target can be usefully examined. Nutritional homeostasis is achieved by regulation of feeding and postingestive utilization of food.

Prior feeding on certain foods may also affect future food selection. Insects may be induced to accept, and even prefer, foods that were originally unacceptable. Nutrient deficiencies in certain foods may result in aversion learning of these foods. Bernays examines both associative and nonassociative forms of learning in relation to feeding and food choice behavior in Chapter 10. Unlike vertebrates, insects show changes in sensitivity of peripheral chemoreceptors due to previous food experience. Such modifications together with synaptic changes in neural pathways and metabolic and hormonal feedbacks are discussed as possible mechanisms underlying plasticity in feeding behavior.

Changes in food requirements with ontogeny of the insect reflect changes in nutritional demands. The needs are different during growth, reproduction, and the preparation for diapause. Barton Browne in Chapter 11 discusses the regulatory mechanisms underlying these developmental stage-related changes in feeding behavior. Food intake is modified in two ways: a change in rate of total food intake or in relative amounts of various foods consumed. Two distinct categories of physiological mechanisms depending on their causal pathway activating feeding behavior are recognized. Demand-mediated mechanisms are based on nutrient feedback in which hemolymph nutrient concentrations directly affect feeding behavior. This results in a match between the rate of uptake of nutrients and the nutritional demands of the tissues (nutritional compensation). Non-demand-mediated mechanisms do not involve hemolymph nutrient concentrations but are based on a variety of neural and hormonal signals which modify feeding behavior. The distinction between these two categories of physiological mechanisms and the concept of nutritional homeostasis as outlined in Chapter 9 are used as the background to discuss the long term regulation of feeding.

Regulation of Feeding: Practical Aspects

The use of a variety of foods by insects has many negative effects on humans. The two most important consequences of feeding activity are (a) the competition with humans for food of adequate quality and quantity and (b) the attack of insects on humans. Old questions such as how to deter or stop insect feeding or how to stimulate feeding on pesticide-treated materials constantly demand new answers due to the development of resistance to current insecticides and the restrictions in usage by environmental concerns. Understanding the regulatory mechanisms underlying insect feeding is a prerequisite for the development of more effective and environmentally safe control methods of pest insects. Therefore, the insights into regulation of insect feeding presented in the previous chapters provide the proper background for an in-depth discussion of improving existing and developing new pest control methods.

The successful use of baits depends on a thorough knowledge of phagostimulation. Pesticides may affect insect feeding behavior, possibly preventing ingestion of lethal doses of a pesticide. The use of feeding stimulatory compounds in pesticide formulations is discussed by Avé.

The most obvious control methods involve knowledge of chemical regulatory factors. Plant secondary compounds have potential value as antifeedants per se, as well as in the development of resistant cultivars of crops both by normal breeding methods and by genetic engineering techniques. Frazier and Chyb in Chapter 13 discuss the various inhibitors of feeding based on their physiological mechanisms and their advantages and limitations for use in insect control.

References

Bernays, E. A. (1985). Regulation of feeding behavior. In: Kerkut, G. A., and Gilbert, L. I. (eds), *Comprehensive Insect Physiology, Biochemistry and Pharmacology,* vol. 4. Pergamon Press, Oxford, pp. 1–32.

Bernays, E. A., and Chapman, R. F. (1994). *Host-Plant Selection by Phytophagous Insects.* Chapman and Hall, New York.

Brues, C. T. (1946). *Insect Dietary. An Account of the Food Habits of Insects.* Harvard University Press, Cambridge.

Chapman, R. F. (1982). *The Insects. Structure and Function.* Harvard University Press, Cambridge, MA.

Dethier, V. G. (1976). *The Hungry Fly.* Harvard University Press, Cambridge, MA.

Hsiao, T. H. (1985). Feeding behavior. In: Kerkut, G. A., and Gilbert, L. I. (eds.), *Comprehensive Insect Physiology, Biochemistry and Pharmacology* vol. 9. Pergamon Press, Oxford, pp. 471–512.

Montllor, C. B. (1991). The influence of plant chemistry on aphid-feeding behavior. In: Bernays, E. A. (ed.), *Insect–Plant Interactions.* vol. 3. CRC Press, Boca Raton, FL, pp. 125–173.

Morgan, E. D., and Warthen, J. D., Jr. (1990). Insect feeding deterrents (1980–1987). In: Morgan, E. D., and Mandava, N. B. (eds.), *Handbook of Natural Pesticides,* vol. 6. CRC Press, Boca Raton, FL, pp. 83–134.

Städler, E. (1992). Behavioral responses of insects to plant secondary compounds. In: Rosenthal, G. A., and Berenbaum, M. R. (eds.), *Herbivores. Their Interactions with Secondary Plant Metabolites,* vol. 2. Pergamon Press, San Diego, pp. 45–88.

Strong, D. R., Lawton, J. H., and Southwood, R. (1984). *Insects on Plants. Community Patterns and Mechanisms.* Harvard University Press, Cambridge, MA.

Warthen, J. D., Jr. (1990). Insect feeding deterrents (1976–1980). In: Morgan, E. D., and Mandava, N. B. (eds.), *Handbook of Natural Pesticides,* vol. 6. CRC Press, Boca Raton, FL, pp. 23–82.

Mechanics of Feeding

1

Mechanics of Food Handling by Chewing Insects

R. F. Chapman

1.1. Introduction

Food handling in chewing insects consists of two separate functions: cutting the food into fragments that can be ingested, and passing these fragments back into the foregut. The processes of prey capture in predaceous insects or manipulation of the food prior to feeding in phytophagous insects are not considered here.

The mouthparts in chewing insects are considered to be retained in a more primitive condition than those in sucking insects (Chapter 2); that is, chewing insects have separate mandibles, maxillae, and labium, retaining some semblance of the ancestral paired, segmental appendages. The mandibles are of primary importance in cutting up the food, while they and the other appendages are involved with its subsequent manipulation and movement to the mouth and into the gut.

1.2. Mandibles and Their Role

Each mandible is a short, stout appendage in which the inner face, that which abuts onto the other mandible when they are closed, is developed into a series of cusps. By analogy with mammalian anatomy, the more distal part of the mandible, with a cutting function, is called the *incisor region* and the proximal part, often with a grinding function, is the *molar region* (Fig. 1.1). Because the cusps of the two sides fit into each other, the mandibles of the two sides are asymmetrical. The cuticle of the whole mandible is heavily sclerotized, and in the region of the cusps it is usually black and even harder than elsewhere. Each mandible is hollow and the cavity is in direct continuity with the hemocoel of the head, although much of the volume is occupied by air sacs.

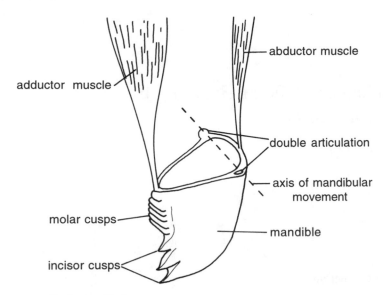

Figure 1.1.   Basic mandibular structure.

1.2.1. Hardness of Mandibles

Hillerton et al. (1982) determined the hardness of different parts of the mandible of *Locusta migratoria*. In this species the left mandible overlaps the right so that the inner surface of the left shears along and across the outer face of the right mandible. Hillerton et al. (1982) found that the outer face of the left incisor region and the inner face of the right incisor region were much harder than other parts of the mandibular cuticle, including the two shearing surfaces (Fig. 1.2). No other part of the body, except the dorsal mesothorax, was as hard. These authors suggest that the hard surfaces form the cutting edges that bite off leaf fragments, and, because of the manner in which the mandibles move, they are

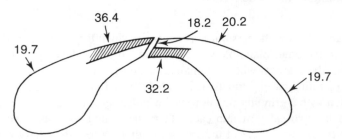

Figure 1.2.   Diagrammatic cross section of the mandibles of *Locusta migratoria* showing the hardness values (in kg · mm^{-2}) for different parts. Hatched areas represent the two areas of hardened cuticle. (Based on data in Hillerton et al., 1982.)

self-sharpening. Zinc is present in the cutting edges, but not in the shearing faces or in any other part of the mandibles (Hillerton and Vincent, 1982).

Zinc is also found in the cutting edges of the mandibles of several species of grasshopper, phasmid, caterpillar (Fig. 1.3), and some larval and adult beetles, as well as in the larva of a seed-eating wasp, *Megastigmus spermatotrophus* (Table 1.1) (Fontaine et al., 1991; Hillerton and Vincent, 1982). Zinc and/or manganese were present in 56 of 61 stored products beetles examined by Hillerton and Vincent (1982) and Hillerton et al. (1984). In some, zinc predominated, while in others manganese was more abundant. The distribution among these beetles appears to be related to phylogeny, because there is no case where different species within a genus have different predominant metals. There are cases where neither metal was found in the larva although manganese was present in the adult. Of the five species in which neither zinc nor manganese was found, three occur in stored products (*Anthrenus sarnicus*, *Anthrenus verbasci*, and *Sitophagus hololeptoides*), one is herbivorous (*Melolontha melolontha*) and the last, *Carabus violaceous*, was the only predator examined by Hillerton and Vincent (1982).

Both zinc and manganese were present in the mandibles of ants, but neither was found in significant amounts in the mandibles of four species of cockroach

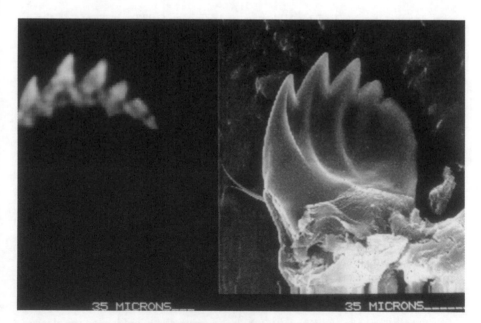

Figure 1.3. Zinc in the mandible of the caterpillar of *Barbara colfaxiana*. Right: Scanning electron micrograph of the inner surface of the right mandible. Left: Zinc x-ray image of the same mandible. (After Fontaine et al., 1991.)

Table 1.1. Numbers of species with zinc (Zn), manganese (Mn), or iron (Fe) in the cutting surfaces of the mandibles[a]

Order	Food habit	Number of species with				
		Zn	Mn	Zn + Mn	Fe	None[b]
Orthoptera	Herbivorous	5				
Phasmida	Herbivorous	2				
Dictyoptera	Omnivorous				1	3
Dermaptera	Omnivorous				1	
Lepidoptera	Herbivorous	11 (L)[c]				
Coleoptera	Herbivorous		4 (L,A)		1 (L)	3 (L,A)
	Wood borers	1 (L,A)	1 (A)	3 (L)		
	Stored products	18 (A)	33 (A)			3 (A)
	Predaceous					1 (A)
Hymenoptera	Wood borer	1 (L)				
	Ants (workers)			5		

[a]Data from Fontaine et al. (1991), Hillerton and Vincent, (1982), and Hillerton et al. (1984).

[b]None or very small amounts.

[c]L, larva; A, adult.

or a dermapteran studied by Hillerton and Vincent (1982). In some of these cases, iron was detected.

Clearly, it is usual for the mandibles of chewing insects to contain zinc, manganese, or, occasionally, iron in the cutting edges. However, the mandibles of some insects possess none of these. If the metals really do confer hardness, as seems likely, it appears that hardness can also be achieved in other ways.

The amounts of these metals in the cutting edges are considerable. In the ants the average concentrations of zinc and manganese in the cutting edges were 4% and 0.4% dry mass, respectively (Schofield et al., 1988). In the larva of *Helicoverpa zea*, zinc comprised 1.3% of the dry weight of the entire mandibular cuticle; in *L. migratoria* it was about 0.5%. However, because these are the amounts in the whole mandible, it is obvious that the concentrations in the cutting edges will be much higher.

The implication is that these metals confer hardness to the cutting edges, but Schofield and Lefevre (1989) suggest that the two metals confer different properties on the cuticle and are not both concerned with hardness. This suggestion arises from their work on the fangs and claws of spiders.

The hard cuticle of the mandibles of the grasshopper, *Chortoicetes terminifera*, is structurally different from normal cuticle which forms other parts of the mandibles (Gardiner and Khan, 1979). It is nonlamellate and is electron-dense, perforated by electron-lucent patches. According to Gardiner and Khan (1979), the electron-dense rods are longitudinally arranged, that is, parallel with the surface of the cuticle. Cuticle with a similar appearance is depicted in the mandible of the larva of *Pieris brassicae* by Ma (1972, Plate 18) and may be characteristic of hard cuticle on the mandibles of all chewing insects.

1.2.2. Sensilla on the Mandibles

In all the species that have been critically examined, the mandibular cusps are innervated by small numbers of scolopidia. They have been described in *Locusta migratoria* by Le Berre and Louveaux (1969), in the larva of the beetle, *Speophyes lucidulus*, by Corbière-Tichané (1971b), in the larva of *Pieris brassicae* by Ma (1972), and in elaterid larvae by Zacharuk (1962). These sensilla are present at the tips of incisor cusps or between molar cusps (Fig. 1.4a, arrows). Similar sensilla are also found in the lacinea of *L. migratoria* (Louveaux, 1972) and *S. lucidulus* (Corbière-Tichané, 1971a), where these structures are heavily sclerotized and toothlike.

Each sensillum consists of a pair of dendrites extending through a canal in the cuticle and ending close beneath the surface. Corbière-Tichané (1971b) and Zacharuk and Albert (1978) demonstrated that these canals did not reach the surface; the belief that they did so had caused most earlier authors to think that these were chemoreceptors. In their distal parts, the two dendrites are joined by a tight junction extending longitudinally along their lengths. More proximally the dendrites are surrounded by a scolopale, which is typical of a scolopidium (Fig. 1.4b).

These sensilla are in positions such that even small amounts of wear must expose the tips of the dendrites. Presumably the tip of the cuticular canal in which the dendrites occur is continually replugged, but nothing is known of this process.

Zacharuk and Albert (1978) showed electrophysiologically that these sensilla on the mandibles of an elaterid beetle larva responded to pressure on the cuticle, but they did not respond to any of the chemicals they applied. Perhaps these receptors help to monitor the hardness of food and modulate the power output of the adductor muscles (see below).

In addition to these scolopidia, insect mandibles have small numbers of campaniform and trichoid mechanoreceptors (Fig. 1.4a). The campaniform sensilla occur primarily in groups adjacent to the hinge (Chapman, 1966; Thomas, 1966), and their axons extend back to the subesophageal ganglion. Baker (1982) records campaniform sensilla near the tip of the mandibles of *Grylloblatta campodeiformis*. Perhaps in this predaceous species they monitor the pressure exerted at the tip. No instances are known of chemoreceptors on the mandibles (Zacharuk, 1985).

1.2.3. Movement of Mandibles

In all mandibulate insects except Archaeognatha, each mandible is articulated with the head capsule at two points (condyles). This limits the movement of the

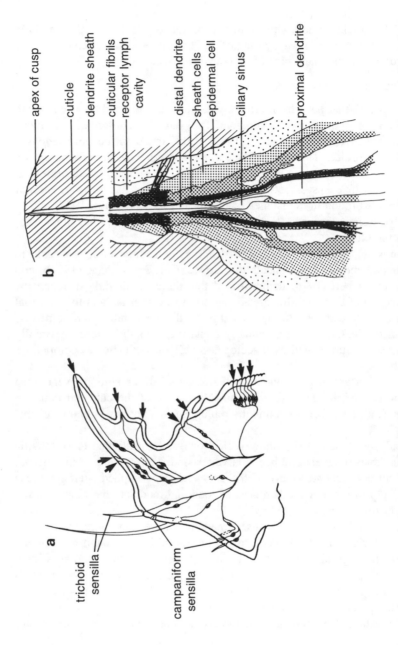

apex of cusp

cuticle

dendrite sheath

cuticular fibrils

receptor lymph cavity

distal dendrite

sheath cells

epidermal cell

ciliary sinus

proximal dendrite

b

trichoid sensilla

campaniform sensilla

a

Figure 1.4. (**a**) Sensory receptors on the mandible of a larva of the beetle, *Speophyes lucidulus*. Arrows indicate the positions of scolopidial organs. (After Corbière-Tichané, 1971.) (**b**) Diagram of a mandibular scolopidium from the beetle *Ctenicera destructor*. (After Zacharuk and Albert, 1978.)

mandibles to one plane (Fig. 1.1), although it is probable that the movement of the mandibles over each other as they close causes some slight displacement from this plane. In grasshoppers, the asymmetry of the mandibles is such that the left always closes over the right, but in saturniid and sphingid caterpillars the left and right mandibles frequently alternate in overlapping each other (Bernays and Janzen, 1988). The extent to which the mandibles overlap appears to be greater in predaceous species.

Movement is produced by a pair of muscles inserted onto apodemes on either side of the line of articulation. The muscle on the side nearest the mouth closes the mandible (adductor muscle), and that inserted on the lateral margin of the mandible opens it (abductor muscle) (Fig. 1.1). Both muscles arise in the upper parts of the head capsule if the insect is hypognathous, like most Orthoptera and caterpillars, or posteriorly if the insect is prognathous, like most beetles. They occupy the greater part of the head capsule, and each muscle is comprised of several units. The adductor muscle is the larger of the two because it is the power-producing muscle, and, at least in grasshoppers and carabid beetles, the cross-sectional area of the adductor muscles is proportional to head size (Bernays and Hamai, 1987; Wheater and Evans, 1989).

In phytophagous insects eating hard food, the head capsule is enlarged to accommodate the large adductor muscles. This is illustrated amongst the grasshoppers and caterpillars by the bigger head capsules, relative to body mass, of grass-feeding species, compared with species feeding on soft, broad-leaved plants (Fig. 1.5a) (Bernays, 1986; Bernays and Hamai, 1987). Amongst caterpillars, head mass as a proportion of body mass is greater in the earlier instars. This is most marked in saturniid larvae, which feed on tough, mature leaves from the time of hatching. Small Saturniidae have a relative head mass about twice that of sphingid larvae, with a similar body mass (Fig. 1.5b). The sphingids, especially during the early instars, feed on plants with relatively soft leaves (Bernays and Janzen, 1988).

Changes in head size may even occur during ontogeny as a consequence of feeding on harder foods. Bernays (1986) showed that caterpillars of *Pseudaletia unipuncta* developed bigger head capsules when they were given hard food from the time of hatching (Fig. 1.6); and the surface area of the adductor muscle apodemes was also increased, indicating that the muscle had a greater cross-sectional area. It was found that insects with bigger head capsules were able to feed more rapidly on tough grass.

Wheater and Evans (1989) measured the forces exerted at the tips of the mandibles in a number of predaceous beetles. They found that, in general, mandible tip force was correlated with the cross-sectional area of the adductor muscles and the mechanical advantage of these muscles (calculated as the basal width of the mandible/the distance from the articulation to the tip of the mandible) (Fig. 1.7). However, some beetles produced less force than expected from their muscle sizes. Wheater and Evans (1989) suggest that the beetles are optimizing

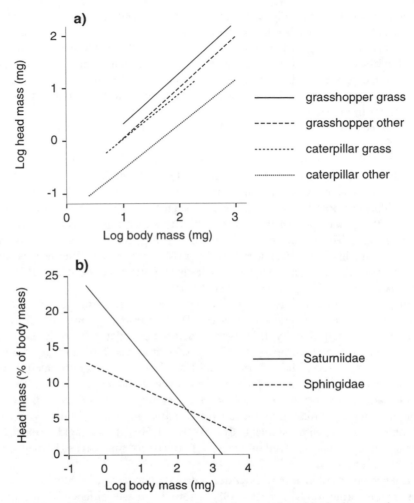

Figure 1.5. (a) Relationship between dry head mass and total body dry mass in grass specialists and other foliage feeders among grasshoppers and caterpillars. (After Bernays, 1986). (b) Relationship between relative head mass and body mass for larval Saturniidae and Sphingidae. (After Bernays and Janzen, 1988.)

the *pressure* exerted at the tips of the mandibles which are of different tip diameters; in other words, the beetles only exert such force (bite hard enough) as is necessary to break through the cuticle of the prey. They argue that, because the cuticles of different prey species have rather similar safety factors, the pressures necessary to break through these cuticles also tend to be uniform.

The opening and closing of the mandibles was studied in *Schistocerca gregaria* by Blaney and Chapman (1970) and by Seath (1977a). Opening was faster than

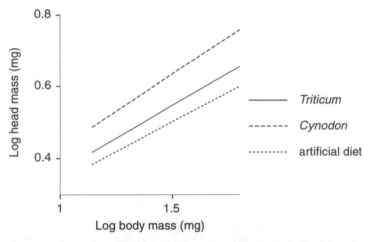

Figure 1.6.   Relationship between dry head mass and total dry body mass in fifth instar caterpillars of *Pseudaletia unipuncta* reared from hatching on three different foods varying in hardness. *Cynodon* was the hardest, and artificial diet was the least hard. (After Bernays, 1986.)

closing, sometimes about twice as fast. The initial phase of closing was also fast, but then the rate of closure declined, presumably as the mandibular cusps encountered the food. Blaney and Simmonds (1987) illustrate a similar pattern in *Locusta migratoria,* but with a final rapid phase of closure immediately before the mandibles start to open again. Perhaps this occurs when the mandibles

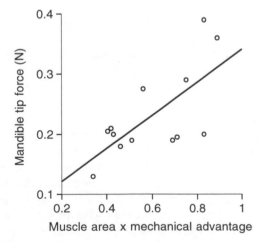

Figure 1.7.   Relationship between mandibular force and the cross-sectional area of adductor muscles × their mechanical advantage in predaceous beetles. Each point is derived from one species. (After Wheater and Evans, 1989.)

suddenly cut through the leaf tissue, although in their figure it corresponds with a period when the abductor muscle was active. The reversal of movement, from opening to closing and vice versa, occurs rapidly, without any pause.

Using film analysis of fifth instar *Schistocerca gregaria* feeding on the grass *Poa annua,* Blaney and Chapman (1970) showed that the duration of a complete cycle of opening and closing varied from 0.27 to 0.55 sec, but was relatively constant within a sequence. Seath (1977a), using electromyograms from adults of the same species feeding on seedling wheat, gives the "normal" chewing frequency as about 1 Hz. Blaney and Simmonds (1987), using the same method with fifth instar nymphs of *Locusta migratoria,* also depict a bite frequency of about 1 Hz.

Bowdan (1988a) recorded the mandibular activity of fifth-instar larvae of *Manduca sexta* by a method analogous to that used for aphids (see Chapter 7). In this case, one electrode was inserted into the food and the other entered the hemolymph via the terminal horn of the larva. She found that the rate of mandibular movement was higher in larvae that had been deprived of food, even for as little as 1 hr. The rate of biting was, again, about 1 Hz, and it increased slightly with the size of the larvae (Bowdan, 1988b). Because this method does not directly record mandibular movement, it is uncertain whether these differences arise from differences in the rate of movement or from pauses between bites. In the caterpillar of *Euxoa messoria* (Noctuidae) feeding on cabbage, Devitt and Smith (1985) found that a sequence of five bites was completed in about 3.2 sec. They depict the cycle of opening and closing as taking about 0.4 sec, followed by an interval of about 0.3 sec during which the mandibles remain closed before they start to open again.

1.2.4. Control of Mandibular Movement

The control of mandibular movement ultimately resides in the mandibular neuromere of the subesophageal ganglion. It is to this ganglion that most of the sensory input from the receptors on the mouthparts projects (Altmann and Kien, 1987; Kent and Hildebrand, 1987), and it is here that the motor neurons of the mandibular muscles reside. Each adductor muscle in the locust is innervated by 12 motor neurons. In the larva of *Manduca sexta* there are 12 adductor motor neurons and 8 abductor motor neurons (Griss, 1990).

The motor pattern controlling chewing movements of the mouthparts is generated in the subesophageal ganglion because normal mandibular movements are elicited even when the subesophageal ganglion is totally isolated from the rest of the central nervous system by cutting the connectives from the thoracic ganglia and those to the brain. In the larva of *Manduca sexta,* chewing movements are normally inhibited by input from receptors in the wall of the thoracic segments. If the ventral connectives are cut behind the subesophageal ganglion, the insect makes spontaneous chewing movements and continues to do so for at least 48 hr (Griss et al., 1991). A similar result is achieved if the nerves connecting the body wall to the thoracic ganglia are cut (Rowell and Simpson, 1992). Probably, mechanoreceptors in or on the wall of the thoracic segments are normally the source of the inhibition which is removed by these nerve-cutting experiments, although Rowell and Simpson do not exclude the possibility that some unknown chemoreceptors are involved. They suggest that the level of this thoracic feedback determines the threshold which must be exceeded by input from sensory receptors on the mouthparts before feeding begins. Cutting the ventral connectives behind the subesophageal ganglion in locusts does not remove the inhibition of chewing activities.

Chewing movements are induced by mechanical stimulation of the labium in larvae of *M. sexta* in which the subesophageal ganglion is isolated from the rest of the nervous system (Griss et al., 1991). Mechanical stimulation of the labrum or stimulation with plant material (chemical?) also increases the rate of chewing, provided that the circumesophageal connectives are intact. In *Locusta migratoria*, Sinoir (1969) showed that mechanical stimulation of the labrum resulted in mandibular abduction. In a normally feeding grasshopper, it is usual for the surface of the food to be contacted first by the chemoreceptors on the tips of the maxillary and labial palps. If the food is acceptable, the insect lowers its head and attempts to bite. It is not clear if biting is only induced by mechanical stimulation, or if chemical stimulation may also produce this effect.

In the larvae of some carabid beetles that are predaceous on Collembola and therefore must have a very fast reaction, mechanosensory input alone is almost certainly responsible for mandible closure. In *Nebria* and *Notiophilus* spp., once the prey is detected, the larva turns to face it with mandibles held wide open. It then lunges at the prey and the mandibles snap shut. Mandible closure is apparently stimulated by contact of the prey with specialized mechanoreceptors carried at the tips of rostral horns which project in front of the head (Fig. 1.8). Mandibular closure may be complete in 7 msec, suggesting a direct connection between the sensilla and the adductor muscle motorneurons (Altner and Bauer, 1982; Spence and Sutcliffe, 1982).

An even faster response occurs in the trap jaw ant, *Odontomachus* sp., where closure of the mandibles may be completed in 0.33 msec. These ants hunt with the mandibles wide open and when they encounter a prey item the mandibles snap shut. It is suggested that, by analogy with the jumping mechanism of a grasshopper, the mandibles are held open by the co-contraction of antagonistic muscles. The adductor muscles are active before, but not during, the strike, implying that they distort some element of the cuticle which then, functionally, resembles a cocked spring. Release of the spring causes the mandibles to snap shut faster than would be possible by direct muscle action. The closure is triggered by contact of the prey with a long mechanosensitive hair on a mandible. There are two such hairs on each mandible. The axons of the sensory neurons are 15–20 μm in diameter, this large size suggesting a high rate of conduction. The axons

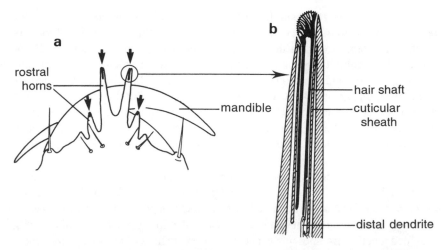

Figure 1.8. Rostral horn receptors in the beetle *Notiophilus biguttatus*. (**a**) Rostrum with horns. Arrows show the positions of receptors. (**b**) Detail of one of the median horns and its receptor. Ventral to the left. (After Altner and Bauer, 1982.)

arborize in the subesophageal ganglion, one from each side having contralateral branches which presumably serve to coordinate the release of the two mandibles. The speed of the response strongly indicates a monosynaptic pathway, but the mechanism of the release is not yet known (Gronenberg et al., 1993).

The sensory input necessary to maintain and change mandibular activity was investigated in *Schistocerca gregaria* by Seath (1977a,b). The mandibles of the two sides normally move synchronously. Artificially driving the mandibles with a mechanical device produced activity in the adductor muscles that was often initiated during the final phases of opening. The activity in the muscles of the two sides was not independent because if the two mandibles were driven at different rates, the activity of the adductor muscles was still synchronized and occurred at the rate of the faster driven mandible. However, the duration of the burst of muscle activity was determined by the position of the slower-moving mandible. This suggests the presence of some position receptor in the mandibles, although the nature of this receptor is unknown. Campaniform sensilla close to the point of articulation on the outside of the mandible are possible candidates. However, the activity of the two muscles is not tightly linked because driving one mandible did not produce activity in the contralateral muscle unless food was present between the mandibles. Probably the scolopidia in the cuticle of the cusps (see above) monitor the presence of food because when these were cauterized in the free mandible, its movements were no longer synchronized with those of the driven mandible. If the mandibles were held in the open position, the adductor muscles were active in a series of short bursts; but if the campaniform sensilla at the bases of the adductor apodeme were cauterized, a continuous train of activity occurred in the muscle. This suggests that the input from this group of campaniform sensilla may serve to inhibit the muscle. Under normal conditions this might occur if the insect was biting on a very hard substrate, in order to avoid damage to the muscles. The scheme suggested by Seath for the control of mandibular movement in *S. gregaria* is shown in Fig. 1.9.

All the muscle fibers of the adductor muscles are of the "fast" type (Griss, 1990). This implies that changes in the forces applied by the mandibles is achieved by changes in the number of muscle units that are active. Griss (1990) found no evidence of inhibitory neurons in *M. sexta*.

Changes in power output may also be achieved by neuromodulators. Serotonin-

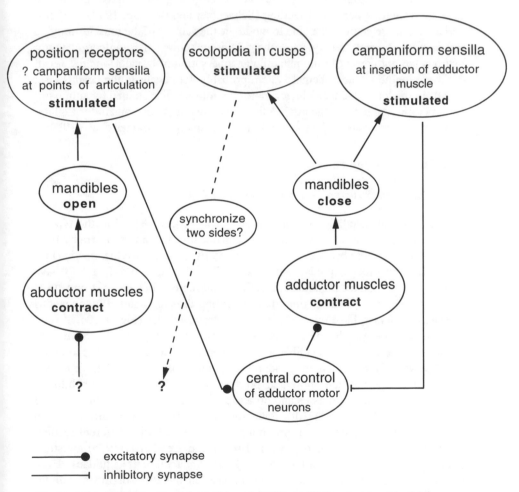

Figure 1.9. Diagram summarizing the control of mandibular movement in *Schistocerca gregaria*. (Based on Seath, 1977a.)

reactive axons have been found in association with the adductor muscles of a locust, a cricket, and a cockroach. Because these axons occur on the outside of muscle fibers, but were not seen to innervate them, it is suggested that they have a neurosecretory function. Application of serotonin while the muscle is active results in increases in the amplitude and rates of contraction and relaxation (Baines et al., 1990a). Davis (1987) describes a neurohemal system in the head of *Periplaneta americana*. A group of two or three nerve cell bodies lies near the root of the mandibular nerve on each side of the ganglion. Axons from these cells are present in the nerves of all the mouthparts, and they branch to form an extensive neurohemal organ at the surface of the nerves. The cells and their axons stain for serotonin. Similar neurons in *Locusta migratoria* have been shown to be active during periods of feeding (Schachtner and Bräunig, 1993). It seems certain that these neurons are able to modulate the activity of the muscles moving the mouthparts, but exactly what they do is unknown. Serotoninergic motor neurons are not present in the mandibular innervation of larval *Manduca sexta*, but a serotoninergic neurohemal system probably does occur in the head (Griss, 1990). In the locust, some adductor motor neurons are also stained by antibodies for proctolin. Acting in concert with the conventional motor transmitter L-glutamate, proctolin enhances the force of muscle contractions (Baines et al., 1990b).

1.3. Cutting Up the Food

1.3.1. Mandible Structure in Relation to Food Type

The structure of mandibles within a group of insects is adapted to the type of food on which the insects feed. This has been documented in Orthoptera by Isely (1944) and in carabid beetles, most recently by Evans and Forsythe (1985) (Fig. 1.10). In both groups, predaceous species (Fig. 1.10a, 1.10d–f, 1.10j) have relatively long, pointed mandibles, and the molar region often bears strong cusps or a ridge which is probably important in tearing the prey apart. Phytophagous insects, in contrast, have relatively much shorter mandibles (Fig. 1.10c, 1.10g–i). Amongst the grasshoppers, species that eat grass (Fig. 1.10h) have long incisor cusps on the left mandible and a ridged molar area; species feeding on broad-leaved plants, which are generally less fibrous than grasses, have incisor and molar cusps that are short and pointed (Fig. 1.10g). Species habitually feeding on a mixture of grasses and broad-leaved plants have mandibles that are intermediate in form (Fig. 1.10i). Patterson (1984) demonstrated differences in the shapes of the mandibles of grasshoppers (Acrididae) with different feeding habits by using morphometric ratios. Brown and Dewhurst (1975) illustrate mandibular forms of caterpillars in the genus *Spodoptera* (Noctuidae). Even within this genus there are distinct differences in the form of the mandibles depending on the food of the species. Grass-feeding species, such as *S. exempta*, have long incisor cusps similar to those depicted in Fig. 1.10h for a grass-

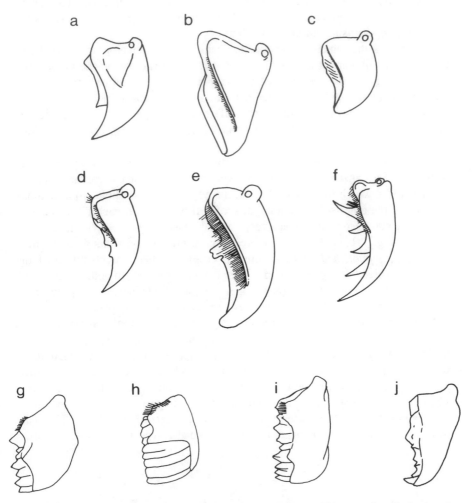

Figure 1.10. Mandibles of some carabid beetles (right mandible seen from below) and some Orthoptera that feed on different foods (left mandible seen from in front). (**a–f**) Carabids. (After Evans and Forsythe, 1985.) (**g–j**) Orthoptera. (After Isely, 1944.) (**a**) *Notiophilus biguttatus*—predaceous, fragments of prey. (**b**) *Diplocheila major*—predaceous on snails. (**c**) *Amara aulica*—plant fragments. (**d**) *Elephrus cupreus*—predaceous, semifluid food, fragments. (**e**) *Carabus problematicus*—predaceous, pre-orally digested. (**f**) *Cicindela campestris*—predaceous, pre-orally digested. (**g**) *Brachystola magna*—phytophagous, forbivorous. (**h**) *Mermiria maculipennis*—phytophagous, graminivorous. (**i**) *Melanoplus differentialis*—phytophagous, mixed plants. (**j**) *Pediodectes haldemanii*—predaceous.

17

feeding grasshopper, whereas polyphagous species, like *S. littoralis,* which feeds predominantly on broad-leaved plants, has sharply pointed cusps, resembling the grasshoppers using similar food plants.

Caterpillars that change their feeding habits during development may have different mandibular forms in different instars. For example, in the notodontid *Heterocampa obliqua,* the first instar larva gouges out leaf tissue from between the veinlets on the underside of the leaf. It has toothed mandibles (Fig. 1.11a). Later instars cut segments out of the leaf from the edge of the leaf blade. In these, the edge of the mandible forms a continuous ridge (Fig. 1.11b) (Godfrey et al., 1989).

The functional importance of mandibular form is demonstrated by the work of Bernays and Janzen (1988). They showed that in saturniid caterpillars of many species, all of which fed on hard leaves, the edge of one mandible works against the face of the other so that a small disc of leaf is cut off at each bite. The leaf fragments cut by one insect are very regular in size with low coefficients of variation (Fig. 1.12a). This is also true amongst grasshoppers that feed on grass (Chapman, *unpublished data*), although occasionally these insects bite off long fragments which become folded up as they enter the foregut. In both the saturniid caterpillars (Fig. 1.13) and the grasshoppers, fragment size increases in proportion to the size of the insect and is relatively similar in different species.

By contrast, Bernays and Janzen (1988) found that sphingid larvae, most of which fed on soft leaves, bite off fragments that are small and vary markedly in size in any one insect (Fig. 1.12b). This variation is produced by the mandibles which bear a complex array of teeth and ridges which are envisaged as tearing away fragments of leaf rather than cleanly cutting them off as in the Saturniidae. They describe the sphingid gut as containing a slurry of fine cellular material. The fragments do not increase in size as the insect gets bigger (Fig. 1.13). The foregut contents of grasshoppers feeding on broad-leaved plants have a similar

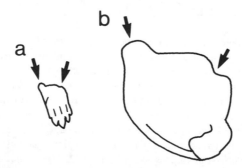

Figure 1.11. Left mandible of the caterpillar of *Heterocampa obliqua.* (**a**) First-instar larva. (**b**) Third-instar larva (drawn to same scale). Arrows show points of articulation with the head capsule. (After Godfrey et al., 1989).

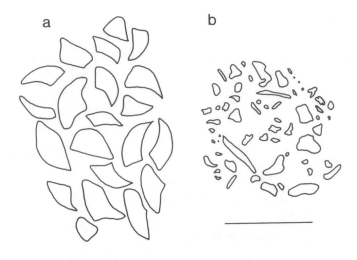

Figure 1.12. Drawings of food particles from the midgut of (**a**) a saturniid, *Rothschildia lebeau,* instar 5 (scale line 2 mm) and (**b**) a sphingid, *Pachylia ficus,* instar 5 (scale line 1 mm). (After Bernays and Janzen, 1988.)

appearance (Chapman, *unpublished data*), although critical analyses have not been carried out.

As further evidence that mandibular form affects the size of the fragments bitten off during feeding, Bernays and Janzen (1988) found that one species of sphingid, *Enyo ocypete,* which habitually feeds on tough leaves had mandibles similar in form to those normally associated with saturniids. And like the saturniids, the fragments in the foregut were large and regular in size (see Fig. 1.13).

That the form of the mandibles is important is further indicated by the fact that *Bootettix argentatus,* a grasshopper feeding specifically on creosote bush, was unable to feed on grass because it was unable to cut through the fibers of the leaf tissue (Chapman et al., 1988).

The implication that arises from these differences in the way that food is processed by the mandibles is that they will result in differences in the efficiency with which enzymes penetrate the leaf tissue and extract the contents. This presupposes that enzyme entry occurs only around the periphery or where the tissues are damaged. Köhler et al. (1991) found a correlation between the numbers of teeth on the pectinate lamellae of the mandibles of different species of millipedes and the rate at which the food was assimilated. Animals with more teeth also had smaller fragments of food in the feces.

1.3.2. Wear of Mouthparts

If Hillerton and Vincent (1982) are correct, the sharpness of mandibles depends on wear of the shearing surfaces. Some degree of wear to the mandibles has

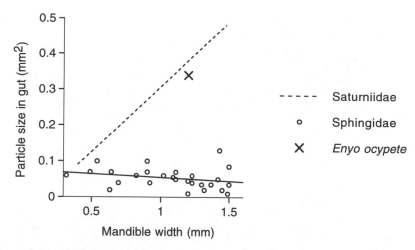

Figure 1.13. Relationship between size of chewed food particles and mandible width in saturniid and sphingid caterpillars. *Enyo ocypete* is a sphingid that feeds on hard leaves. (After Bernays and Janzen, 1988.)

been recorded in several insect groups. Amongst herbivorous insects it has been recorded in grasshoppers (Chapman, 1964; Gangwere, 1965; Gangwere et al., 1976; Kaufmann, 1971), caterpillars (Djamin and Pathak, 1967) and beetles (Raupp, 1985). Amongst fungivores it has been recorded in Collembola (Welton, 1988) and termites (Skaife, 1955) and it is also found in carnivorous beetles (Butterfield, 1986; Houston, 1981; Wallin, 1988). Damage to mandibles and other mouthpart structures is widespread amongst aquatic insect larvae that graze on algae (some mayflies, stoneflies, caddisflies, and larval Diptera) (Arens, 1990). It is probably a widespread phenomenon in longer-lived mandibulate insects.

The progressive change in mandible form has been documented in the laboratory in grasshoppers (Chapman, 1964), a herbivorous beetle, *Plagiodera versicolora* (Raupp, 1985), and a carnivorous beetle, *Pterostichus melanarius* (Wallin, 1988), usually by measuring the length of the incisor cusps relative to the absolute size of the mandible, as indicated by the distance between the condyles (Fig. 1.14a). Fig. 1.14b shows the change in the length of the left incisor cusps of a population of adult *Schistocerca gregaria* over a period of 12 weeks. These insects were fed from the time of the final molt on relatively soft temperate grasses. A similar change occurred in the length of the right incisor cusps. Butterfield (1986) documents the seasonal increase in wear in mandibles of *Carabus problematicus* collected in the field. In this species, wear may be produced by burrowing in the soil as well as by feeding, as Wallin (1988) found in *Pterostichus melanarius*.

In an extensive study of several aquatic grazing species, Arens (1990) illus-

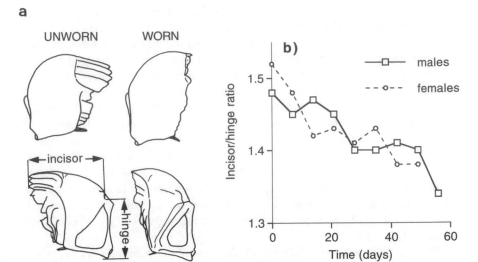

Figure 1.14. (a) The left mandible of the grasshoppper *Austeniella cylindrica* unworn and worn. The figure also shows the measurements from which the incisor hinge ratio in (b) are calculated. (b) Wear of the left mandibles of *Schistocerca gregaria* over a period of days in the laboratory. (After Chapman, 1964.)

trated extensive wear on any of the structures employed in removing algae from the surface. He also demonstrated experimentally that in the mayfly, *Epeorus,* wear is more extreme when the insect feeds from a rougher, more abrasive surface. In these insects the grinding surfaces of the mandibles are also subject to wear, in addition to the parts used to remove the algae from the substrate. In most species this affects the molar areas of the mandibles, but in the stratiomyid larva, *Oxycera,* where crushing is performed in the pharynx, both the tooth and the mortar may be almost completely worn away.

While it is probably true that some degree of wear maintains the sharpness of the incisor cusps, there are examples where the wear is excessive and the original microstructure of the mandibles is completely worn away (Fig. 1.14a). Micrographs, mainly with the scanning electron microscope, demonstrate this point in a caterpillar (Djamin and Pathak, 1967), beetles (Raupp, 1985; Wallin, 1988), and aquatic grazing insects (Arens, 1990). Amongst the herbivores, this excessive wear has sometimes been associated with feeding on unusually tough foods (Djamin and Pathak, 1967; Raupp, 1985). Assuming that the original microstructure of the mandibles has real importance, one is forced to the conclusion that extreme mandibular wear must reduce feeding efficiency. This has been specifically investigated in two cases.

Chapman (1964) fed adult *Locusta migratoria* from the time of emergence a powdered diet to which he added carborundum powder to produce abrasion of the mandibles. At intervals over 6 weeks the insects were timed while they ate fragments of grass leaf blade approximately 100 mm^2 in area. At the same times, samples were taken to determine that the carborundum treatment did increase the amount of wear of the mandibles compared with insects on the diet without carborundum powder. Insects receiving the diet with added carborundum consistently took longer to eat 100 mm^2 of grass leaf compared with insects on plain diet. Even by day 9 of the experiment, the mandibles of insects feeding on the carborundum diet showed evidence of wear, and perhaps this small amount of wear was sufficient to reduce the rate of feeding. However, there was no progressive increase in the differences between the treatments in the time to ingest 100 mm^2 as the mandibles of the insects feeding on the carborundum diet became progressively more worn. As Chapman (1964) suggests, it is possible that the carborundum powder adversely affected the insects in some way unrelated to mandibular wear. These experiments offer no clear evidence of an effect of wear on feeding.

Raupp (1985) produced differences in the wear of the mandibles of *Plagiodera versicolora* by feeding them on soft or hard willow leaves for 3 weeks. The rate of feeding on young leaves was then determined for both groups of beetles. Those that had eaten soft leaves fed at an average rate of 0.250 ± 0.086 mm^2/min, whereas those previously on hard leaves ate more slowly, 0.173 ± 0.018 mm^2/min. This difference is statistically significant ($t = 3.23$, $p < 0.01$) and appears to be an unequivocal demonstration of the effects of mandibular wear. Raupp went on to show that the number of eggs produced was proportional to the amount eaten per day, but he did not demonstrate that the reduced rate of feeding by the beetles with more worn mandibles necessarily resulted in them eating less in the course of a day; they might have fed for longer to compensate for the reduced feeding rate. Thus these experiments provide no evidence that mandibular wear adversely affects food intake over a period.

Wear of the mandibles is also sometimes apparent in larval forms, but presumably fully developed mandibles are produced anew at each molt. It has been suggested by Houston (1981) with respect to carabid beetles and by Arens (1990) with respect to aquatic grazing insects that frequent molting which is seen in these insects may be an adaptation to offset mandibular wear in the larvae.

1.4. Ingesting the Food

The mechanism of ingesting food that has been cut by the mandibles is only poorly studied. In a majority of chewing insects where the maxillae are well-developed, they may play an important part. The lacineae are commonly pointed and strongly sclerotized with the same types of sensilla as the mandibles (see above).

In grasshoppers the movements of the mouthparts are dominated by the movements of the mandibles, but active movements of the other mouthparts occur. As the mandibles swing open, the labrum moves forward and up and the maxillae are forced outward. As the mandibles close, the labrum swings back because of the elasticity of the labroclypeal hinge, and the maxillae make a rapid closing movement. The hairs in extensive tracts on the epipharyngeal surface of the labrum and on the hypopharynx point upward, toward the mouth, so that fragments of food are pushed upward as the labrum moves back. The brushes of long hairs on the mandibles will also move particles toward the mouth during adduction. The hair tracts of the epipharynx are hydrophilic, whereas the rest of the surface of the cuticle of the mouthparts is hydrophobic, and it is envisaged that fluids from the crushed food flow along these tracts to groups of chemoreceptors outside the mouth (Cook, 1977). Amongst Carabidae it is believed that hairs on the epipharynx and the maxillae facilitate the flow of fluids *away* from the mouth during the process of extraoral digestion (Forsythe, 1982).

Evans (1964) grouped predaceous beetles in five categories depending on how they dealt with the food:

1. The prey is dismembered by the mandibles, and further fragmentation is performed by the proventricular teeth (see Section 1.5). The carabid, *Notiophilus,* and many dytiscid beetles are in this category.

2. The prey is dismembered by the mandibles, but the proventriculus acts only as a valve or filter; no further trituration occurs. Examples are *Carabus* and some dytiscids.

3 and 4. The prey is macerated by the mandibles and various degrees of extraoral digestion occur. Only liquids and small particles of food are ingested; the undigested remains of the body are discarded. Many Staphylinidae and larval forms are in this category.

5. Maceration is not important. Enzymes are injected into the prey through channels in the mandibles and the digested remains are sucked into the foregut. *Dytiscus* and *Lampyris* larvae are examples.

Amongst the carabid beetles that fragment their prey with the mandibles, the maxillae are in antiphase with the mandibles, closing as the mandibles open. The lacinea bears rows of large hairs which serve to rake the food toward the mouth; and when the mandibles close, their basal regions push the food further into the cibarium toward the mouth. Movement of the food through the mouth and into the pharynx is then presumably achieved by changes in the shape of the cibarium due to muscular activity. Larger fragments of food are prevented from entering the cibarium by anteriorly directed spines on the epipharyngeal surface of the labrum and at the base of the labium. These fragments drop down on to the labium where anteriorly directed microtrichia direct them forward. They are prevented from falling out by the hairs at the front of the labium and, probably, by the large hairs on the labial palps. From the labium they are recirculated to the lacineae and the mandibles (Fig. 1.15) (Evans and Forsythe, 1985; Forsythe, 1982).

1.4.1. Control of Ingestion

Maintained ingestion requires the frontal ganglion to be connected to the subesophageal ganglion in *Manduca sexta* (Griss et al., 1991), although Bell (1984) found that a majority of frontal ganglionectomized larvae increased weight and pupated. He concluded that "the frontal ganglion is not absolutely essential for

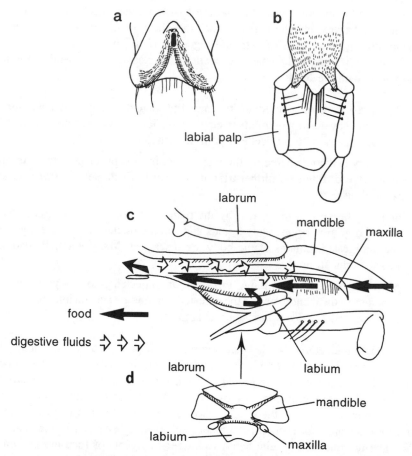

Figure 1.15. Mouthparts of *Carabus* sp. (After Forsythe, 1982.) (**a**) labrum. (**b**) labium. (**c**) Diagram of a sagittal section through the mouthparts showing the movements of food and digestive fluids. (**d**) Section through the mouthparts (c) at the position shown by the arrow.

feeding." Removal of the protocerebrum did not, however, prevent ingestion (Griss et al., 1991). In grasshoppers, removal of the frontal ganglion does not prevent ingestion over the long term (see Penzlin, 1985), but Bernays and Chapman (1973) found that cutting the anterior and median pharyngeal nerves resulted in slow feeding and regurgitation of gut fluids. These nerves innervate the pharynx immediately behind the mouth and presumably control swallowing. Both Clarke and Langley (1963) and Dogra and Ewen (1971), working with *Locusta migratoria* and *Melanoplus sanguinipes*, respectively, record that food intake was unaffected "once the effect of the operation had worn off." Taken together, these various studies indicate that the frontal ganglion does have some role in regulating food intake in both caterpillars and grasshoppers, but that over a period of days the insects can compensate for damage to this system.

While movement of food through the mouth appears to be a mechanical process due to the activity of the mouthparts, Hamamura et al. (1962) suggested that there were separate biting and swallowing factors for silkworm larvae, *Bombyx mori*. They suggested that cellulose was the swallowing factor because the addition of cellulose powder to an artificial diet in agar considerably enhance the amounts eaten of a 32-hr period. It was shown that increasing proportions of cellulose increased the amounts of artificial diet eaten by *Schistocerca gregaria* (Dadd, 1960a) and resulted in better growth and survival (Dadd, 1960b). In neither case were detailed observations made on behavior, and it is possible that the increased food intake occurred over a period of time rather than in the course of one feed, with the increased bulk contributed by the cellulose affecting the rate of movement of food through the gut. It is also probable that the lower concentrations of nutrients in the diets containing cellulose resulted in reduced feedback from the hemolymph so that the insects ate more over short periods (see Chapter 9). It appears that there is currently no good evidence for the existence of a distinct swallowing factor.

1.5. Manipulation of Food in the Foregut

Once inside the mouth, the food is presumably passed back to the crop and midgut by peristaltic activity of the pharynx and esophagus. The foregut of grasshoppers is armed with small backwardly directed spines which are apparently concerned with maintaining the movement of food back through the gut. There is no evidence that they effect further trituration of the food after it is ingested, because no reduction in the size of food particles was found in the midgut compared with the foregut (Chapman, *unpublished data*). Amongst phytophagous insects it also appears to be true that fragments in the fecal pellets are similar in size to those found in the foregut. Bernays and Janzen (1988) comment that "the fecal pellets of saturniid larvae are simply tightly packed wads of almost morphologically intact leaf discs." The same seems to be true of grass-feeding grasshoppers.

Some insects, however, have a gizzard or proventriculus at the posterior end of the foregut armed with large, heavily sclerotized teeth. These occur in cockroaches, crickets, and some beetles. It is assumed that a primary function of these strongly armed gizzards is to effect further trituration of the food. This has been specifically examined in some carabids that were fed on adult *Drosophila*. In all six species of beetle examined, the fragments of *Drosophila* cuticle were smaller in the gut posterior to the proventriculus than they were in the crop (Fig. 1.16). This was attributed to mechanical grinding, not to the activity of enzymes (Cheeseman and Pritchard, 1984). Reports on *Blatta orientalis* and *Ips radiatae* suggest that the proventriculus also has a grinding function in these insects, although the data were not quantified (Eaton, 1942; Eidmann, 1924).

The foregut is innervated and its movements controlled by the stomatogastric nervous system (see Penzlin, 1985, for details of the anatomy). Bernays and Chapman (1973) showed that cutting the esophageal nerves from the hypocerebral ganglion of *Locusta migratoria* resulted in reduced meal sizes because the food did not pass back along the foregut.

1.6. Conclusion

The mechanics of food handling by chewing insects is still poorly understood. Much of our present understanding is based on the morphology of the mouthparts and foregut, with little rigorous analysis of the mechanical processes involved. Similarly, our knowledge of the physiological mechanisms which regulate food handling is extremely fragmentary.

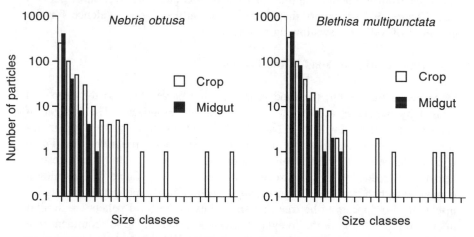

Figure 1.16. Distribution of particle sizes in the crop and in the gut posterior to the proventriculus in two beetles, *Nebria obtusa* and *Blethisa multipunctata*. (After Cheeseman and Pritchard, 1984.)

References

Altmann, J. S., and Kien, J. (1987). Functional organization of the subesophageal ganglion in arthropods. In: Gupta, A. P. (ed.), *Arthropod Brain*. Wiley, New York, pp. 265–301.

Altner, H., and Bauer, T. (1982). Ultrastructure of a specialized, thrust-sensitive, insect mechanoreceptor: stimulus-transmitting structures and sensory apparatus in the rostral horns of *Notiophilus biguttatus*. *Cell Tissue Res.* 226, 337–354.

Arens, W. (1990). Wear and tear of mouthparts: a critical problem in stream animals feeding on epilithic algae. *Can. J. Zool.* 68, 1896–1914.

Baines, R. A., Tyrer, N. M., and Downer, R. G. H. (1990a). Serotinergic innervation of the locust mandibular closer muscle modulates contractions through the elevation of cyclic adenosine monophosphate. *J. Comp. Neurol.* 294, 623–632.

Baines, R. A., Lange, A. B., and Downer, R. G. H. (1990b). Proctolin in the innervation of the locust mandibular closer muscle modulates contractions through the elevation of inositol trisphosphate. *J. Comp. Neurol.* 297, 479–486.

Baker, G. T. (1982). Sensory receptors on the mandibles and labrum of *Grylloblatta campodeiformis* Walker. *Zool. Anz. Jena* 5, 341–344.

Bell, R. A. (1984). Role of the frontal ganglion in lepidopterous insects. In: Bořkovec, A. B., and Kelly, T. J. (eds.) *Insect Neurochemistry and Neurophysiology*. Plenum Press, New York, pp. 321–324.

Bernays, E. A. (1986). Diet-induced head allometry among foliage-chewing insects and its importance for graminivores. *Science* 231, 495–497.

Bernays, E. A., and Chapman, R. F. (1973). The regulation of feeding in *Locusta migratoria:* internal inhibitory mechanisms. *Entomol Exp. Appl.* 16, 329–342.

Bernays, E. A., and Hamai, J. (1987). Head size and shape in relation to grass feeding in Acridoidea (Orthoptera). *Int. J. Insect Morphol. Embryol.* 16, 323–330.

Bernays, E. A., and Janzen, D. H. (1988). Saturniid and sphingid caterpillars: two ways to eat leaves. *Ecology* 69, 1153–1160.

Blaney, W. M., and Chapman, R. F. (1970). The functions of the maxillary palps of Acrididae (Orthoptera). *Entomol. Exp. Appl.* 13, 363–376.

Blaney, W. M., and Simmonds, M. S. J. (1987). Control of mouthparts by the subesophageal ganglion. In: Gupta, A. P. (ed.), *Arthropod Brain*. Wiley, New York, pp. 303–322.

Bowdan, E. (1988a). The effect of deprivation on the microstructure of feeding by the tobacco hornworm. *J. Insect Behav.* 1, 31–50.

Bowdan, E. (1988b). Microstructure of feeding by tobacco hornworm caterpillars, *Manduca sexta*. *Entomol. Exp. Appl.* 47, 127–136.

Brown, E. S., and Dewhurst, C. (1975). The genus *Spodoptera* (Lepidoptera, Noctuidae) in Africa and the Near East. *Bull. Entomol. Res.* 65, 221–262.

Butterfield, J. E. L. (1986). Changes in life-cycle strategies of *Carabus problematicus* over a range of altitudes in northern England. *Ecol. Entomol.* 11, 17–26.

Chapman, R. F. (1964). The structure and wear of the mandibles in some African grasshoppers. *Proc. Zool. Soc. London* 143, 305–320.

Chapman, R. F. (1966). The mouthparts of *Xenocheila zarudnyi* (Orthoptera, Acrididae). *J. Zool. London* 148, 277–288.

Chapman, R. F., Bernays, E. A., and Wyatt, T. (1988). Chemical aspects of host-plant specificity in three *Larrea*-feeding grasshoppers. *J. Chem. Ecol.* 14, 557–575.

Cheeseman, M. T., and Pritchard, G. (1984). Proventricular trituration in adult carabid beetles (Coleoptera: Carabidae). *J. Insect Physiol.* 30, 203–209.

Clarke, K. U., and Langley, P. A. (1963). Studies on the initiation of growth and moulting in *Locusta migratoria migratorioides* R. & F. II. The role of the stomatogastric nervous system. *J. Insect Physiol.* 9, 363–373.

Cook, A. G. (1977). The anatomy of the clypeo-labrum of *Locusta migratoria* (L.) (Orthoptera: Acrididae). *Acrida* 6, 287–306.

Corbière-Tichané, G. (1971a). Ultrastructure du système sensoriel de la maxille chez la larva du Coléoptère cavernicole *Speophyes lucidulus* Delar. (Bathysciinae). *J. Ultrastruct. Res.* 36, 318–341.

Corbière-Tichané, G. (1971b). Ultrastructure de l'équipement sensoriel de la mandibule chez la larve du *Speophyes lucidulus* Delar. (Coléoptère cavernicole de la sous-famille des Bathysciinae). *Z. Zellforsch.* 112, 129–138.

Dadd, R. H. (1960a). Observations on the palatability and utilisation of food by locusts, with particular reference to interpretation of performances in growth trials on synthetic diets. *Entomol. Exp. Appl.* 3, 283–304.

Dadd, R. H. (1960b). The nutritional requirements of locusts. I. Development of synthetic diets and lipid requirements. *J. Insect Physiol.* 4, 319–347.

Davis, N. T. (1987). Neurosecretory neurons and their projections to the serotonin neurohemal system of the cockroach *Periplaneta americana* (L.), and identification of mandibular and maxillary motor neurons associated with this system. *J. Comp. Neurol.* 259, 604–621.

Devitt, B. D., and Smith, J. J. B. (1985). Action of mouthparts during feeding in the dark-sided cutworm, *Euxoa messoria* (Lepidoptera: Noctuidae). *Can. Entomol.* 117, 343–349.

Djamin, A., and Pathak, M. D. (1967). Role of silica in resistance to Asiatic rice borer, *Chilo suppressalis* (Walker), in rice varieties. *J. Econ. Entomol.* 60, 347–351.

Dogra, G. S., and Ewen, A. B. (1971). Effects of severance of stomatogastric nerves on egg-laying, feeding, and the neuroendocrine system in *Melanoplus sanquinipes*. *J. Insect Physiol.* 17, 483–489.

Eaton, C. B. (1942). The anatomy and histology of the proventriculus of *Ips radiatae* Hopkins (Coleoptera: Scolytidae). *Ann. Entomol. Soc. Am.* 35, 41–49.

Eidmann, H. (1924). Untersuchungen über die Morphologie und Physiologie des Kaumagens von *Periplaneta orientalis* L. *Z. Wiss. Zool.* A122, 281–307.

Evans, M. E. G. (1964). A comparative account of the feeding methods of the beetles

Nebria brevicollis (F.) (Carabidae) and *Philonthus decorus* (Grav.) (Staphylinidae). *Trans. R. Soc. Edinb.* 66, 91–109.

Evans, M. E. G., and Forsythe, T. G. (1985). Feeding mechanisms, and their variation in form, of some adult ground-beetles (Coleoptera: Carabidae). *J. Zool. London* 206, 113–143.

Fontaine, A. R., Olsen, N. Ring, R. A., and Singla, C. L. (1991). Cuticular metal hardening of mouthparts and claws of some forest insects of British Columbia. *J. Entomol. Soc. Brit. Columbia* 88, 45–55.

Forsythe, T. G. (1982). Feeding mechanisms of certain ground beetles (Coleoptera: Carabidae). *Coleopt. Bull.* 36, 26–73.

Gangwere, S. K. (1965). Food selection in the oedipodine grasshopper *Arphia sulphurea* (Fabricius). *Am. Midl. Nat.* 74, 67–75.

Gangwere, S. K., Evans, F. F., and Nelson, M. L. (1976). The food-habits and biology of Acrididae in an old-field community in southeastern Michigan. *Great Lakes Entomol.* 9, 83–123.

Gardiner, B. G., and Khan, M. F. (1979). A new form of insect cuticle. *Zool. J. Linn. Soc.* 66, 91–94.

Godfrey, G. L., Miller, J. S., and Carter, D. J. (1989). Two mouthpart modifications in larval Notodontidae (Lepidoptera): their taxonomic distributions and putative functions. *J. New York Entomol. Soc.* 97, 455–470.

Griss, C. (1990). Mandibular motor neurons of the caterpillar of the hawk moth *Manduca sexta*. *J. Comp. Neurol.* 296, 393–402.

Griss, C., Simpson, S. J., Rohrbacher, J., and Rowell, C. H. F. (1991). Localization in the central nervous system of larval *Manduca sexta* (Lepidoptera: Sphingidae) of areas responsible for aspects of feeding behaviour. *J. Insect Physiol.* 37, 477–482.

Gronenberg, W., Tautz, J., and Hölldobler, B. (1993). Fast trap jaws and giant neurons in the ant *Odontomachus*. *Science* 262, 561–563.

Hamamura, Y., Hayashiya, K., Naito, K.-I., Matsuura, K., and Nishida, J. (1962). Food selection by silkworm larvae. *Nature (London)* 194, 754–755.

Hillerton, J. E., and Vincent, J. F. V. (1982). The specific location of zinc in insect mandibles. *J. Exp. Biol.* 101, 333–336.

Hillerton, J. E., Reynolds, S. E., and Vincent, J. F. V. (1982). On the indentation hardness of insect cuticle. *J. Exp. Biol.* 96, 45–52.

Hillerton, J. E., Robertson, B., and Vincent, J. F. V. (1984). The presence of zinc or manganese as the predominant metal in the mandibles of adult, stored-product beetles. *J. Stored Product Res.* 20, 133–137.

Houston, W. W. K. (1981). The life cycles and age of *Carabus glabratus* Paykull and *C. problematicus* Herbst (Col.: Carabidae) on moorland in northern England. *Ecol. Entomol.* 6, 263–271.

Isely, F. B. (1944). Correlation between mandibular morphology and food specificity in grasshoppers. *Ann. Entomol. Soc. Am.* 37, 47–67.

Kaufmann, T. (1971). Biology and Ecology of *Melanoplus borealis* (Orthoptera: Acrididae) in Fairbanks, Alaska with special reference to feeing habits. *Michigan Entomol.* 4, 3–13.

Kent, K. S., and Hildebrand, J. G. (1987). Cephalic sensory pathways in the central nervous system of larval *Manduca sexta* (Lepidoptera: Sphingidae). *Philos. Trans. R. Soc.* 315B, 1–36.

Köhler, H.-R., Alberti, G., and Storch, V. (1991). The influence of the mandibles of Diplopoda on the food—a dependence of fine structure and assimilation efficiency. *Pedobiologia* 35, 108–116.

Le Berre, J. R., and Louveaux, A. (1969). Equipementes sensoriels des mandibules de la larve du premier stade de *Locusta migratoria* L. *C. R. Acad. Sci. Paris* 268, 2907–2910.

Louveaux, A. (1972). Equipementes sensoriels et système nerveux périphérique des pièces buccales de *Locusta migratoria* L. *Insectes Sociaux* 19, 359–368.

Ma, W. C. (1972). Dynamics of feeding responses in *Pieris brassicae* Linn. as a function of chemosensory input: a behavioral, ultrastructural and electrophysiological study. *Meded. Lanbouwhogesch. Wageningen* 72–11, 162 pp.

Patterson, B. D. (1984). Correlation between mandibular morphology and specific diet of some desert grassland Acrididae (Orthoptera). *Am. Midl. Natl.* 111, 296–303.

Penzlin, H. (1985). Stomatogastric nervous system. In: Kerkut, G. A., and Gilbert, L. I. (eds.) *Comprehensive Insect Physiology, Biochemistry and Pharmacology*, vol. 5. Pergamon Press, Oxford. pp. 371–406.

Raupp, M. J. (1985). Effects of leaf toughness on mandibular wear of the leaf beetle, *Plagiodera versicolora*. *Ecol. Entomol.* 10, 73–79.

Rowell, C. H. F., and Simpson, S. J. (1992). A peripheral input of thoracic origin inhibits chewing in the larva of *Manduca sexta*. *J. Insect Physiol.* 38, 475–483.

Schachtner, J., and Bräunig, P. (1993). The activity pattern of identified neurosecretory cells during feeding behaviour in the locust. *J. Exp. Biol.* 185, 287–303.

Schofield, R., and Lefevre, H. (1989). High concentrations of zinc in the fangs and manganese in the teeth of spiders. *J. Exp. Biol.* 144, 577–581.

Schofield, R. M. S., Lefevre, H. W., Overley, J. C., and MacDonald, J. D. (1988). X-ray microanalytical surveys of minor element concentrations in unsectioned biological samples. *Nucl Instrum Methods Physics Res* B30, 398–403.

Seath, I. (1977a). Sensory feedback in the control of mouthpart movements in the desert locust *Schistocerca gregaria*. *Physiol. Entomol.* 2, 147–156.

Seath, I. (1977b). The effects of increasing load on electrical activity in the mandibular closer muscles during feeding in the desert locust, *Schistocerca gregaria*. *Physiol. Entomol.* 2, 237–240.

Sinoir, Y. (1969). Le rôle des palpes et du labre dans le comportement de prise de nourriture chez la larve du criquet migrateur. *Ann. Nutrit. Aliment.* 23, 167–194.

Skaife, S. H. (1955). *Dwellers in Darkness*. Longmans, London.

Spence, J. R., and Sutcliffe, J. F. (1982). Structure and function of feeding in larvae of *Nebria* (Coleoptera: Carabidae). *Can. J. Zool.* 60, 2382–2394.

Thomas, J. G. (1966). The sense organs on the mouth parts of the desert locust (*Schistocerca gregaria*). *J. Zool.* 148, 420–448.

Wallin, H. (1988). Mandible wear in the carabid beetle *Pterostichus melanarius* in relation to diet and burrowing behaviour. *Entomol. Exp. Appl.* 48, 43–50.

Welton, M. N. (1988). A morphometric analysis of mandibular teeth in *Folsomia* (Collembola: Isotomidae). *Zool. J. Linn. Soc.* 94, 99–109.

Wheater, C. P., and Evans, M. E. G. (1989). The mandibular forces and pressures of some predacious Coleoptera. *J. Insect Physiol.* 35, 815–820.

Zacharuk, R. Y. (1962). Sense organs of the head of larvae of some Elateridae (Coleoptera): Their distribution, structure and innervation. *J. Morphol.* 111, 1–22.

Zacharuk, R. Y. (1985). Antennae and sensilla. In: Kerkut, G. A., and Gilbert, L. I. (eds.) *Comprehensive Insect Physiology, Biochemistry and Pharmacology*, vol. 6. Pergamon Press, Oxford, pp. 1–69.

Zacharuk, R. Y., and Albert, P. J. (1978). Ultrastructure and function of scolopophorous sensilla in the mandible of an elaterid larva (Coleoptera). *Can. J. Zool.* 56, 246–259.

2

Mechanics of Food Handling by Fluid-Feeding Insects

J. G. Kingsolver and T. L. Daniel

2.1. Introduction

Many insects meet their nutritional requirements through an entirely liquid diet, by feeding on fluids ranging from plant nectars and phloem to mammalian blood. A rich tradition in comparative anatomy and behavior has detailed (a) the diversity of mouthparts found in nectar-, blood-, and phloem-feeding insects and (b) the variety of behaviors used during the feeding process. This tradition has placed less emphasis on the common features in fluid feeding that are likely to result from the basic physical processes of moving fluid from the external environment to the inside of the insect.

An understanding of fluid feeding will require knowledge of both (a) the fundamental physical processes involved in fluid feeding and (b) the morphological, behavioral, and resource diversity exhibited by various insect taxa. In this review, we shall first take an engineering perspective on fluid feeding that analyzes the mechanics of fluid flow through insect mouthparts. Because most of the basic physics involved are well understood, we now have the beginnings of a quantitative theoretical framework for fluid feeding. We shall emphasize those key aspects of fluids, morphology, and behavior that determine fluid and nutrient uptake rates during feeding. We shall then consider in some detail the mechanics of feeding in several major groups of insects feeding on nectar, blood, or phloem. Here we shall emphasize which aspects of the bewildering diversity found in mouthparts are likely to be of functional significance, and we will evaluate the adequacy of the engineering models in those few study systems in which sufficient quantitative data are available. Finally, we shall identify areas for future research on fluid feeding that appear particularly interesting and fruitful, including (a) the fluid mechanics of disease transmission of blood parasites by blood-feeding insects and (b) the mechanics of piercing mammalian skin and plant tissues.

2.2. Mechanical Aspects of Fluid Feeding

Every organism that feeds on a fluid must rely on some force that drives the motion of fluid through some feeding structure. The flow rate of that fluid is determined by three principal features: (1) the physical properties of the moving fluid, (2) the geometry of the feeding structures, and (3) the nature of the force produced. To understand how the design of any feeding structure may affect the nutrient uptake rate for any organism, we must first examine the physical rules that govern each of these features. In turn, we can establish approximate models for analyzing fluid intake rates and, accordingly, nutrient acquisition rates.

2.2.1. The Moving Fluid: Physical Properties of Liquids

Insects feed on a wide range of liquids that may be simple solutions of sugars such as nectar or more physically enigmatic fluids that consist of protein solutions (e.g., saliva and tears) or suspensions of particles (e.g., blood). All of these fluids are classically characterized as substances with a particular density that deform continuously under an applied force. The rate at which they deform is directly proportional to their viscosity, a measure of the tendency of fluids to resist shearing (deformation). Table 2.1 shows some values for the viscosity and density of fluids that are relevant to insect feeders. Physical factors (such as temperature) that affect the viscosity of these fluids may play important roles in

Table 2.1. Physical properties of fluids encountered by "liquivorous" insects

Fluid	Temperature (°C)	Density (kg·m^{-3})	Viscosity $(\text{kg·m}^{-1}\text{·sec}^{-1})$	Surface tension (N·m^{-1})
Water	10	999.7	0.0013	0.074
	20	998.2	0.0010	0.073
	30	995.7	0.0008	0.071
	40	992.2	0.0006	0.070
"Nectars"				
20% sucrose	15	1082	0.0025	
	20	1081	0.0019	
	25	1079	0.0017	
40% sucrose	15	1178	0.0075	
	20	1176	0.0062	
	25	1174	0.0051	
60% Sucrose	15	1289	0.0795	
	20	1286	0.0585	
	25	1284	0.0440	
Xylem	20	1000	0.0010	
Phloem	20	1010	0.002	
Blood				
0.4 hematocrit	37	1025	0.001–0.003	

determining the rate of nutrient uptake by insect feeders. For example, for any liquid, an increase in its temperature will yield a decrease in its viscosity. For a biologically reasonable change in temperature, the magnitude of viscosity change is not likely to be much greater than 50%: thus the viscosity of water varies from 1.7 cP at 0°C to 0.66 cP at 40°C. Similar temperature dependence is seen in most aqueous solutions.

The concentration of dissolved or suspended nutrients also affects the viscosity of a liquid. The sensitivity of fluid viscosity to nutrient (solute) concentration is often quite dramatic (see Table 2.1), usually showing a strong exponential relationship. This exponential dependence of viscosity on concentration holds for a variety of biological liquids, including nectar, blood, and phloem, and has important implications for liquid feeding (see below).

For so-called Newtonian fluids, such as nectar and phloem, the viscosity depends only on temperature and concentration. In contrast, for non-Newtonian fluids such as blood, viscosity also depends upon the shearing they experience. In this case, viscosity is some function of how fast the fluid is deformed, or, more precisely, its rate of shear (the spatial gradient of its velocity). To illustrate some of the complexities of non-Newtonian fluids, a brief description of the viscosity of vertebrate blood is useful.

From a mechanical viewpoint, blood is a suspension of small particles (primarily red blood cells) in a dilute aqueous (proteinaceous) solution (plasma). The viscosity of vertebrate blood decreases with increasing shear rates (Fung, 1981; Goldsmith et al., 1989). This behavior arises from the way in which blood cells move within a vessel or some feeding apparatus. Blood cells tend to aggregate and adhere to one another (Fahraeus, 1929; Fung, 1981; Goldsmith et al., 1989). The less aggregated the cells are, the more easily the suspension of particles can move. This aggregation is easily broken up by shears in flowing blood. Thus, the faster the flow and the greater the shear, the smaller the aggregations will be, and the lower the observed viscosity of the blood. At very low shears, these aggregations become so pronounced that blood nearly solidifies, yielding a very high viscosity. Below a certain minimum rate of shearing (the yield stress), there is no fluid movement at all: here the blood acts like a solid rather than a liquid.

The viscosity of blood can also vary with the size of the tube through which it flows (Fig. 2.1). For tubes less than 100 μm [$\sim$6 ln (μm)] in diameter, viscosity of human blood decreases with decreasing tube diameter, reaching a minimum at tube diameters of about 6 μm [$\sim$2 ln (μm)]. As tube diameter decreases below 6 μm, viscosity rises dramatically. This remarkable mechanical behavior, known as the *Fahraeus–Lindqvist effect* (see Goldsmith et al., 1989), results from the migration of the flexible red cells to the center of the tube, which reduces the average viscosity of the blood for tubes between 6 and 100 μm. Tube diameters below 6 μm approach the size of a single red cell, resulting in large increases in blood viscosity. Thus, when an insect with a tube diameter similar to the diameter of vertebrate red blood cells feeds on blood, much of the

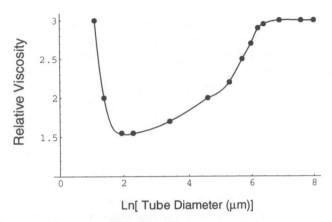

Figure 2.1. The relative viscosity of blood (relative to water, with relative viscosity = 1.0) plotted against the natural log of the tube diameter in which it flows using data from Fung (1981) and Goldsmith et al. (1989). A hematocrit of 0.4 is assumed for this plot.

stress generated by feeding activity is devoted to significant deformation of the cells themselves (Fig. 2.1). The stresses associated with fluid flow per se are considerably smaller than these deformation stresses. These effects of tube size and shearing rate on blood viscosity have some interesting implications for blood-feeding insects (see Section 2.4).

2.2.2. The Moving Fluid: Geometric Considerations

Armed with an understanding of the viscosity of relevant fluids, we must now determine how that parameter, along with the geometry of the feeding structures and the forces generated by the feeding process, affects rates of nutrient uptake. The simplest situation involves feeding with a hollow tube, much like a straw. Engineering principles (Batchelor, 1967; Vogel, 1981) show that, for Newtonian fluids, there is a distinct relationship between the volume flow rate of fluid (Q) and the pressure difference generated by the feeding process (ΔP), the dimensions of the feeding apparatus (radius r and length L), and the viscosity of the fluid (μ):

$$Q = \frac{\pi r^4 \, \Delta P}{8\mu L} \tag{1}$$

This equation, known as the *Hagen–Poiseuille relationship*, shows that, for a given driving force (ΔP), flow rate is inversely proportional to the viscosity of the fluid and the length of the food canal. Note that flow rate is also extremely sensitive (a fourth-power dependence) to the radius of the food canal. Thus, simple isometric scaling arguments suggest that larger organisms will feed at significantly greater flow rates.

While the Hagen–Poiseuille relation provides a simple way to quantify the flow rates that result from a particular driving force and a particular morphology, its applicability is limited by four key assumptions that underlie its derivation: (1) The flow is laminar, (2) the flow is "fully developed," (3) the fluid velocity is constant over time at any point in the tube, and (4) the flow is Newtonian. Any attempt to quantify or model flow rates must first ask whether any of these assumptions are violated. We introduce a few additional relationships that provide indices for the extent to which each of these assumptions apply.

The first assumption asserts that the flow is laminar. Thus turbulent phenomena such as eddies and whirls in the fluid are presumed to be absent and the fluid moves in an orderly manner, following the geometry of the tube. Eddies are inertial processes that are triggered by instabilities in the flow, irregularities in the surfaces over which fluids move (bumps, knobs, rough surfaces), and abrupt changes in tube geometry (expansions, contractions, curves). These are dissipated by the action of viscosity. Thus, in extremely viscous flows, such flow structures are rarely seen. An index for the likelihood of the presence of turbulence is called the *Reynolds number* (Re), which measures the importance of inertial forces relative to viscous forces in a moving fluid:

$$Re = \rho U D / \mu \qquad\qquad [2]$$

where ρ is the density of the fluid, U is the average velocity; D is the diameter of the tube, and μ is the viscosity of the fluid. When the Reynolds number is much less than 1, viscous forces dominate and inertial events are negligible. When the Reynolds number is much greater than 1, the flow may well be turbulent. For flow in straight smooth tubes, Reynolds numbers in excess of about 2000 show turbulent events. Any of the above geometric irregularities will decrease the Reynolds number at which turbulence will be seen.

For all the insect feeders we see, the Reynolds numbers are quite small. For example, a fairly large butterfly such as *Danaus* feeds at about 1 μl·sec^{-1} with a food canal diameter of about 70 μm. Using water for the lowest estimate of fluid viscosity yields a Reynolds number of about 10, far lower than the critical Reynolds numbers for turbulence. Smaller creatures, those that feed on more viscous liquids, or those that feed more slowly will all have lower Reynolds numbers.

Such low values for the Reynolds numbers have two additional implications. First, they indicate that viscous forces are so great that surface irregularities or tube curvatures have little effect on the flow, because they are incapable of introducing eddies or other deviations from the assumption of laminarity. Accordingly, the Hagen–Poiseuille relationship should hold for the diversity of tubular mouth parts we encounter. Second, rapid changes in the diameter of the feeding apparatus will have little qualitative effect on the flow. Thus, as fluid enters the feeding apparatus, it rapidly assumes the flow seen through the entire length of the tube. Thus, the second assumption that the flow is fully developed is likely to be satisfied.

The third assumption asserts that the fluid velocity at any point in the tube is at steady state (i.e., constant velocity), and, accordingly, any forces required for *acceleration* of the fluid are absent. Clearly, feeding among insects is a cyclic process, with periodic filling and emptying of the feeding apparatus. As such, the velocity cannot be constant. The crucial question, however, is: how large are the forces associated with fluid accelerations relative to the forces associated with fluid shearing? The Reynolds number provides only a partial answer to this question: we know that at low Reynolds numbers, inertial processes are negligible. But, if the time scale of velocity changes is very rapid, then even at low Reynolds numbers, accelerational forces may be important. We use the Strouhal number (St) to assess the relative importance of such forces and the applicability of the Hagen–Poiseuille relationship:

$$St = \omega D/U \tag{3}$$

where ω is the frequency of velocity change and D and U are as above. This index measures two velocity scales: one associated with steady flow (U) and the other associated with periodic flow (ωD). When the Strouhal number is much less than 1, accelerational forces are negligible; forces associated with steady flow dominate. From estimates of the frequency and volumes associated with lepidopteran feeders, we find that Strouhal numbers are generally less than 0.01 (Daniel et al., 1989).

The very low values for the Strouhal and Reynolds numbers suggest that the Hagen–Poiseuille relationship will provide an accurate relationship between morphology, forcing, and fluid properties for feeding structures with simple geometries (e.g., smooth hollow tubes). Even for insects with mouthparts that are far more complicated, we shall see that such a relationship can provide useful qualitative insights.

Finally, for insects feeding on non-Newtonian fluids such as vertebrate blood, the relationship between pressure, flow rate, tube morphology, and fluid properties is more complex (see e.g., Fung, 1981), and a derivation of this relationship is beyond the scope of this chapter (but see Section 2.4).

2.2.3. Moving the Fluid: Physical Forcing

The pressure difference ΔP in equation [1] represents the processes that drive the fluid through and around the myriad of mouthparts encountered. What are the mechanical determinants of this driving force? There are three principal mechanisms that can yield a driving force: (1) active muscle contraction, (2) capillary forces, and (3) external pressures. It is possible that all three may play a role in feeding. Indeed, the relative importance of these has not been fully explored. Below, we describe basic rules that govern each of these processes.

2.2.3.1. MUSCLES

Many insects use some sort of muscular pump to drive fluid through their mouthparts. In sucking insects, contraction of skeletal muscles within the head capsule actively expands an internal cavity. The pressure drop (ΔP) resulting from this action is simply the muscle force divided by the cross-sectional area of muscle. One might imagine that the muscles might produce some constant force, and thus a constant pressure difference, up to some maximum force that could be generated by the relevant muscles (e.g., Kingsolver and Daniel, 1979).

However, this approach ignores an essential feature about contracting muscle: the contractile force is inversely related to the speed of shortening. This inverse relationship between force and velocity is given explicitly by Hill's equation (named after A. V. Hill: see McMahon, 1984) simplified below as:

$$F = \frac{T_0(V_{max}-V)}{V_{max}+4V} \qquad [4]$$

where V is the shortening velocity, T_0 is the maximum isometric tension, and V_{max} is the maximum shortening velocity. Both T_0 and V_{max} vary considerably among different organisms and among different muscle types within any organism. While specific values for these parameters are not available for those muscles that drive the feeding process, we can use values for insect skeletal muscle as an approximation: one length/sec for the maximum shortening velocity and a maximum isometric stress (force per unit area) of 80 kPa (Prosser, 1973; McMahon, 1984).

The above relationship tells us that the force driving flow through the feeding apparatus depends upon the speed of movement of internal structures. At the same time, we know from the Hagen–Poiseuille relationship above that the speed of fluid motion is directly related to the applied force. These two conditions must be simultaneously satisfied in order for the predicted flows and feeding rates to be both physically and physiologically correct.

2.2.3.2. CAPILLARY FORCES

Surface tension or capillarity may play a crucial role in forcing fluid through or around the feeding structures of a variety of insects. In butterflies, for example, the initial immersion of the food canal into a nectar source presents an interface between the liquid, air, and the solid surface of the food canal. Just as with capillary tubes immersed in water, a force arises that may draw the fluid up the tube. Bees and flies may immerse tongues of complex morphology into nectar or other fluid food sources. The interstices created by the presence of cuticular hairs and the like present a surface over which capillary forces may draw fluid up the structure.

Some discussion of the physical basis of capillarity may be useful for understanding its potential importance for fluid feeding. Capillarity results from cohesive forces between the molecules at the interface of two fluids. The surface of the interface takes on a shape that minimizes the total surface energy of the system and balances all other forces acting on the system. This surface energy results in a pressure difference across the interface. For the simple case of an air-filled hollow tube immersed at one end in water, the pressure difference draws fluid up the tube. The magnitude of that pressure difference depends upon the geometry of the tube (its radius, r) and a surface tension coefficient (γ) that depends on the pair of fluids (here, air and nectar or water):

$$\Delta P = 2\gamma \cos \theta / r \qquad\qquad [5]$$

where θ is the contact angle between the fluid and the surface of the tube.

The pressure difference increases in direct proportion to the surface tension coefficient, γ. Water and simple sugar solutions at an interface with air have very large γ relative to most liquids (for example: 0.073 $N \cdot m^{-1}$ for air–water interfaces; 0.03 $N \cdot m^{-1}$ for air–oil interfaces; 0.02 $N \cdot m^{-1}$ for air–alcohol interfaces). However, in contrast to viscosity, γ does not depend strongly on sucrose concentration. For example, increasing sucrose concentration from 0% to 50% increases γ by only about 5%. The physical units of γ [$N \cdot m^{-1}$ (i.e., force per length)] are also instructive: they imply that the force resulting from capillarity will be directly proportional to some length in the system. For a hollow tube, that length is its inner radius; for a hairy glossa, it is the length along which there is contact between the free surface of nectar and the solid surface of the tongue.

While the capillary forces increase in direct proportion to some critical length scale, equation [5] shows that the pressure difference driving the flow declines with increasing tube radius (more generally, the radius of curvature of the structure). Indeed, because the pressure is defined as a force acting over some area (length squared), this inverse dependence is always seen. While ever smaller tube radii lead to ever greater forcing, the Hagen–Poiseuille relationship shows that flows will still decrease dramatically.

Thus, organisms that use capillary action for driving the relevant flows face an intriguing tradeoff: capillarity requires small radii for effective forcing, but the flow rate and removal of liquid requires large radii to offer effectively low resistance. The consequences of such tradeoffs have not been fully explored for such insects as euglossine bees (see Section 2.3) or other insect feeders relying on surface tension [but see Kingsolver and Daniel (1983), for a discussion of such consequences for hummingbird feeding).

The contact angle (θ) depends on the wettability of the solid surface upon which fluid adheres. Highly wettable surfaces have very low contact angles ($<$ $\pi/2$), whereas nonwettable (or hydrophobic) surfaces may have large contact angles. It is interesting to note that insect cuticle is commonly viewed as a surface covered with highly hydrophobic organic material (various lipids and waxes). Such hydrophobicity would counteract any effective use of capillarity in feeding on liquids. We suspect that those insects utilizing capillarity for feeding must have cuticular surface modifications that augment their wettability. While there is, at present, no information about the organic makeup of the extracuticular covering of mouthparts, we predict that the relevant structures of such feeders are coated with polyanionic proteins to augment their wettability.

2.2.3.3. EXTERNAL PRESSURE

While sucking insects may actively generate the necessary pressure differences (ΔP) by muscular activity, the food source itself may be positively or negatively pressurized. To understand whether such pressurization is important in the feeding process, one can use the Hagen–Poiseuille relationship along with observed feeding rates and morphologies to calculate the pressure differential. For example, Daniel and Kingsolver's (1983) calculations show that pressure differential for three blood-feeding insects is well in excess of the capillary blood pressures [~3 kPa (Farrell, 1991)] of their hosts (the mosquito, *Aedes aegypti*, ~8 kPa; the bedbug, *Cimex lectularius*, ~80 kPa; and the louse, *Pediculus humanus*, ~20 kPa). With the possible exception of *A. aegypti* (and perhaps other mosquitoes), the pressurization of the food source is unlikely to be an important determinant of the blood-feeding process.

Unlike pressures in blood supplies, plant fluids may be under extremely high pressures. Phloem, for example, may be positively pressurized with values ranging from 0.2 to 1 MPa. As we shall see, these external pressures contribute importantly to feeding rates in phloem-feeding insects. Xylem, on the other hand, may be under strong negative pressures, often exceeding −1 MPa (Scolander et al., 1965). The implications of these large negative pressures to the mechanics of feeding are poorly understood: indeed, some novel physiological adaptations may be required for effective xylem feeding (see Section 2.5).

2.2.4. Putting the Pieces Together

The physical principles discussed above provide the basic building blocks for developing quantitative models of fluid feeding. The main purpose of such models is to make predictions about rates of nutrient uptake during feeding, as a means of exploring how fluid properties and aspects of the feeding structures determine nutrient uptake rates. Here we illustrate how the general physical principles outlined above may be used to develop models for specific types of insect feeders.

Nutrient uptake rate can be examined over several time scales. For the time during which there is bulk motion of fluid into the animal, the instantaneous nutrient uptake rate is the product of the volume flow rate and the nutrient concentration. But, for a variety of reasons, flow rate may vary in time, and it is therefore important to identify the average nutrient uptake rate for some relevant time scale.

Our discussion above shows that there are several interacting processes that determine feeding rates (Fig. 2.2): (1) the mechanics of fluid flow through the feeding structure and (2) the mechanics of the driving forces that produce the fluid movement. The balance between driving forces and the resulting fluid flow can be specified, and, using principles of conservation of mass, momentum, and energy, a mathematical model predicting nutrient intake rates can be derived. To illustrate this general approach, we derive a specific model for suction feeding in Lepidoptera in the Appendix. The spirit of this analysis is simply one of

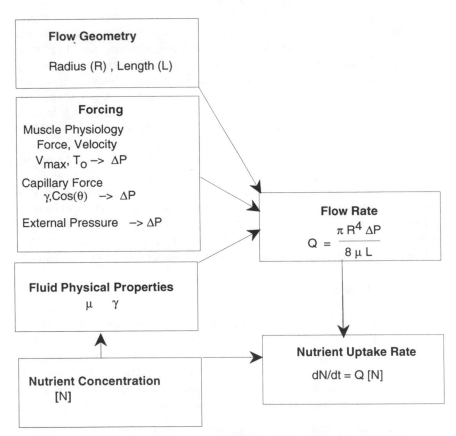

Figure 2.2. This flow diagram shows how the various parameters and physical relationships interact to determine nutrient uptake rate by fluid feeders. The model in the Appendix uses this approach in predicting feeding rates for nectar feeders.

substituting all appropriate relationships into a single expression for nutrient uptake rate.

2.2.5. Mechanical Access to Food

One final issue about the mechanics of feeding concerns access to the food source. For nectar-feeding insects, that access is limited solely by the length of the feeding structures relative to the length of the corolla in which nectar may be located. For those insects that obtain food from *inside* plant or animal structures, the force required to penetrate these structures may limit their access. For example, Hoffman and McEvoy (1986) showed that food access for spittlebugs is limited by the mechanical barriers of their host plants. The strength or toughness of such barriers is clearly a limitation to access.

While there is too little information about the mechanics of penetrating host tissues, we can develop a few indicators of how the structure of the feeding apparatus may limit access. One such limitation is the possibility that the proboscis may buckle in response to the force required for penetration. Buckling occurs when the force along the proboscis exceeds its structural strength. This strength is determined by the geometry of the proboscis and the stiffness of its cuticle. In general, the longer and narrower the proboscis and the lower its cuticular stiffness, the more likely it will be to buckle. Details of the mechanics of buckling of tubular structures are discussed in greater detail in Wainwright et al. (1976), who point out that many other factors (end support and specific buckling mode, for example) may strongly affect the likelihood of failure by this scheme. It is interesting to note that while sharp, narrow mouthparts would likely require lower forces for penetration than would blunt, wide ones, they would be more likely to buckle. This tradeoff in mouthpart design has not been examined.

Penetration forces may also be limited by the muscles available to accomplish such activity. Here, again, there is too little information about the recruitment of muscles and their physiological properties to carefully evaluate such a limitation.

2.3. Nectar Feeding

Many insects utilize plant nectars as a food source. In some cases (e.g., mosquitoes, muscid flies, blowflies), nectar represents one of a variety of resources used (e.g., Dethier and Rhoades, 1954; also see Chapter 8). Other insect groups obtain energy almost exclusively from plant nectars. Here we shall focus our discussion on two groups of specialized nectivores that use quite different mechanisms for nectar feeding: butterflies and bees.

2.3.1. Lepidoptera: Simple Suction Tubes

Adult Lepidoptera obtain energy almost entirely by feeding on plant nectars, plant saps, rotting fruit, and similar resources. Because plant nectars are primarily a mixture of simple sugars and water (Baker and Baker, 1982), the nectar-feeding success of butterflies and moths can be readily quantified in terms of rates of energy intake. Because a single flower generally contains a small amount of nectar relative to the storage capacity of an insect, it is useful here to distinguish two levels of analysis of the mechanics and energetics of nectar feeding: (1) the determinants of fluid and energy intake rates during feeding on a single nectar source (e.g., an individual flower or floret) and (2) the determinants of foraging success during a single foraging bout or meal.

The mouthparts of a butterfly or moth are highly modified relative to more generalized insects, with the mandibles, labium, and maxillary palps greatly reduced. The proboscis itself consists of two elongated maxillary galeae that are connected both ventrally and dorsally to form the food canal (Fig. 2.3) (Snod-grass, 1935; Eastham and Eassa, 1955). During feeding, the proboscis is uncoiled and its tip is extended into the nectar source [see Krenn (1990) for a description of the mechanism of uncoiling], and nectar is drawn through the food canal by the action of a muscular suction pump.

The suction pump in Lepidoptera is a large, muscularized cavity that is a combination of the cibariopharyngeal and buccopharyngeal cavities (Eastham and Eassa, 1955); for simplicity we shall use the term *cibarium* to refer to this compound structure in the discussion below. In *Pieris rapae,* for example, the cibarium is approximately cylindrical, with the long axis oriented dorsoventrally in the head capsule (Fig. 2.3) (Eastham and Eassa, 1955; Daniel et al., 1989). The floor of the cibarium is heavily sclerotized, whereas the roof is flexible and invested with various muscles that expand and contract the cibarium and control the flow of nectar through the system. Contraction of the four dilator muscles that insert on the cibarium roof is primarily responsible for generating the suction that drives movement of the nectar into the cibarium. Compressor muscles force nectar from the cibarium into the esophagous to complete a single sucking cycle.

Using the general modeling approach outlined earlier, one can develop a model describing nectar intake during a single sucking cycle that has two main parts: (1) the fluid mechanics of nectar flow through the food canal and (2) the mechanics of muscle contraction producing cibarial expansion (see the Appendix for a derivation of the model). Using measurements of the geometry of the food canal, the cibarium, and the cibarial muscles for *Pieris rapae* butterflies, along with information on nectar viscosity and density, we can use the model to predict how nectar (sucrose) concentration, morphology of the feeding apparatus, and physiological characteristics of the muscles affect rates of fluid and energy intake during sucking.

Several general results from such model simulations are of interest here. First, at any nectar concentration, increasing the food canal radius increases the flow and energy intake rates (Fig. 2.4). Note that the relationship between radius and intake rate is not simply predicted by Hagen–Poiseuille's equation (equation [1]),

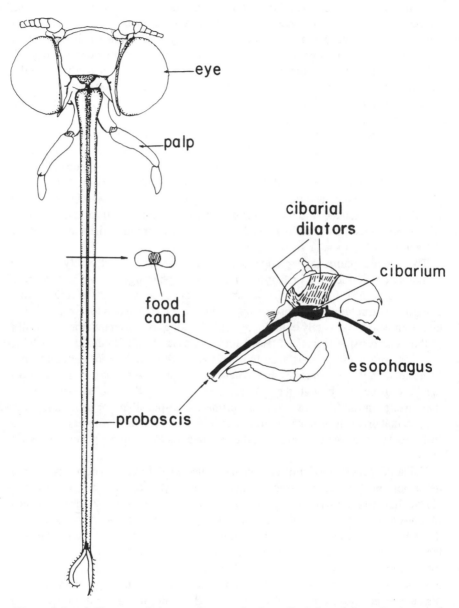

Figure 2.3. A diagram of the mouthparts of a butterfly (adapted from Eastham and Eassa, 1955) showing the path followed by nectar (in black). Muscles in the head capsule attach to the top of the cibarial cavity. Their contraction drives nectar through the food canal.

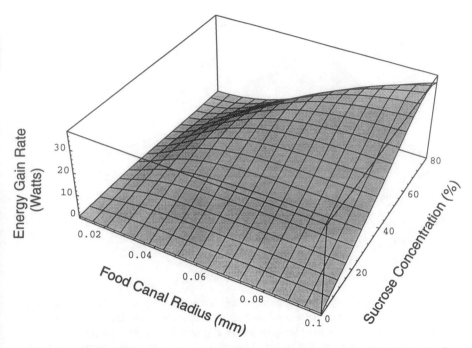

Figure 2.4. Model predictions for the energy gain rate of a nectar-feeding butterfly plotted against food canal radius and sucrose concentration. We used the data in Daniel et al. (1989) for the morphological and physiological parameters for *Pieris rapae* in generating these predictions. Note that as the food canal radius increases, the nectar sugar concentration that maximizes energy gain rates also increases.

and that the relationship differs at different concentrations. This result occurs because the pressure difference generated by the muscles is not constant, as assumed by the Hagen-Poiseuille equation. Second, for any given geometry of the feeding apparatus (food canal, cibarium, and cibarial muscles), the rate of energy intake is maximal at some intermediate concentration (Figs. 2.4 and 2.5); for convenience we shall term this concentration the *optimal concentration*. Note that the location of this optimal concentration varies markedly as the geometry of the system [e.g., food canal radius (Fig. 2.4); radius of the cibarium (Fig. 2.5)] changes. Fourth, for any given food canal radius and muscle cross-sectional area, there is an intermediate cibarial size that yields the maximal rate of energy intake; this maximum always occurs at sucrose concentrations of 35–45% (Fig. 2.5). This overall optimum occurs at the point at which the power output of the cibarial muscles is maximal.

These model predictions emphasize the important interaction between fluid mechanics and muscle mechanics in determining the relationships among morphology, fluid properties, and intake rate. Increasing the food canal radius,

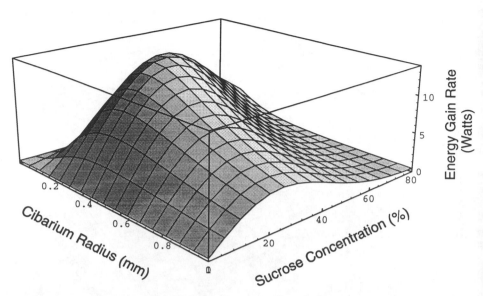

Figure 2.5. Model predictions for the energy gain rate of a nectar-feeding butterfly plotted against cibarium radius and sucrose concentration. We used the data in Daniel et al. (1989) for the morphological and physiological parameters for *Pieris rapae* in generating these predictions. Here, the sucrose concentration that maximizes energy gain rate decreases with increasing cibarium radius. Moreover, there is a unique cibarium radius and sucrose concentration that leads to a maximal energy gain rate.

cibarial radius, or cibarial muscle area each increases the energy intake rate, but the optimal nectar concentration also changes. To achieve the highest rates of energy intake, the relative sizes of the food canal, the cibarium, and the cibarial muscles must be "in tune"; for butterflies and moths with such "well-tuned" geometries, the optimal nectar concentration occurs at 35–45% sucrose. Thus, identifying "optimal" nectar sources requires quantitative information about feeding morphology, unless one assumes that these nectar feeders are optimally designed in the sense used here.

Empirical laboratory studies of five butterfly species have measured average rates of fluid and energy intake during feeding on artificial or real flowers (May, 1985; Pivnick and McNeil, 1985; Boggs, 1988; Hainsworth et al., 1991). In each case, average energy intake rate was highest at concentrations of ~35–45% sucrose, in agreement with the above model predictions for "well-tuned" feeding geometries. Recent data on hawkmoths (*Manduca sexta*) suggests a slightly lower optimal concentration near 30% (Stevenson, 1992); this is of particular interest because of the relatively long food canal in this nectar feeder. All of these empirical studies indicate that the volume rate of flow decreases monotonically with increasing nectar concentrations.

Some data on the morphological and environmental correlates of feeding performance are also available. For example, in *Vanessa cardui* (Hainsworth et al., 1991) and in *Speyeria mormonia* (Boggs, 1988), body mass is correlated positively with nectar intake rate. On the other hand, larger size (and larger food canal radius) of females relative to males is not associated with higher intake rates (Pivnick and McNeil, 1985; Boggs, 1988). One study (Pivnick and McNeil, 1985) shows that nectar intake rates increase with increasing temperature, as expected because of reduced viscosity at higher temperatures. In sum, the available empirical data suggest that the existence and location of the optimal nectar concentration during nectar intake (35–45% sucrose) is in good agreement with model predictions, but agreement with predictions about morphological determinants of feeding is more equivocal. In particular, sexual differences in food canal size are not related to intake rates, and there are likely other determinants of differences in feeding rates between males and females.

Our discussion has emphasized those factors expected to have important effects on feeding rates during ingestion; but it is also of interest to note aspects of morphology and behavior that are *not* likely to be of functional significance in feeding. For example, the proboscis is generally strongly curved during feeding (Eastham and Eassa, 1955), and the degree of curvature may vary among species and with the nectar source. In addition, the surface of the food canal is strongly ribbed, and such sculpturing may vary interspecifically (e.g., Kingsolver and Daniel, 1979). However, details of curvature and surface texture of the food canal are unlikely to have substantial effects on nectar flow rates, because inertial effects are of little importance at the low Reynolds numbers found in insect mouthparts. Similarly, the acceleration and deceleration of the flow as the pump expands and contracts and the frequency of the sucking cycle are unlikely to affect the flow dynamics (see Section 2.2).

To evaluate our current understanding of the determinants of nectar and energy intake rates during feeding with simple suction tubes, two types of studies are now needed. First, we need quantitative morphometric measurements of the food canal, cibarium, and cibarial muscles for a number of species over the wide range of sizes found in Lepidoptera. Such data, currently available for only a single species, can determine whether the feeding apparatuses of butterflies and moths are "well-tuned" in the sense described above. In conjunction with laboratory data on intake rates for a range of sucrose concentrations, these studies can evaluate whether optimal nectar concentrations during feeding are relatively similar for all Lepidoptera, or whether they vary systematically with size or morphology. Second, measurements of instantaneous flow rates in the food canal and pressures in the cibarium are needed to directly test the predicted relationships between muscle mechanics and flow mechanics. All studies to date have measured only average intake rates and have inferred pressure differences from feeding rates; these studies cannot discriminate between the model described above (Daniel et al., 1989) and an earlier, empirically based model (Pivnick and McNeil, 1985). Instantaneous pressure measurements would also give valuable information on the details of the sucking cycle. Such measurements are now in progress for *Manduca sexta* (R. D. Stevenson, *personal communication*).

Thus far, we have considered energy intake rates during feeding on a single nectar source. Laboratory feeding studies with butterflies on large nectar sources show that the meal size (the volume of nectar ingested before the insect becomes unresponsive to food) varies from ~5 ml to ~50 ml; this depends strongly on body size in *Vanessa cardui* and *Thymelicus lineola,* but not in *Speyeria mormonia* (Pivnick and McNeil, 1985; Hainsworth, 1989; Hainsworth et al., 1991; Boggs and Ross, 1993). Meal size also varies with nectar concentration, and it is greatest at concentrations of 25–40% sucrose in the two species measured to date. These meal sizes are considerably larger than the nectar volumes found in flowers and florets used generally by butterflies in the field (e.g., Watt et al., 1974; May, 1985); a single meal will typically involve feeding from 10–20 separate nectar sources. As a result, the energetics of foraging to obtain a complete meal will depend not only on the energy intake rates during feeding on a single nectar source, but also on the energy profitability of different nectar sources and the energetics of movement among nectar sources.

A detailed discussion of the foraging energetics and foraging behavior in nectar feeders is beyond the scope of the present chapter (for such discussion see Stephens and Krebs, 1986; Hainsworth, 1989; Hainsworth et al., 1991), but three points are relevant here. First, the energy profitability of a nectar source for a butterfly will depend on nectar concentration and nectar volume. In the field, many butterflies utilize nectar sources with concentrations of 15–40% sucrose (Watt et al., 1974; Pyke and Waser, 1981; May, 1985), but at least some species feed frequently on concentrations of 40–65% (Pivnick and McNeil, 1985). While this variation represents a fourfold range of concentrations, nectar volumes often vary over one to two orders of magnitude (Watt et al., 1974; May, 1985). As a result, nectar volume may often be a more important determinant of food choice in the field (May, 1985). Second, as the transit time between nectar sources becomes long relative to the time required to ingest nectar at a single source, feeding on more concentrated nectars will increase the average rate of net energy gain (Heyneman, 1983; Pivnick and McNeil, 1985). Third, behavioral choices by foraging butterflies may maximize not rates of energy intake, but rather net energy per meal (Hainsworth, 1989). Recent lab studies show that when given a choice between nectar sources of 35% and 70% sucrose, *Vanessa cardui* will consistently choose the more concentrated nectar when its volume is greater (Hainsworth and Hamill, *unpublished data*); this choice leads to a lower rate of energy intake during feeding, but a greater energy content per meal. The available evidence is equivocal as to whether the energy content in a single meal increases monotonically with nectar concentration, or is maximal at concentrations of 50–60% sucrose (Pivnick and McNeil, 1985; Hainsworth et al., 1991). These considerations emphasize that the energetics of foraging must consider both the mechanical determinants of feeding rates and the ecological context in which foraging occurs.

2.3.2. Bees (Apoidea): Capillary Mechanisms

The proboscis of a bee represents a composite that is more complex structurally and functionally than that found in Lepidoptera. Here we shall emphasize those aspects of the morphology that are likely to be of functional significance [see e.g., Michener and Brooks (1984) for a more detailed comparative anatomical discussion]. Except in colletids, the bee's tongue, or glossa, is relatively long and slender, and bears transverse rows of distally projecting hairs; in many bees these rings (annulli) of hairs may continue nearly all the way around the glossa, and they occur over much of the glossal surface [see Snodgrass (1956) for figures]. The glossa is sheathed anteriorly by the two overlapping maxillary galeae, and posteriorly by the two labial palpi. The food canal through which nectar moves during feeding is the region bounded by the outer surface of the glossa and the inner surfaces of the galeae and labial palpi. In long-tongued bees, the distal tip of the glossa may possess a spatulate flabellum, which contains the opening through which salivary gland secretions are discharged. Proximally, the food canal empties into the functional mouth leading to the cibariopharyngeal sac (Snodgrass 1956). Several pairs of dilator muscles are inserted on the anterior (upper) surface of the sac, between thick bands of transverse compressor muscles; the posterior "floor" of the sac is heavily sclerotized. Contraction of the dilators expand the sac to generate the suction driving nectar from the food canal through the mouth; the compressors then force nectar posteriorly from the sac into the esophagus.

Unlike the Lepidoptera, a "soda straw" analogy for feeding is not appropriate for bees. Careful observations (Muller, 1888; Snodgrass, 1956) and cinefilm analyses (Harder, 1982) of feeding have revealed the detailed kinematics of feeding for long-tongued bees. During feeding, the maxillary galeae and the labial palpi are held relatively motionless, while the glossa is repeatedly extended into the nectar source and then retracted. During extension, nectar moves onto the glossa, and the rings of hairs on the glossa expand outward. The nectar-laden glossa is then retracted until it is enclosed by the galeae and labial palpi; now the nectar is within the food canal. Nectar is then removed from the glossa, and the glossa is again extended, beginning the next cycle. We term the rapid, repeated extension and retraction of the glossal tongue the *licking cycle*. This description emphasizes that there are three different components to the licking cycle for feeding in long-tongued bees: a loading phase, during which the nectar moves onto the glossa during its extension; a retraction phase, during which the glossa is retracted into the galeae and labial palpi; and an unloading phase, during which nectar is removed from the glossa (Kingsolver and Daniel, 1983b).

Capillary or surface tension phenomena may play two major roles during the loading phase. First, if only the distal portion of the glossa is inserted into the nectar during extension, the pressure difference generated by capillary action will be important in determining the rate at which nectar moves up and onto the glossa (Kingsolver and Daniel, 1983b). This may be of particular importance for bees with very long glossae (e.g., many euglossines) and when feeding on flowers with long corolla tubes. In contrast, when feeding on flowers with shorter corolla tubes, much of the glossa may be inserted into the nectar during extension, so that capillarity will have little affect on the rate of nectar loading. Second, the amount of nectar loaded onto the glossa will be strongly affected by surface tension considerations.

The hairs on the glossa have three important mechanical consequences for nectar feeding in bees. First, as noted above, these hairs contribute substantially to the capillary forces that drive nectar movement onto the glossa during the loading phase, by increasing the contact length at the interface between the nectar and air (see Section 2.2). Second, the hairs on the glossa expand outward as loading proceeds, increasing the amount of nectar loaded onto the tongue. A. Bertsch (*unpublished data*) has directly measured this expansion for *Bombus* species, showing that the effective radius of the glossa (including the hairs) increases by about 50% during expansion. This expansion is the result of capillary forces that tend to decrease the total surface energy of the system at the liquid–air interface. Third, during the unloading phase, the hairs in the food canal increase the resistance to fluid movement that results from the suction provided by the cibarial pump. For the low Reynolds numbers found in this situation, the resistance to flow will be proportional to the total surface area of the hairs in contact with the fluid. Thus, the capillary forces that drive nectar movement and influence the amount of nectar during loading scale with the length (or perimeter) of the hairs; but the viscous forces that resist fluid movement during unloading scale with the surface area of the hairs. As a result, we would expect that some intermediate density and length of hairs on the glossa would yield the highest overall rates of nectar flow, such that neither the loading nor the unloading phase is the dominant rate-limiting step.

We can briefly illustrate how a simple model for the average rate during capillary feeding in bees might be developed, using the Hagen–Poiseuille relationship (equation [1]) and the pressure generated by capillary forces (equation [5]). First we must identify the relevant surfaces over which tensional forces are generated. For bees, the interstices among the hairs generate small gaps in which capillary forces arise. We can use an array of many small, parallel, tubes as a plausible first approximation for an otherwise complicated geometry. We then note that as nectar rises through this array during the loading phase, the length term (L) in the Hagen–Poiseuille equation will change in time. Thus, the flow rate of nectar varies in time during the loading phase. Therefore, our measure of nutrient uptake rates must consider the average flow rate during the loading phase.

Using the capillary pressure term (equation [5]) in the Hagen–Poiseuille relationship and conservation of mass to relate the volume flow rate to the speed of fluid in the array, we arrive at an expression for the average fluid uptake rate during the loading phase for a capillary feeders:

$$Q[S] = n\pi R^2 [R \cos \theta/(4\mu T]^{1/2} \qquad [6]$$

where n is the effective number of tubes (related to the number of glossal hairs), each with radius R, and T is duration of the loading phase.

There are several important differences between this result and that for constant-pressure (or constant-flow-rate) nectar feeders. First, the uptake rate depends on the 5/2 power of the tube radius, rather than fourth power as suggested by the Hagen–Poiseuille relationship. This weaker dependence follows from the fact that the pressure differential is inversely proportional to the tube radius. Second, the uptake rate varies as $\mu^{-1/2}$ rather than μ^{-1}. This difference arises from the fact that the flow rate depends upon the extent to which the tubes are filled. This square root dependence on viscosity increases the sucrose concentration at which uptake rates are maximized. (Note that this result (equation [6]) refers only to average intake rate during a single loading phase; the effects of r and μ on intake rate over the entire licking cycle will also be affected by the interactions with the muscular pump during unloading, as in the case for butterflies described above.)

The third difference is the square-root dependence of average intake rate on the duration of the loading phase (T). This result suggests that ever shorter loading phases lead to ever greater uptake rates. Thus a nectar feeder that rapidly dips and retracts its glossa should experience higher uptake rates than one that holds its glossa in nectar until the glossa is filled. In fact, our analyses of capillary feeding of hummingbirds show that the average intake rate is maximal when the duration of the loading phase is equal to the duration of the unloading phase (Kingsolver and Daniel, 1983a).

Laboratory studies using artificial feeders have measured rates of fluid and energy intake for several long-tongued bees feeding over a range of sugar (sucrose) concentrations. For four species of *Melipona* (Apidae), the mean rates of fluid intake were roughly similar for concentrations up to 30–45% sucrose, but declined markedly with increasing concentrations above 45% (Roubik and Buchmann, 1984). More fine-scale data for several bumblebee species (*Bombus*) showed that mean intake rates were constant for sucrose concentrations of 10–40%, then declined rapidly with further increases in concentration (Harder, 1986). Results for *Apis mellifera* showed a similar pattern (Buchmann and Shipman, *unpublished data*). For *Bombus, Apis,* and *Melipona,* mean energy intake rates were highest at concentrations of 50–60% sucrose. As with the butterfly studies, increasing temperature increased rates of intake at all concentrations in *Apis* (Buchmann and Shipman, *unpublished data*), as predicted by the decreased viscosity at higher temperatures. These studies clearly suggest that sucrose concentrations yielding the highest rates of energy intake are higher for long-tongued bees than for butterflies, as predicted by our simple analysis above. These studies also showed that meal size (proventricular load) was largely independent of temperature and concentration in *Apis* (Buchmann and Shipman, *unpublished data*).

Harder (1986) has examined some of the details of the licking cycle in *Bombus*. His laboratory measurements of feeding on concentrations of 20%, 55%, and 70% sucrose showed that licking frequency was 4–5 Hz and independent of concentration, whereas volume per lick decreased with increasing concentration. The reduction in volume per lick at higher concentrations could result from the decreased capillary flow rate onto the glossa during the loading phase due to higher viscosity, and/or from decreased flow from the glossa during unloading. Harder suggests that the independence of intake rate (and, presumably, volume per lick) at lower concentrations (10–40%) indicates that the tongue is entirely saturated to maximum volume during the loading phase, as a result of partial or total immersion of the tongue during loading. In contrast, in nectar-feeding hummingbirds, volumes per lick decrease continuously with increasing concentration, and the tongue grooves are only partially filled with nectar at any concentration (e.g., Ewald and Williams, 1982; Kingsolver and Daniel, 1983; Tamm and Gass, 1986). Harder's interpretation implies that intake rates for bumblebees are limited by licking frequency, such that the loading phase is the rate-limiting step. It would be useful to know why licking frequencies in bees are so low. In contrast, hummingbirds typically use licking frequencies of 10–20 Hz, close to the predicted "optimal" licking frequency maximizing average intake rates at which neither the loading nor unloading phases are rate-limiting (Kingsolver and Daniel, 1983). Additional data on licking frequencies, volumes per lick, and especially the relative durations of the loading and unloading phases of the licking cycle would be especially valuable.

The length of the proboscis determines the flowers or nectar sources on which a bee is able to feed efficiently. Cinefilm analyses of *Bombus* suggest that the length of the glossa is most relevant here for long-tongued bees (Harder, 1982). Among bumblebees, proboscis length was positively correlated with feeding rates (time required to visit a given flower) on flowers with long corolla tubes; conversely, proboscis length and feeding rate were negatively correlated for feeding on short-corolla flowers (Inouye, 1980). Accordingly, differences in proboscis length among coexisting bumblebee species is associated with differences in flower corolla length in the field, and they may result in resource partitioning of the available nectar sources (Heinrich, 1976; Inouye, 1976; Harder, 1985). Among bees, glossal length is as large as 3–4 cm in some tropical species of euglossines that feed on flowers with extremely long corollas. It is noteworthy that the glossa of such euglossines is considerably less "hairy", both in hair density and hair length, than in bumblebees and honeybees (Michener and Brooks, 1984). Because the resistance to nectar flow during the unloading phase will be proportional to the length of the proboscis, reduced hairiness in these euglossines may be important in reducing the resistance to flow during this phase (see above). More detailed kinematic studies of euglossine feeding would be of particular interest in understanding the morphological determinants of bee feeding mechanics; no such studies are currently available.

Thus far, our discussion of bee feeding has considered only the time on a flower to actually ingest the nectar (ingestion time). But the time required to obtain nectar while on a flower also involves an access time, the time required to enter and leave a flower (Harder, 1983). Lab studies of bumblebees feeding at artificial flowers showed that access time increased linearly with flower (corolla) depth. In contrast, ingestion time was independent of flower depth for flowers shallower than the length of the bee's glossa; for flowers deeper than the glossal length, ingestion time increased rapidly with increasing flower depth (Harder, 1983). This increase in ingestion time is the result of a reduced nectar load per lick while feeding on deep flowers (Harder, 1986) and suggests that the rate and amount of nectar flow onto the glossa during the loading phase can be an important determinant of intake rates in this situation.

2.4. Blood Feeding

The blood-feeding habit has evolved multiple times in several different insect orders. Because of their medical importance as disease vectors, blood-feeding insects have been studied in some detail in terms of functional and comparative anatomy, the feeding process, chemical and physical factors affecting the rates and regulation of feeding, and host attraction. There are a number of comprehensive reviews of this literature (e.g., Snodgrass, 1944; Downes, 1958; Hocking, 1971; Friend and Smith, 1977; and see Chapter 6); and Lehane (1991) provides a fine recent discussion of these topics. Here we shall give a brief overview of some of the common features of the blood-feeding process, including the important interaction between the geometry of flow and the mechanical properties of blood. We will then consider in some detail the blood-feeding process in reduviids and in mosquitoes, for which quantitative information is available on key components of the feeding process. Finally, we shall describe some potential implications of the fluid mechanics of blood feeding for disease transmission.

2.4.1. The Blood-Feeding Process

Blood feeding in insects involves three main steps: puncture and penetration through the epidermis of the skin, location of blood within the host, and the ingestion of blood. The multiple, independent evolution of blood-feeding habits is reflected by the fact that different mouthpart structures are used for these purposes in different insect groups (see Lehane, 1991). Skin penetration requires cutting or puncturing of the skin, usually involving toothed structures on the relevant mouthparts, and anchoring onto the skin to allow penetration. The literature describes various "sawing" and "scissoring" movements used during puncture and penetration for different blood feeders; but to our knowledge there has been no detailed consideration of the mechanics of puncture by blood feeders. Given the substantial medical literature on the mechanical properties of skin (e.g., Fung, 1981), and the very different types of wounds inflicted by various blood feeders (e.g., Lehane, 1991), this seems a fruitful area for future research. The location of blood within the host requires the formation of a blood pool, or (in some insects) the insertion of the feeding parts into an individual blood vessel. Salivary gland secretions play a central role in these blood location processes (see Chapter 3). In all blood feeders, the ingestion of blood involves the sucking of blood through a food canal by means of a muscular pump. In some blood feeders (e.g., assassin bugs), only a single (pharyngeal) pump occurs, whereas in others (e.g., mosquitoes, sucking lice) there are two pumps (cibarial and pharyngeal) (Snodgrass, 1935). To our knowledge, there is no information on functional relations between the two pumps in these latter groups, nor on the functional consequences of one versus two pumps (but see below).

Many blood-feeding insects also feed on sugar solutions from flower nectars, fruit, or other sources (see Chapter 8). These sugars are in many cases essential as an energy source and to sustain normal longevity, especially within the Diptera (Downes, 1985). In mosquitoes, for example, females and males drink both nectar and water; females likely feed more often on nectar than on blood; as in most blood-feeding Diptera, only females utilize blood. While nectar feeding may be essential for longevity, successful reproduction by females depends on obtaining proteins from one or more blood meals. In most mosquitoes, for example, oocyte maturation requires a blood meal, and egg output is directly related to blood meal volume (Gillett, 1972; Edman and Lynn, 1975). Over 75% of the protein in normal human blood is contained in the red cells, with the remainder in the blood plasma (e.g., Daniel and Kingsolver, 1983). Because many blood feeders may be susceptible to mortality by the host during feeding, rapid feeding times and/or rates may be advantageous (Gillett, 1969).

As described earlier in this chapter (see Section 2.2), the presence of the flexible red cells contributes to the distinctive fluid mechanical properties of blood. Blood hematocrit is defined as the volume fraction of the blood occupied by red blood cells. As blood hematocrit increases, the rate at which blood viscosity increases is much slower than for a comparable suspension of rigid particles. At the size scale of insect feeding structures, several additional mechanical features of blood become important (Kingsolver and Daniel, 1993). First, the hematocrit of blood in small blood vessels such as venules and capillaries is lower than that in the circulation as a whole. For example, in humans, hematocrit in the pulmonary arteries is ~0.4, whereas that in small venules and capillaries is 0.2 or less (Goldsmith et al., 1989). Because blood feeders generally obtain blood directly (in vessel feeders) or indirectly (in pool feeders) from these smaller blood vessels, they will usually experience substantially reduced hematocrits. Second, blood viscosity will vary with the food canal diameter of the blood feeder as a result of the Fahraeus–Lindqvist effect (Fig. 2.1), as discussed earlier. Because reduced viscosity could clearly affect blood flow rates during feeding, it is of interest that many mosquitoes and assassin bugs have food canals with diameters around 10 μm, where the viscosity of human blood approaches its minimal value.

The tube size at which the minimum viscosity occurs is determined primarily by the size of a single red cell. Most mammals, including humans, have red cells of similar shape and size, but this is does not hold for vertebrates in general. Thus red cells in birds are more spherical in shape, and they are more than twice as large as those in humans; and red cells in reptiles and amphibians are often larger still. Although no data are currently available, this variation in red cell size and shape will undoubtedly affect the relationship between blood viscosity and tube diameter, with important possible consequences for blood-feeding rates on different vertebrate hosts (Kingsolver and Daniel, 1993).

Some blood feeders such as sucking lice (e.g., *Pediculus*) have food canals with a diameter smaller than that of a single red cell (Tawfik, 1968). In this case the flexible red cells must deform and pass through the food canal in single file, and the effective viscosity of the blood may be quite large (see Section 2.2). As a result, the maximum feeding rates (blood volume/time) for *Pediculus* are 10–100 times lower than those for larger blood feeders (Tawfik, 1968). Based on studies of blood flow in capillaries, the critical tube diameter below which flow of human blood is zero is 3–4 μm (Fung, 1981); this places an absolute lower bound on the size of the food canal for blood feeders. This lower bound is likely to be considerably larger for blood feeders of birds and reptiles, with their larger and less flexible red cells.

As discussed earlier, positive pressures occur within the blood of the host, and our calculations suggest that such pressures could contribute to feeding rates in mosquitoes (see Section 2.2). One interesting consequence of feeding on pressurized food sources is that the nutrient concentration (in this case, blood hematocrit) which maximizes uptake will be lower than that for insects using only muscles to generate the pressure difference.

2.4.2. *Mechanics of Blood Feeding in* Rhodnius

Many aspects of the mechanics, behavior, and regulation of feeding in the blood-sucking reduviid bug *Rhodnius prolixus* have been studied in some detail (e.g., Lehane, 1991). These animals are capable of very rapid blood feeding; for example, fifth-instar bugs can obtain a 300-mg blood meal in about 15 min, for an average flow rate of $0.33 \mu l \cdot sec^{-1}$ (Buxton, 1930). These large flow rates are achieved despite a food canal of only 8–10 μm in diameter. Conservative calculations suggest that *Rhodnius* is apparently capable of generating pressure differences in excess of 1–2 atmospheres during feeding, implying that a negative pressure is produced within the cibarial (pharyngeal) pump (Bennet-Clark, 1963a). Experimental manipulations of ambient (air) and blood pressures support this notion (Bennet-Clark, 1963a,b). The existence of such negative pressures requires that the cavitation ("breaking-up") of the blood be avoided. Bennet-Clark (1963a,b) suggests that this process is analogous to the negative pressures within the water column in the xylem vessels of trees, where the wettability (surface tension properties) of the xylem surfaces is crucial (Preston, 1952). How this operates at much higher flow rates that change rapidly over time remains unclear.

The impressive pressure differences generated by *Rhodnius* during feeding are the result of a cibarial pump that occupies a large proportion of the head [see Bennet-Clark (1963a,b) for figures and description]. The pump consists of (a) a rigid, U-shaped plate attached to the ventral surface of the head and (b) a flexible top that is attached laterally to the base plate by rubbery cuticle. Large muscles run dorsolaterally from the pump's top surface to the dorsal wall of the head; this musculature runs the entire length of the head. The contraction of these cibarial dilator muscles expand the pump and draw blood through the food canal. The pumping rate is approximately 7 cycles/sec (Hz), and the stroke volume (fluid volume/cycle) is ~0.06–0.07 μl (Smith, 1979).

Smith (1979) has measured the feeding and pumping rates for fifth-instar *Rhodnius* feeding on artificial diets with viscosities ranging from 0.8 to 7.0 $g \cdot m^{-1} \cdot sec^{-1}$ (for food canals of 8 to 10 μm diameter, as in *Rhodnius*, blood viscosity is ~2 $g \cdot m^{-1} \cdot sec^{-1}$). As viscosity increases, the pumping frequency, the stroke volume, and the average fluid intake rate all decrease monotonically. Using these data, Smith (1979) estimated the mechanical power output of the pump and found that the power output was lowest at the lowest and highest viscosity considered, with a maximal output at viscosities near 2 $g \cdot m^{-1} \cdot sec^{-1}$. Of particular interest is that decreasing the viscosity from 2.2 to 0.8 $g \cdot m^{-1} \cdot sec^{-1}$ increased the fluid intake rate by only ~20%, far less than that predicted by Poiseuille's equation (a predicted ~267% increase). These calculations suggest that the cibarial pump of *Rhodnius* will generate the maximal power output while feeding on fluid with viscosities similar to blood, leading to rapid blood-feeding rates (see below).

The size of the blood meal in *Rhodnius* is remarkable: a 300-mg meal represents a sixfold increase in body mass during feeding. This is made possible by the rapid extension and expansion of the abdomen during feeding [see Bennet-Clark (1962) for details].

2.4.3. Mechanics of Blood Feeding in Aedes

As an important vector for human disease, *Aedes aegypti* has been intensively studied with respect to blood feeding. Meal size in *Aedes* is ~4.2 μl (Clements,

1963), with an average blood intake rate of ~ 0.016 $\mu l \cdot sec^{-1}$ (Tawfik, 1968)—
an order of magnitude lower than in *Rhodnius*. As in other mosquitoes, there
are two muscular pumps [see Snodgrass (1935) for figures and description of
the feeding structures]. The posterior wall (floor) of the anterior (cibarial) pump
is strongly sclerotized and forms a basin. The dorsal wall (ceiling) of the pump
is flexible; two dilator muscles run from the top of this ceiling dorsally to the
clypeus. The posterior (pharyngeal) pump is elongated, with dilator muscles
acting both laterally and anteriorly. Our histological studies with *Aedes* suggest
that the posterior pump musculature is considerably larger and that it may provide
the primary driving force during feeding (Kingsolver, Daniel, and Meyhofer,
unpublished data).

We have developed models for blood feeding in mosquitoes that are analogous
to those for nectar feeding in butterflies described earlier (see Appendix)
(Kingsolver and Daniel, 1993) and that incorporate both muscle and fluid mechanics. In the current model, the fluid dynamics are described in terms of an approximation based on Hagen–Poiseuille's equation developed by Fung (1981) that
incorporates some of the non-Newtonian behavior of blood. We have used morphological measurements on *Aedes aegypti* in the model, considering only the
pharyngeal pump in our analysis. We can use the model to explore how blood
hematocrit and morphological features affect the rates of intake during feeding.
Because protein is the primary nutrient for female mosquitoes in blood, it is
useful to consider the effects of these factors on the rate of protein intake.

Despite the non-Newtonian nature of blood, our simulation results are qualitatively quite similar to those for nectar-feeding butterflies (see Section 2.3). For
a given set of morphological parameters, the model predicts that the protein
intake rate is maximum at some intermediate blood hematocrit (Fig. 2.6). As
with nectar feeding, this intermediate maximum is the result of the exponential
relationship between blood viscosity and red cell concentration (hematocrit). For
any given food canal radius and muscle cross-sectional area, there is an intermediate pump size that yields the maximal rate of protein intake; this maximal rate
always occurs at hematocrits of 0.6–0.7 (Fig. 2.7). As in nectar feeding, the
interaction between the resistance to fluid movement and the speed and power
output of the muscular pump determines the relationship between intake rates
and hemocrit for blood feeders.

No empirical data on blood intake rates as a function of hematocrit are available
to evaluate these model predictions for *Aedes*. However, the hematocrit of human
blood does not normally exceed 0.45: mosquitoes will not experience hosts with
hematocrits of 0.6–0.7. Note that for the range of hematocrits seen in human
hosts (~ 0.3–0.45), protein intake rates is predicted to increase with increasing
hematocrit. However, these calculations do not include the possible effects of
positive blood pressure on the relationship between hematocrit and intake rate,
which is likely to decrease the hematocrit at which maximum protein intake rates
occur (see above).

The above simulations do not include the Fahraeus–Lindqvist (F–L) effect in

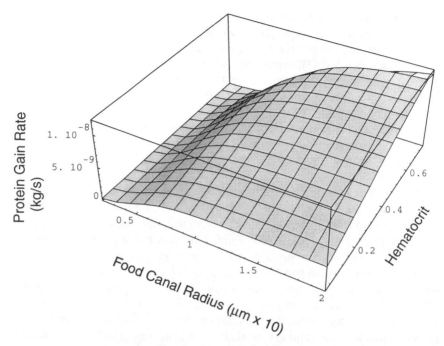

Figure 2.6. Model predictions for the protein intake rate of a blood-feeding mosquito plotted against food canal radius and blood hematocrit. We used measurements of the morphological and physiological parameters for *Aedes aegypti* (J. Kingsolver, T. Daniel, and E. Meyhofer, *unpublished data*) in generating these predictions. Note that as food canal radius increases, the blood hematocrit that maximizes protein gain rate also increases.

computing blood viscosity. Unfortunately, data on this effect are available only for human blood of ~0.4 hematocrit, and the strength of the effect will change with hematocrit. We can, however, incorporate empirical data relating blood viscosity to tube size (e.g., Fig. 2.1) for a hematocrit of 0.4 into the model, to consider the consequences of the F–L effect for protein intake rate for food canals of different size. Our simulation results (Kingsolver and Daniel, *unpublished data*) suggest that the F–L effect will increase protein intake rates substantially for food canals of 10 to 60 μm diameter. However, increasing canal size always increases protein intake rate, even in the range where the F–L effect is greatest: the strong dependence of flow resistance on food canal diameter more than counterbalances the effects of canal diameter on blood viscosity.

One important limitation of these simulations is that they consider only the posterior of the two pumps in mosquitoes. No information is available about the relative importance or the coordination of the two pumps. One important difference between one- and two-pump systems is that with a single pump the flow rate is zero during some part of each pumping cycle, whereas with two pumps

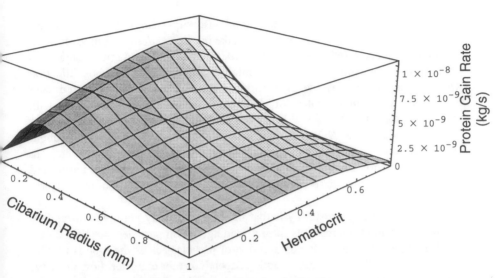

Figure 2.7. Model predictions for the protein intake rate of a blood-feeding mosquito plotted against cibarium radius and blood hematocrit. We used measurements of the morphological and physiological parameters for *Aedes aegypti* (J. Kingsolver, T. Daniel, and T. Meyhofer, *unpublished data*) in generating these predictions. There is a unique cibarium radius and blood hematocrit that leads to a maximal protein gain rate.

a continuous, positive flow can in principle be maintained. This is of interest for blood feeding because of the existence of a yield stress for blood (see Section 2.2). Unlike the situation in Newtonian fluids, a force larger than the yield stress is required to start blood movement from rest. It would be of interest to compare pumping cycles and flow rate in blood feeders with one (e.g., reduviids) and two (e.g., mosquitoes) sucking pumps.

2.4.4. Mechanics of Disease Transmission

All of our discussion of the mechanics and dynamics of blood flow has assumed that we are considering normal, vertebrate blood. We have seen that many of the unusual mechanical properties of blood result from the flexibility and shape of red cells. However, a number of factors, including disease, can alter these characteristics of red cells. For example, when a human red cell is infected by malarial parasites (*Plasmodium falciparum*), the red cell becomes both more spherical and more rigid (Cranston et al., 1984; Taraschi et al., 1986). Experimental studies of blood show that increasing the sphericity or rigidity of red cells will increase blood viscosity and will largely eliminate the Fahraeus–Lindqvist effect (e.g., Fung, 1981). This suggests that parasites within red blood cells could substantially increase blood viscosity for blood feeders. On the other hand, the anemia commonly associated with malarial infection will decrease the hematocrit level, perhaps reducing blood viscosity (Kingsolver and Daniel, 1993). One preliminary study of blood intake rates of *Aedes* mosquitoes feeding on malarial-infected (*Plasmodium yoelii yoelii*) and uninfected mice suggests an interesting relationship between blood intake rates and the level of host infection (parasitemia, defined as the proportion of the red blood cells infected with parasites). Blood intake rates for mosquitoes feeding on infected hosts with parasitemias of 10–30% were 20–25% higher than those for mosquitoes feeding on uninfected hosts (J. A. Wolff, *unpublished data*). Above parasitemias of about 30%, however, blood intake rates decreased rapidly with increased parasitemia; at parasitemias >50%, blood intake rates were 30–35% below those for uninfected hosts. More study of these relationships and their mechanical bases is clearly indicated.

Another intriguing possibility is that blood feeders may not obtain a random "sample" of blood components during feeding. When particles suspended in a fluid flow through a tube, the distribution of particles across the tube is usually not uniform; and this particle distribution depends on tube diameter, flow velocity, and particle size and shape. For example, in tubes between 6- and 100-μm diameter, the red cells in normal blood migrate away from the tube walls toward the center of the tube, owing to the shape and flexibility of the red cells (see Section 2.2). Experimental studies of particle flow show that more spherical and more rigid particles do not concentrate as greatly toward the center of the tube (Fung, 1981). As noted above, red cells infected with blood parasites are more spherical and rigid relative to normal cells. For vessel feeders such as mosquitoes and reduviids, the positioning of the food canal in the blood vessel may affect the encounter rate with infected and noninfected red cells. Entry effects on the movement of particles with different shape and mechanical properties (i.e. infected and noninfected red cells) into the distal end of the food canal may also be important. These factors suggest that the relative abundance of infected and noninfected red cells obtained by the blood feeder during vessel feeding may differ from their relative abundances within the host blood, with consequences for the probability of disease transmission during blood feeding (e.g., Aron and May, 1982). At this point, neither the fluid mechanics of mixtures of rigid and flexible particles, nor relevant studies with blood feeders, have been conducted. The mechanical consequences of parasites for blood feeding and for parasite transmission to the vector remain intriguing and unexplored issues.

2.5. Phloem and Xylem Feeding

Many phytophagous insects obtain nutrients directly from the liquids contained within plants, primarily from phloem, xylem, and mesophyllic cell cytoplasm. In contrast to nectar and blood as discussed above, the viscosity of phloem and xylem is relatively low because of their low nutrient content. The low viscosity does not mean that mechanical considerations are necessarily unimportant, how-ever: many phloem and xylem feeders process large volumes of liquid (relative to their body size) through food canals of small diameter to obtain sufficient nutrients.

Because some phloem feeders are important agricultural pests and vectors of plant disease, aspects of the feeding mechanics, feeding behavior, and nutrition have been studied for some major groups of phloem feeders, especially aphids. Here we shall summarize the functional anatomy and behavior of feeding in Hemiptera and Thysanoptera, the means by which insects penetrate plant tissues, and the determinants of fluid and nutrient uptake rates.

2.5.1. Anatomy and Behavior of Feeding

The primary feeding structure in Hemiptera (Homoptera + Heteroptera) is a long and relatively slender beak consisting principally of the labium [see Snodgrass (1935) for figures and description of the feeding structures]. The two mandibles and two maxillary palpi form a set (bundle) of long, slender bristles or stylets. The inner surfaces of the two maxillary stylets form both the food canal (anteriorly) and the salivary canal (posteriorly) during feeding. In phytoliquivorous Hemiptera, the mandibular stylets are the primary organs used in piercing plant tissues; penetration is achieved by a series of short, alternating thrusts of these two stylets. The food canal empties into a single, large cibarial cavity. The posterior and lateral walls of the cibarium are strongly sclerotized; on the flexible anterior wall are attached the dilator muscles that contract to generate suction in the cibarium.

Salivary secretions play an important role in Hemipteran feeding (Miles, 1973; also see below and Chapter 3). For example, during penetration of the plant tissues all Homoptera (and some Heteroptera) discharge saliva wherever the stylets penetrate, to form a tubular lining—the stylet sheath—along the entire path of penetration. Flow through the stylet sheath may proceed even when the stylets themselves are withdrawn (Miles et al., 1964; Kinsey and McLean, 1967; Miles, 1973). Most Hemiptera utilizing plant liquids feed directly from vascular structures such as phloem or xylem elements, or directly from leaf mesophyll cells.

The feeding apparatus and method of feeding in thrips (Thysanoptera) is unique among insects [see Snodgrass (1935) for figures and description]. The mouthparts form a broad cone defined by the clypeus and labrum in front, the labium behind, and the galeal parts of the maxillae at the sides. Within this cone are the piercing stylets, consisting of a pair of maxillary stylets, the single (left) mandible, and (in some species) the hypopharynx (Lewis, 1973). The margins of the two maxillary stylets fit together to form the food canal. The single cibarial pump is quite similar in general form to that in Hemiptera (Snodgrass, 1935). Many thrips suck juices from leaves or flowers of vascular plants, but a variety of other food sources are used, including pollen grains, fungal hyphae, and young shoots. During leaf feeding, the mandibular stylet penetrates the epidermal cells of the leaf, and the mandibular and/or maxillary stylets are used to rupture cell walls within the mesophyll; the cell contents of the mesophyll are then sucked through the food canal. The vascular tissues are rarely penetrated during feeding by thrips (Lewis, 1973).

2.5.2. Penetrating Plant Tissues

Many plant tissues present a considerable physical barrier to potential phytoliqui-vores. Some aspects of plant penetration have been studied, particularly in aphids. Because many plant feeders are known to transmit viral and other plant diseases near the tips of the stylets during probing and penetrating (Pollard, 1973), the path traversed by the stylets during penetration has been a topic of special interest. In most aphids, for example, the stylet path is frequently intercellular, terminating in individual phloem cells (sieve tubes); but intracellular penetrations also occur, as well as intramural paths where stylets lie between cell walls and plasmalemmas (Kimmins and Tjallingii, 1985; Spiller et al., 1985; Kimmins, 1986). An intercel-lular track likely represents the path of least resistance, though quantitative verification of this is not available. Many Coccids, whose stylet bundles are frequently less than 3 μm in diameter, utilize intracellular stylet paths. The remarkable ability of Hemiptera to penetrate hard plant tissues with slender mouthparts has inspired many detailed descriptions on behavior and mouthpart actions occurring during penetration, though there is little quantitative information on the solid mechanics of penetration. Cell walls may be punctured by direct pressure, by sawing and twisting of the barbed mandibular stylets, or by forcibly injecting watery saliva (Pollard, 1973; Miles, 1973). The salivary secretions of aphids contain pectinesterase and other enzymes that degrade plant cell walls (Adams and McAllan, 1958; Ma et al., 1990) and facilitate intercellular penetra-tion (Adams and McAllan, 1958; Ma et al., 1990). Indeed, aphid-resistant varie-ties of some host plants possess pectins that can inhibit pectinesterase activity, resulting in reduced probing and stylet penetration (Campbell and Dreyer, 1985; Campbell et al., 1986).

The depth of tissue penetration can be readily determined for most Hemipterans by histological examination of the stylet sheath. In many cases the depth of penetration can exceed 75–80% of the length of the stylets; in some Coccids, the penetration depth may reach 10^3 times the diameter of the stylets (Pollard, 1973). Not surprisingly, penetration depth depends on tissue type, the depth of vascular elements, and tissue hardness, with important consequences for the distribution of feeding sites. For example, spittlebugs may be restricted to feeding sites near the terminal bud as a result of the greater hardness of lignified tissues away from the bud (Hoffman and McEvoy, 1986).

While biochemical aspects of penetration have received increasing attention, quantitative mechanical analyses of plant penetration remain largely absent. Systematic studies of the consequences of stylet length and diameter, stylet fine structure (e.g., barbs), and stylet movements for the rate and depth of penetration in a variety of plant tissues would be particularly valuable.

2.5.3. Mechanical Determinants of Nutrient Uptake

In contrast to plant penetration, the mechanical determinants of nutrient uptake in some phytoliquivores have been examined in some quantitative detail. Here we summarize current understanding of phloem feeding in aphids, provided largely by the pioneering studies of T. Mittler and his collaborators.

Many aphid species insert their stylet tips directly into phloem sieve tubes for feeding. Kennedy and Mittler (1953) first demonstrated that once the stylets are in place in the sieve tubes, the aphid may be severed from the stylets, and phloem sap will exude continuously from the severed stylets. This allows an indirect assessment of the contribution of the sucking cibarial pump to fluid flow (see Chapter 7). For example, the willow-feeding aphid, *Tuberolachnus salignus*, has a food canal that is only 1.2 μm in diameter at the distal end and 1.8 mm in length, yet it achieves a flow rate of 1–2 $\mu l \cdot hr^{-1}$ while feeding on phloem (Mittler, 1957). Application of Hagen–Poiseuille's equation (equation [1]) indicates that a pressure differential of more than 20 atmospheres is required to generate such flow rates. However, flow rates through severed stylets and during feeding by intact aphids were similar, both during nymphal stages and during adult stages (though flow rates increased consistently with stage). Thus, the turgor pressure in the host plant's phloem can largely account for the rapid flow rates observed during aphid feeding. While the aphid must actively swallow, its cibarial pump may contribute relatively little to intake rate during normal feeding (Mittler, 1957) (but see below). Other lines of evidence confirm the importance of turgor pressure in the rapid feeding rates of aphid. For example, in wilting or water-stressed plants, rates of honeydew production and of reproduction by aphids are reduced (Mittler, 1957; Kennedy et al., 1958; Weatherley et al., 1959; Wearing, 1967; Wearing and Van Emden, 1967).

An aphid must process large volumes of phloem to obtain sufficient nutrients, and consequently it excretes substantial quantities of honeydew. During extended feeding, a balance between phloem intake and honeydew output is obtained; hence by monitoring phloem and honeydew flow rates and concentrations, a quantitative accounting of mass and nutrients is possible. Sugar concentrations (primarily sucrose) in the phloem of actively photosynthesizing plants rarely exceed 5–10%, and they may be substantially less in dormant plants. Honeydew contains a mixture of sucrose, glucose, and fructose and does not differ from phloem sap in total sugar concentration by more than about 5% (Mittler, 1958a), a very low absorption efficiency. In terms of nitrogen balance, 10 or more different amino acids typically are found in phloem sap, with total nitrogen concentrations generally less than 0.5%; comparisons with honeydew suggest that aphids absorb at least 55% of the nitrogen in phloem from actively growing plants, with little evidence of differential absorption of different amino acids (Mittler, 1958b). Greater nitrogen absorption efficiencies are found for aphids feeding on mature or senescent foliage, where phloem nitrogen concentrations are lower (Mittler, 1958b). The low absorption efficiencies for sugar may be related to the osmoregulatory challenge resulting from the rapid rate of flow through the aphid's gut (Mittler, 1958b).

Although plant turgor pressure contributes importantly to the feeding rate of aphids, this does not mean that aphids are incapable of active sucking using their cibarial pump to feed effectively. This is demonstrated by feeding studies of aphids on artificial diets at atmospheric pressure, which have explored systematically the effects of sugar and amino acid concentrations on feeding rates, growth, and survival (Mittler and Dadd, 1963a,b; Mittler, 1967a,b). For example, for *Myzus persicae* feeding on sucrose solutions from 0% to 40% (Mittler, 1967a), the volume rate of fluid intake was much greater on 5% sucrose than on pure water (0%), as a result of increased feeding stimulation. For concentrations greater than 5% sucrose, fluid intake rate decreased continuously with increasing concentration. The rate of sugar intake is maximal at intermediate concentrations of 25–30%. Calculations using Hagen–Poiseuille's equation suggest that the pressure differential generated by the aphid's cibarial pump increases continuously with increasing sucrose concentration, and is highest at the highest concentration considered (40% sucrose). These results are generally consistent with the model predictions for suction feeding nectarivores discussed above (see Section 2.3), although the maximum rate of sucrose intake occurred at somewhat lower concentrations than that predicted (and observed) for butterflies. In aphids, this may occur because of the high resistance provided by the very small food canal diameter. Analogous measurements of amino acid solutions show that the rate of amino acid intake increases continuously with amino acid concentrations from 0% to 5%. Note that intake rates of amino acids and sucrose are not directly proportional to growth and survival for aphids (Mittler and Dadd, 1963a,b; Mittler, 1967a); for example, for *Myzus persicae,* larval growth rate is maximal on 25–30% sucrose. Amino acids rather than sucrose may generally be the limiting nutrient for growth and development in phloem-feeding aphids (Cartier, 1967).

These studies with artificial diets demonstrate that aphids can feed effectively on diets at atmospheric pressures by use of their cibarial pump. However, aphids normally feed and excrete honeydew at greater rates on turgid plants than on artificial diets (Dadd and Mitler, 1965). Experimental increases in pressure applied to appropriate artificial diets have been shown to increase longevity (and presumably feeding rates) in *Myzus persicae* (Wearing, 1968). The relative importance of turgor pressure, active sucking, and swallowing rate in determining feeding rates is likely to depend on both plant condition and the nutritional requirements of the aphids.

Some quantitative measurements of feeding by thrips have been done for *Limothrips cerealium* feeding on plant cell contents of wheat (Chisholm and Lewis, 1984). These thrips were able to ingest whole chloroplasts up to 6 μm in diameter, even though the diameter of the maxillary tube was only about 1 μm. The cibarium pumps at a rate of 2–6 cycles/sec, resulting in intake rates of 8.5×10^{-5} μl·min^{-1}, equivalent to 12.5% of their body mass per hour. Feeding rates on dilute sucrose solutions through artificial membranes were several times greater than rates on plant materials for *Limothrips ceralium* (Chisholm and Lewis, 1984) and for *Thrips tabaci* (Day and Krzykiewicz, 1954).

In contrast to phloem feeding, quantitative information on mechanical determinants of rates of xylem feeding is largely lacking. Nutrient concentrations are much lower in xylem than in phloem, with amino acid concentrations generally below 0.08% and sugar concentrations of only 0.005% (Horsfield, 1978). Feeding and excretion rates in the xylem-feeding sharpshooter *Hordnia circellata* (Hemiptera: Cicadellidae) were 10–100 times total wet body mass per day (Mittler, 1967b), much higher (on a body mass basis) than for comparable phloem feeders (e.g., Auclair, 1963). These higher feeding rates for xylem feeders are likely needed to maintain adequate rates of nutrient intake and absorption for the insects. These high feeding rates are achieved even though the insect must apparently use its cibarial pump to generate suction pressures that overcome the substantial negative pressures found in xylem elements (Mittler, 1967b). For example, even in forest herbs and shrubs not under water stress, xylem pressures of -1.0 to -2.0 MPa (-10 to -20 atmospheres) are common (Scolander et al., 1965). How xylem feeders generate such pressures, and how they avoid cavitation of the liquid when the xylem is initially penetrated, remains unknown. Indeed, the maximum isometric stress reported for contracting insect skeletal muscle (80 kPa) appears far too low for xylem feeding. One may therefore expect some novel physiological adaptations for effective xylem feeding that have not been previously explored.

2.6. Future Prospects

We have tried throughout our discussion to point out important questions that would benefit from further study. Five general issues seem of special interest at this point.

First, in only one system (*Rhodnius*) do we approach the level of detailed information needed to understand the *dynamical* aspects of fluid feeding in insects. Quantitative data on the rates of cibarial pumping and stroke volume, instantaneous rates of flow, the pressures generated during muscle contraction, licking frequencies, and lick volume for bees and other insects are almost entirely lacking. Such data are essential to adequately test (a) our current understanding of insect feeding and (b) the predictions of current mathematical models.

Second, more detailed studies of blood feeding in mosquitoes would be particularly profitable. Experimental and observational studies of the effects of blood pressure, hematocrit levels, blood parasites, and the Fahraeus–Lindqvist effect on rates of blood intake may provide important insights about determinants of blood feeding, host choice, and disease transmission in blood feeders. Our mechanical analyses suggest potentially important roles for these factors, but data are entirely lacking.

Third, the functional morphology of bee tongues in relation to feeding rates seems ripe for study. Comparative studies demonstrate the morphological diversity in these structures, but current mathematical analyses can provide little guidance as to the functional significance of this diversity. Measurements of the rate and volume of nectar loading as functions of features of the glossae and glossal hairs would be particularly enlightening.

Fourth, quantitative analyses of the solid mechanics of penetrating plant tissues by phloem and xylem feeders are largely lacking. Such studies could contribute importantly to our understanding of herbivore resistance and to plant disease transmission in phytoliquivores. Similarly, quantitative analyses of skin penetration by blood feeders would be valuable, drawing on the extensive biomedical literature on mechanical properties of skin.

Finally, we have tried to demonstrate that one's understanding of the mechanics of fluid feeding benefits substantially from a quantitative engineering approach to the problem. We hope that this may encourage larval and nymphal entomologists interested in the study of feeding to include basic mechanical engineering as part of their training.

Acknowledgments

H. Bennet-Clark, C. Boggs, S. Buchmann, F. R. Hainsworth, L. Harder, M. Lehane, and D. Roubik provided valuable comments on earlier versions of the manuscript. This work was supported in part by NSF grant BSR-8908131 to JGK.

Appendix: A Simple Model for Lepidopteran Sucking

During feeding on nectars, a butterfly or moth uses a tubular proboscis to draw nectar through its food canal by contracting muscles attached to an internal cavity, the cibarium (Fig. 2.3) (for details see Section 2.3). In this simple case, the mechanics of fluid flow can be described by the Hagen–Poiseuille relationship (equation [1]), and the driving force generated by the contracting muscles can be described by Hill's equation (equation [4]). The key to the analysis is that the power or momentum causing fluid movement is precisely that generated by the contracting cibarial muscles.

For nectar feeding, we are interested in the determinants of the rate of sugar uptake (here described in terms of sucrose). Using equation [1], the sucrose uptake rate is simply

$$\frac{dS}{dt} = Q[S] = \frac{\pi r^4 \Delta P[S]}{8\mu L} \qquad \text{[A.1]}$$

where dS/dt is the sucrose uptake rate and $[S]$ is the concentration of sucrose in nectar. Recall that nectar viscosity is a function of sucrose concentration, rising approximately exponentially with increasing $[S]$. Thus equation [A.1] predicts that dS/dt has an intriguing dependence of sucrose concentration. At very low concentrations, uptake rates are necessarily low because of the relative lack of any sucrose; at very high concentrations, uptake rates are low because of the relatively high viscosity of the solution. At intermediate sucrose concentrations, the uptake rate is maximized. Indeed, because viscosity is related exponentially to concentration for many types of liquids (see above), we expect that nutrient uptake rates will in general be maximal at intermediate nutrient concentrations (see Section 2.4).

The pressure difference in equation [A.1] is determined by the mechanics of the muscles that drive the flow. There are two approaches one can use here: (1) We can use conservation of energy to assert that the power output of the muscles equals the power associated with the flow, or (2) we can use conservation of momentum to assert that the force produced by the muscles is equal to the force associated with the flow. Here we use the latter approach. Thus, the pressure produced by contracting muscles must necessarily equal the pressure drop across the proboscis. Because the pressure is simply the contractile force divided by the cross-sectional area of the cibarium (in this case equal to muscle cross-sectional area), we can directly substitute Hill's equation into equation [4] above:

$$\Delta P = \frac{F_{muscle}}{A_{cib}} = \frac{T_0(V_{max} - V)}{A_{cib}(V_{max} + 4V)} \qquad \text{[A.2]}$$

The final step in the analysis invokes conservation of mass, asserting that the flow rate of nectar into the cibarium is equal to the flow rate of nectar through the food canal. The rate of change in cibarial volume must therefore equal that flow rate. With this step, we may relate the velocity of muscle shortening to the flow rate of nectar. For our hypothetical butterfly, the change in cibarial volume is equal to the product of the muscle contraction velocity and the cross-sectional area of the cibarium:

$$Q = VA_{cib} \qquad \text{[A.3]}$$

After all of the substitutions are made (Hill's equation, conservation of mass and momentum), we arrive at an equation for the flow rate of nectar that relates all of the relevant morphological, physiological, and physical parameters:

$$Q = \frac{\pi R^4 \{T_0(V_{max} - Q/A_{cib})\}}{8\mu L \{A_{cib}(V_{max} + 4Q/A_{cib})\}} \qquad [A.4]$$

The sucrose uptake rate ($dS/dt = Q[S]$) is given by the solution to the above quadratic equation for Q [see Daniel et al. (1989) for the rather untidy solution]. We can use this solution to explore how the physiology of muscle, the morphology of the feeding apparatus, and the physics of flow interact to affect the behavior of the nutrient uptake rate (see Section 2.3 for a discussion of these model results).

Similar approaches may be developed to account for nutrient uptake rates for insects feeding on non-Newtonian fluids such as blood. In this case, the uptake rate of protein (rather than sucrose) is the relevant measure of feeding performance. While the Hagen–Poiseuille relationship no longer applies, a similar one may be developed to account for the relationship between pressure differential and flow rate (Kingsolver and Daniel, 1993; Daniel and Kingsolver, 1983) for fluids whose viscosity may vary with flow. The results of such analyses parallel those seen for nectar feeders in many important ways (for discussion see Section 2.4).

References

Adams, J. B., and McAllan, J. W. (1958). Pectinase in certain insects. *Can. J. Zool.* 36, 305–308.

Aron, J. L., and May, R. M. (1982). The population dynamics of malaria. In: Anderson, R. M. (ed.), *Population Dynamics of Infectious Diseases*. Chapman & Hall, New York, pp. 139–179.

Auclair, J. L. (1963). Aphid feeding nutrition. *Annu. Rev. Entomol.* 8, 439–490.

Baker, I., and Baker, H. G. (1982). Some constituents of floral nectars of *Erythrina* in relation to pollinators and systematics. *Alertonia* 4, 25–37.

Batchelor, G. K. (1967). *An Introduction to Fluid Dynamics*. Cambridge University Press, New York.

Bennet-Clark, H. C. (1962). Active control of the mechanical properties of insect cuticle. *J. Insect Physiol.* 8, 627–633.

Bennet-Clark, H. C. (1963a). Negative pressures produced in the pharyngeal pump of the blood-sucking bug, *Rhodnius prolixus*. *J. Exp. Biol.* 40, 223–229.

Bennet-Clark, H. C. (1963b). The control of meal size in the blood sucking bug, *Rhodnius prolixus*. *J. Exp. Biol.* 40, 741–750.

Boggs, C. L. (1988). Rates of nectar feeding in butterflies: effects of sex, size, age, and nectar concentration. *Functional Ecol.* 2, 289–295.

Boggs, C. L., and Ross, C. L. (1993). The effects of adult food limitation on life history traits in *Speyeria mormonia* (Lepidoptra: Nymphalidae). *Ecology* 74, 433–441.

Buxton, P. A. (1930). The biology of a blood-sucking bug, *Rhodnius prolixus*. *Trans. R. Entomol. Soc. Lond.* 78, 227–236.

Campbell, B. C., and Dreyer, D. L. (1985). Host-plant resistance of sorghum: differential hydrolysis of sorghum pectic substances by polysaccharases of greenbug biotypes (*Schizaphis graminum*, Homoptera: Aphididae). *Arch. Insect Biochem. Physiol.* 2, 203–215.

Campbell, B. C., Jones, K. C., and Dreyer, D. L. (1986). Discriminative behavioral responses by aphids to various plant matrix polysaccharides. *Entomol. Exp. Appl.* 41, 17–24.

Cartier, J. J. (1967). Factors of host plant specificity and artificial diets. *Ann. Entomol. Soc. Am.* 60, 18–21.

Chisholm, I. F., and Lewis, T. (1984). A new look at thrips (Thysanoptera) mouthparts, their action and effects of feeding on plant tissue. *Bull. Entomol. Res.* 74, 663–675.

Clements, A. N. (1963). *The Physiology of Mosquitoes*. Pergamon Press, New York.

Cranston, H. A., Boylan, C. W., Carroll, G. L., Sutera, S. P., Williamson, J. R., Gluzman, I. Y., and Krogstad, D. J. (1984). *Plasmodium falciparum* maturation abolishes physiologic red cell deformability. *Science* 223, 400–403.

Dadd, R. H., and Mittler, T. E. (1965). Studies on the artificial feeding of the aphid *Myzus persicae* (Sulzer). III. Some major nutritional requirements. *J. Insect Physiol.* 11, 714–743.

Daniel, T. L., and Kingsolver, J. G. (1983). Feeding strategy and the mechanics of blood sucking in insects. *J. Theor. Biol.* 105, 661–672.

Daniel, T. L., Kingsolver, J. G., and Meyhofer, E. (1989). Mechanical determinants of nectar-feeding energetics in butterflies: muscle mechanics, feeding geometry, and functional significance. *Oecologia (Berlin)* 79, 66–75.

Day, M. F., and Krzykiewicz, H. (1954). Physiological studies on thrips in relation to transmission of tomato spotted wilt virus. *Aust. J. Biol. Sci.* 7, 274–281.

Dethier, V. G., and Rhoades, M. V. (1954). Sugar preference-aversion functions for the blowfly. *J. Exp. Zool.* 126, 177–203.

Downes, J. A. (1958). The feeding habits of biting flies and their significance in classification. *Annu. Rev. Entomol.* 3, 249–266.

Eastham, L. E. S., and Eassa, Y. E. E. (1955). The feeding mechanism of the butterfly *Pieris brassicae* L. *Philos. Trans. R. Soc. London Ser B* 239, 1–43.

Edman, J. D., and Lynn, H. C. (1975). Relationship between bloodmeal volume and ovarian development in *Culex nigripalpus* (Diptera: Culicidae). *Entomol. Exp. Appl.* 18, 492–496.

Ewald, P. W., and Williams, W. A. (1982). Function of the bill and tongue in nectar uptake by hummingbirds. *Auk* 99, 573–576.

Fahraeus, R. (1929). The suspension stability of the blood. *Physiol. Rev.* 9, 241–274.

Farrell, A. P. (1991). Circulation of body fluids. In: Prosser, C. L. (ed.), *Environmental and Metabolic Animal Physiology*. Wiley–Liss, New York, pp. 509–558.

Friend, W. G., and Smith, J. J. B. (1977). Factors affecting feeding by bloodsucking insects. *Annu. Rev. Entomol.* 23, 309–331.

Fung, Y. C. (1981). *Biomechanics*. Springer-Verlag, New York.

Gillett, J. D. (1969). Natural selection and feeding speed in a blood-sucking insect. *Proc. R. Soc. London Ser B* 167, 316–329.

Gillett, J. D. (1972). *The Mosquito*. Doubleday, New York.

Goldsmith, H. L., Cokelet, G. R., and Gaehtgens, P. (1989). Robin Fahraeus: evolution of his concepts in cardiovascular physiology. *Am. J. Physiol.* 257, 1005–1015.

Hainsworth, F. R. (1989). "Fast food" vs. "haute cuisine": painted ladies, *Vanessa cardui* (L.), select food to maximize net meal energy. *Funct. Ecol.* 3, 701–707.

Hainsworth, F. R., Precup, E., and Hamill, T. (1991). Feeding, energy processing rates and egg production in painted lady butterflies. *J. Exp. Biol.* 156, 249–265.

Harder, L. D. (1982). Measurement and estimation of functional proboscis length in bumblebees (Hymenoptera: Apidae). *Can J. Zool.* 60, 1073–1079.

Harder, L. D. (1983). Flower handling efficiency of bumble bees: morphological aspects of probing time. *Oecologia (Berlin)* 57, 274–280.

Harder, L. D. (1985). Morphology as a predictor of flower choice by bumble bees. *Ecology* 66, 198–210.

Harder, L. D. (1986). Effects of nectar concentration and flower depth on flower handling efficiency of bumble bees. *Oecologia (Berlin)* 69, 309–315.

Heinrich, B. (1976). Resource partitioning among some eusocial insects: bumblebees. *Ecology* 57, 874–889.

Heyneman, A. (1983). Optimal sugar concentrations of floral nectars—dependence on nectar energy flux and pollinator foraging costs. *Oecologia (Berlin)* 60, 198–213.

Hocking, B. (1971). Blood-sucking behavior of terrestrial arthropods. *Annu. Rev. Entomol.* 16, 1–26.

Hoffman, G. D., and McEvoy, P. B. (1986). Mechanical limitations on feeding by meadow spittlebugs *Philaenus spumarius* (Homoptera: Cercopidae) on wild and cultivated host plants. *Ecol. Entomol.* 11, 415–426.

Horsfield, D. (1978). Evidence for xylem feeding by *Philaenus spumarius* (L.) (Homoptera: Cercopidae). *Entomol. Exp. Appl.* 24, 95–99.

Inouye, D. W. (1976). Resource partitioning and community structure: a study of bumblebees in the Colorado Rocky Mountains. Ph.D. Dissertation, University of North Carolina, Chapel Hill.

Inouye, D. W. (1980). The effect of proboscis and corolla tube lengths on patterns and rates of flower visitation by bumblebees. *Oecologia (Berlin)* 45, 197–201.

Kennedy, J. S., and Mittler, T. E. (1953). A method of obtaining phloem sap via the mouthparts of aphids. *Nature* 171, 258.

Kennedy, J. S., Lamb, K. P., and Booth, C. O. (1958). Responses of *Aphis fabae* (Scop.) to water shortage in host plants in pots. *Entomol. Exp. Appl.* 1, 274–291.

Kimmins, F. M. (1986). Ultrastructure of the stylet pathway of *Brevicoryne brassicae* in host plant tissue, *Brassica oleracea*. *Entomol. Exp. Appl.* 41, 283–290.

Kimmins, F. M., and Tjallingii, W. F. (1985). Ultrastructure of sieve penetration by aphid stylets during electrical recording. *Entomol. Exp. Appl.* 39, 135–141.

Kingsolver, J. G., and Daniel, T. L. (1979). On the mechanics and energetics of nectar feeding in butterflies. *J. Theor. Biol.* 76, 167–179.

Kingsolver, J. G., and Daniel, T. L. (1983a). Mechanical determinants of nectar feeding strategy in hummingbirds: energetics, tongue morphology, and licking behavior. *Oecologia (Berlin)* 60, 214–226.

Kingsolver, J. G., and Daniel, T. L. (1993). The mechanics of fluid feeding in insects. *Proc. Thomas Say Publ. Entomol.* 1, 149–162.

Kinsey, M. G., and McLean, D. L. (1967). Additional evidence that aphids ingest through an open stylet sheath. *Ann. Entomol. Soc. Am.* 60, 1263–1265.

Krenn, H. W. (1990). Functional morphology and movements of the proboscis of Lepidoptera (Insecta). *Zoomorphology* 110, 105–114.

Lehane, M. J. (1991). *Biology of Blood-Sucking Insects*. Harper Collins Academic, London.

Lewis, T. R. (1973). *Thrips: Their Biology, Ecology, and Economic Importance*. Academic Press, New York.

Ma, R., Reese, J. C., Black, W. C., and Bramel-Cox, P. (1990). Detection of pectinesterase and polygalacturonase from salivary secretions of living green bugs, *Schizaphis graminum* (Homoptera: Aphididae). *J. Insect Physiol.* 36, 507–512.

May, P. G. (1985). Nectar uptake rates and optimal nectar concentration of two butterfly species. *Oecologia (Berlin)* 66, 381–386.

McMahon, T. A. (1984). *Muscles, Reflexes, and Locomotion*. Princeton University Press, Princeton, NJ.

Michener, C. D., and Brooks, R. W. (1984). Comparative study of the glossae of bees (Apoidea). *Contrib. Am. Entomol. Inst.* 22, 1–73.

Miles, P. W. (1973). The saliva of Hemiptera. *Adv. Insect Physiol.* 9, 183–255.

Miles, P. W., McLean, D. L., and Kinsey, M. G. (1964). Evidence that two species of aphid ingest food through an open stylet sheath. *Experientia* 20, 582.

Mittler, T. E. (1957). Studies on the feeding and nutrition of *Tuberolachnus salignus* (Gmelin) (Homoptera: Ahididae). I. The uptake of phloem sap. *J. Exp. Biol.* 34, 334–341.

Mittler, T. E. (1958a). Studies on the feeding and nutrition of *Tuberolachnus salignus* (Gmelin) (Homoptera: Aphididae). II. The nitrogen and sugar composition of ingested phloem sap and excreted honeydew. *J. Exp. Biol.* 35, 74–84.

Mittler, T. E. (1958b). Studies on the feeding and nutrition of *Tuberolachnus salignus* (Gmelin) (Homoptera: Aphididae). III. The nitrogen economy. *J. Exp. Biol.* 35, 626–638.

Mittler, T. E. (1967a). Effect of amino acid and sugar concentrations on the food uptake of the aphid *Myzus persicae*. *Entomologia Exp. Appl.* 10, 39–51.

Mittler, T. E. (1967b). Water tensions in plants—an entomological approach. *Ann. Entomol. Soc. Am.* 60, 1074–1076.

Mittler, T. E., and Dadd, R. H. (1963a). Studies on the artificial feeding of the aphid *Myzus persicae* (Sulzer). I. Relative uptake of water and sucrose solutions. *J. Insect Physiol.* 9, 623–645.

Mittler, T. E., and Dadd, R. H. (1963b). Studies on the artificial feeding of the aphid *Myzus persicae* (Sulzer). II. Relative survival, development, and larvipostion on different diets. *J. Insect Physiol.* 9, 741–757.

Muller, H. (1888). On the fertilisation of flowers by insects and on the reciprocal adaptations of both. *Nature* 8, 187–189.

Pivnick, K. A., and McNeil, J. N. (1985). Effects of nectar concentration on butterfly feeding: measured feeding rates for *Thymelicus lincola* (Lepidoptera: Hespiridae) and a general feeding model for adult Lepidoptera. *Oecologia (Berlin)* 66, 226–237.

Pollard, D. G. (1973). Plant penetration by feeding aphids (Hemiptera, Aphidoidea): a review. *Bull. Entomol. Res.* 62, 631–714.

Preston, R. D. (1952). Movement of water in higher plants. In Frey-Wyssling, A. (ed.), *Deformation and Flow in Biological Systems*. Amsterdam.

Prosser, C. L. (1973). *Comparative Animal Physiology*. W. B. Saunders, Philadelphia.

Pyke, G. H., and Waser, N. M. (1981). On the production of dilute nectars by hummingbird and honeyeater flowers. *Biotropica* 13, 260–270.

Roubik, D. W., and Buchmann, S. L. (1984). Nectar selection by *Melipona* and *Apis mellifera* (Hymenoptera: Apidae) and the ecology of nectar intake by bee colonies in a tropical forest. *Oecologia (Berlin)* 61, 1–10.

Scolander, P. F., Hammel, H. T., Bradstreet, E. D., and Hemmingsen, E. A. (1965). Sap pressure in vascular plants. *Science* 148, 339–346.

Smith, J. J. B. (1979). Effect of diet viscosity on the operation of the pharyngeal pump in the blood-feeding bug *Rhodnius prolixus*. *J. Exp. Biol.* 82, 93–104.

Snodgrass, R. E. (1935). *Principles of Insect Morphology*. McGraw–Hill, New York.

Snodgrass, R. E. (1944). The feeding apparatus of biting and sucking insects affecting man and animals. *Smithson. Misc. Collect.* 104, 1–113.

Snodgrass, R. E. (1956). *Anatomy of the Honey Bee*. Comstock Publications, Ithaca, New York.

Spiller, N. J., Kimmins, F. M., and Llewellyn, M. (1985). Fine structure of aphid stylet pathways and its use in host plant resistance studies. *Entomol. Exp. Appl.* 38, 293–295.

Stephens, D. W. and Krebs, J. R. (1986). *Foraging Theory*. Princeton University Press, Princeton, NJ.

Stevenson, R. D. (1992). Feeding rates of the tobacco hawkmoth *Maduca sexta* at artificial flowers. *Am. Zool.* 31, 57A.

Tamm, S., and Gass, C. L. (1986). Energy intake rates and nectar concentration preferences by hummingbirds. *Oecologia (Berlin)* 70, 20–23.

Taraschi, T. F., Parashar, A., Hooks, M., and Rubin, H. (1986). Perturbation of red cell membrane structure during intracellular maturation of *Plasmodium falciparum*. *Science* 232, 102–105.

Tawfik, M. S. (1968). Feeding mechanisms and the forces involved in some blood-sucking insects. *Quaest. Entomol.* 4, 92–111.

Vogel, S. (1981). *Life in Moving Fluids*. Willard Grant Press, New York.

Wainwright, S. A., Biggs, W. D., Currey, J. D., and Gosline, J. M. (1976). *Mechanical Design in Organisms*. Edward Arnold, London.

Watt, W. B., Hoch, P. C., and Mills, S. G. (1974). Nectar resource use by *Colias* butterflies. *Oecologia (Berlin)* 14, 353–374.

Wearing, C. H. (1967). Studies on the relations of insect and host plant. II. Effects of water stress in host plants in pots on the fecundity of *Myzus persicae* (Sulz.) and *Brevicoryne brassicae* (L.). *Nature* 213, 1052–1053.

Wearing, C. H. (1968). Responses of aphids to pressure applied to liquid diet behind parafilm membrane: longevity and larviposition of *Myzus persicae* (Sulz.) and *Brevicoryne brassicae* (L.) (Homoptera: Aphididae) feeding on sucrose and sinigrin solutions. *N. Z. J. Sci.* 11, 105–121.

Wearing, C. H., and Van Emden, H. F. (1967). Studies on the relations of insect and host plant. I. Effects of water stress in host plants in pots on infestation by *Aphis fabae* Scop., *Myzus persicae* (Sulz.) and *Brevicoryne brassicae* (L.). *Nature* 213, 1051–1052.

Weatherley, P. E., Peel, A. J., and Hill, G. P. (1959). The physiology of the sieve tube. Preliminary experiments using aphid mouthparts. *J. Exp. Bot.* 10, 1–16.

3

Insect Saliva: Function, Biochemistry, and Physiology

J. M. C. Ribeiro

3.1. Introduction

Insects utilize a diverse array of feeding strategies and hosts, and thus they encounter problems associated with the mechanical or biochemical properties of their food. Part of the solution to these problems is provided by the secretion of saliva. Often these glands are the largest exocrine organs in insects and other arthropods, and they may be essential for successful feeding. They may serve a *physical* role by solubilizing or adding a liquid vehicle to solid foods, as in most chewing insects, or they may serve a *physiological* role by actually making food available to ingestion as in most stylet-feeding insects.

Many diseases of medical, veterinary, and agricultural interest are transmitted by insect or tick vectors to their hosts; in many cases, pathogen inoculation is via saliva. Diseases of vertebrates such as malaria, babesiosis, and many viral diseases are transmitted this way. Plants can become infected with *Phytomonas* protozoa or with viral diseases by inoculation of the disease agent through vector saliva (Freymuller et al., 1990). Many aphids, for example, do their damage to agricultural crops mostly by the viruses they transmit (Miles, 1972). Accordingly, salivary glands of vector arthropods are a subject of interest for parasitologists and arbovirologists.

3.2. Salivary Gland Structures

Insect salivary glands are composed of a unicellular layer of ectodermal epithelium organized in three general structural types: (1) *tubular* glands, where the cells are organized around a central duct which may be lined by cuticle, and most of the proteinaceous secretion is stored in apical vesicles within the cells (as seen in most Diptera); (2) *alveolar* (or acinar) glands, where the cells are organized like grapes around a branching duct structure (as in grasshoppers and ticks); and (3) *reservoir* glands, where the cells are organized around a large bladder-like cavity that contains most of the secretory material (as in many Heteroptera) (Fig. 3.1).

Reservoir-type salivary glands may also be bound by a layer of smooth muscle cells that help in the emptying of the glands during feeding (Ammar, 1984; Baptist, 1941). In the other gland types, secretion is accomplished through a net water flow through the epithelium. These structural types may also be organized in several lobes, often with different cellular types, adding another dimension to the gland's structure. Accessory salivary glands may also be found in many insects and probably reflect a division of labor regarding a fluid-rich versus an enzyme-rich salivary demand (Goodchild, 1966). Sometimes intermediate forms are also seen, such as the tubular glands of anopheline mosquitoes that contain enlarged distal lobes lacking a cuticular lining, but with secretory material stored extracellularly, as in reservoir glands.

Salivary glands will often have distinct cell types, associated with different types of secretion. Thus, the anterior lobe of Heteroptera is associated with the production of sheath materials, and the posterior lobe is associated with enzyme-rich secretion (Miles, 1972). In adult female mosquitoes, the anterior lobes are associated with sugar feeding, and the posterior lobes are associated with blood feeding (Marinotti et al., 1990).

These basic structural types may indicate the demands on the glands during feeding. Alveolar glands represent a structure that maximizes cell surface in contact with the hemocoel, and they reflect modes of salivation that require a large and sustained secretion of watery saliva (as in grasshoppers and ticks). A reservoir type may indicate the needs of an insect that feeds occasionally and intensely, quickly discharging a large amount of saliva but requiring a time period for recharge of the glands that may take days or weeks (as in triatomine bugs).

In most insects the common salivary duct discharges into a pump structure, located in the head posterior to the base of the feeding stylets. This structure is morphologically similar to the pharyngeal feeding pump, but is of much smaller size. Its function is to pump the salivary fluids into the meal or the host. In many cases (as in aphids), the sap is present at several atmospheres of pressure, and saliva has to be ejected against these enormous pressure gradients.

3.3. Physical Function of Saliva

3.3.1. Solubilizer or Carrier of Solid Foods

Terrestrial insects feeding on solid food require the mixing of food with water to allow proper presentation of phagostimulants to receptors and also to allow the digestive enzymes to act. Although there are gut structures that may secrete

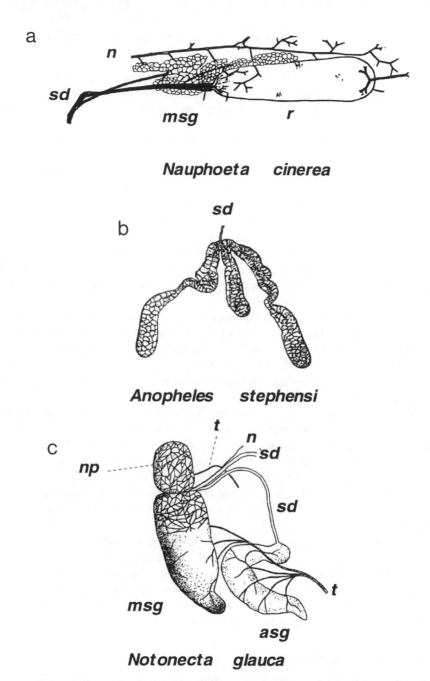

Figure 3.1. Salivary glands. (**a**) Alveolar (or acinar) salivary glands of the cockroach, *Nauphoeta cinerea*. Notice that this gland has a reservoir which is separate from the glandular tissue. (Modified from Bowser-Riley, 1978.) (**b**) Tubular salivary glands of the mosquito, *Anopheles stephensi*. (Modified from Wright, 1969.) (**c**) Reservoir salivary glands of the water bug, *Notonecta*. (Modified from Baptist 1941.) Abbreviations: asg, accessory salivary gland; msg, main salivary gland; n, nerve; np, nervous plexus; r, reservoir; sn, salivary nerve; sd, salivary duct; t, trachea.

water to help in this process, saliva is universally added to the meal. In chewing insects, saliva holds together the small solid pieces of food before ingestion and initiates the solubilization of these particles. Additionally, saliva may be an important component of the food-harvesting process itself. Solid foods, like dried honeydew or sugar, are dissolved by alternating egestion of saliva and suction of the externally dissolved food (Eliason, 1963). In the housefly, saliva wets the spongy mouthparts which are applied onto food surfaces for food gathering. In some fluid-feeding phytophagous insects (leafhoppers, for instance), a copious amount of salivary fluid is ejected while the feeding stylets rapidly move into the plant tissues, followed by withdrawal of this mixture of saliva and cell debris into the food canal (Miles, 1972).

In many aquatic insects, the salivary glands may produce a viscous material that spreads over special mouthpart structures and that aids in the capture of food particles, as in the Simuliidae and Chironomidae (Ross and Craig, 1980). Accordingly, nets of filamentous material may be produced which serve as a sieving device for food harvest. Synthesis of these proteinaceous products may be intense, but these are ingested and recycled with the food. The salivary glands of such insects usually display polytenic chromosomes as a means of amplifying the message required for producing vast amounts of a small number of proteins.

3.3.2. Sealing Around Feeding Mouthparts

In many fluid-feeding insects and acari, the mouthparts may be embedded for long periods of time in the host tissues during feeding. In the case of aphids feeding on plant sap, fluid in the sieve tubes is under high pressure. A viscous salivary secretion is often produced by such insects during penetration. This specialized saliva hardens and seals the mouthparts in place, forming a "stylet sheath" that prevents removal or leakage of sap. Upon removal, a proteinaceous cone-shaped structure is left behind (Fig. 3.2). Oxidizing enzymes are secreted with the sheath material and are probably associated with the hardening of the structure. Indeed, removal of oxygen prevents hardening when salivary gland homogenates are mixed *in vitro* (Miles, 1972).

Insects, such as seed-feeding Heteroptera, produce a hardening flange on the surface of the food substrate. This flange enables the insect to firmly attach the labium to the surface and prevents it from sliding sideways during penetration (Miles, 1972). Similarly, ixodid ticks produce a salivary cement substance that fixes these acari in place during the feeding process (Kemp et al., 1982). The hardening process of these cement substances is poorly known, but an understanding could lead to novel water-based cement substances of practical use.

3.3.3. Maintenance of the Feeding Mouthparts

It has been observed that mosquitoes whose salivary glands have been destroyed by arboviral infections developed lesions on the mouthparts characterized by the

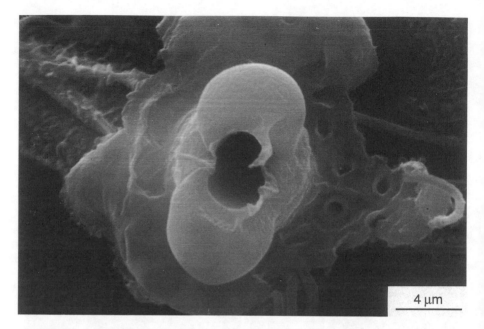

Figure 3.2. Scanning electron micrograph of the salivary flange produced by the plant-hopper, *Peregrinus maidis,* when feeding on *Sorghum.* The hole in the center was formed when the stylets were removed from the plant substrate. The small holes to the right on the flange surface were made by chemoreceptors at the tip of the labium. (Photograph courtesy of R. Chapman.)

separation of the feeding stylets (Grimstad et al., 1980). Perhaps a continuous low level of salivary secretion is needed for the maintenance (keeping proper moisture content) of the feeding mouthparts in mosquitoes and other insects.

3.4. Physiological Function of Saliva

3.4.1. Preingestive Food Treatment

Insects with chewing mouthparts mechanically reduce the food to small particles before ingestion. Saliva aids this process by initiating the solubilization of the food and by contributing enzymes, such as amylase in cockroaches. Indeed, in *Periplaneta* species, most (if not all) gut amylase activity is of salivary origin (Agrawal and Bahadur, 1978, 1981). Similarly, larval blowflies produce a saliva that is rich in proteases, aminopeptidases, and phosphatases (Anderson, 1982), and termites have salivary glands rich in cellulase, cellobiase, amylase, and invertase (Veivers et al., 1982). Insects with piercing mouthparts rely very heavily on their saliva to make food available for ingestion, either by having compounds that counteract host defense responses and/or by producing most of the digestive enzymes for external digestion and liquefaction of the food before ingestion can take place. Clearly, insects with plant-feeding, predacious, or blood-feeding behaviors developed very specialized salivary biochemical tools to deal with their particular feeding needs.

3.4.1.1. PHYTOPHAGOUS INSECTS

In the case of many Hemiptera feeding on plant sap, salivary secretions provide a sealant around the mouthparts and may additionally be important in preventing host defense responses that could interrupt feeding. Indeed, salivary peroxidases and catechol oxidases found in aphids inactivate toxic phytochemicals found in host plants (Miles, 1972; Miles and Peng 1989; Peng and Miles, 1988). Thus saliva of these insects not only serves a physical role, but can counteract plant defense reactions that would otherwise make the tissue unsuitable for feeding. This role is similar to the function of saliva in blood-feeding arthropods, which blocks host substances that arrest blood flow after vessel injury (see below).

Pectinase and other cell wall digestive enzymes, as well as amylase, proteases, and lipases, have been described in the saliva of plant-feeding Hemiptera (Laurema et al., 1985; Ma et al. 1990; Miles, 1972). These enzymes are important in aphids to aid the penetration of the feeding stylets through the hard plant tissues before the sap can be tapped (McAllan and Adams, 1961).

Aphids can also secrete salivary modified amino acids, such as indoleacetic acid (IAA), which are plant growth hormones. These compounds are involved in gall formation. Galls are hypertrophic plant structures that form a shape that protects the feeding insect or the feeding colony. For an excellent review on the saliva of Hemiptera, the reader should consult Miles (1972).

Other plant-feeding insects, such as leaf hoppers or seed-eating Hemiptera, will feed directly on plant tissues, using piercing mouthparts. A mode of feeding described as "lacerate and flush" by Miles (1972) aptly describes the process whereby the flexible maxillae quickly move inside the plant tissues, destroying and breaking the cellular tissue, while at the same time copious amounts of saliva are released. This saliva solubilizes the tissue, allowing suction of this new liquefied tissue.

3.4.1.2. PREDACIOUS INSECTS

Predacious Heteroptera have specialized salivary glands with large activity of digestive enzymes. These insects often prey on other insects of smaller or similar size, but giant water bugs may also prey upon tadpoles, adult amphibians and reptiles, small fish, and even birds. Proteases, aminopeptidases, phosphatases, phospholipases, esterases, and hyaluronidases are injected into the prey through the fast moving maxillae (Cohen, 1993; Edwards, 1961; Rees and Offord, 1969). Additionally, neurotoxins may also be injected that will rapidly paralyze the prey

(Edwards, 1961). After injection of this cocktail of enzymes, the predator often rests with the prey still attached to its mouthparts, while the enzymes work their way into the prey tissues. After enzymatic liquefaction of these tissues, the predator sucks its meal which has been partially digested. The digestive process continues in the gut.

Research on the salivary component of predacious bugs, like those of plant-feeding insects, has been very sporadic and limited to superficial characterization of enzymes and toxins. None has been purified to homogeneity. It is possible that many novel enzymes and neurotoxins could be isolated from these insects, particularly with the present advances in biochemical microtechniques and molecular biology, which have led to cloning and expression of salivary gland-specific genes of insect vectors of human disease (James et al., 1989, 1991; Lerner and Shoemaker, 1992).

3.4.1.3. BLOOD-FEEDING INSECTS

When feeding on blood, insects have to counteract the hemostatic process (i.e., defenses against blood loss) in their host (Fig. 3.3). Hemostasis is a highly redundant phenomenon, where platelet aggregation, blood coagulation, and vasoconstriction act together to prevent blood loss from injured tissues. Each of these three processes also exhibits redundancy, having different pathways that can act simultaneously toward the final common result. As a consequence, blood-feeding insects have evolved a sophisticated array of salivary antihemostatic components that are necessary for the process of finding blood during probing or that keep the blood flowing during feeding (Law et al., 1992; Ribeiro, 1987).

3.4.1.3.1 The Hemostatic Process

Initiation of platelet aggregation occurs through three different pathways triggered by adenosine diphosphate (ADP), collagen, or thrombin. ADP, normally in submicromolar concentrations in the extracellular fluid, is found in millimolar concentrations within cells. When cells are ruptured, they release their nucleotides in the extracellular milieu, and this induces a very rapid (within 5 sec) platelet aggregation. Platelets also have receptors for collagen. Collagen fibrils, present in subendothelial structures, will be available for platelets when the endothelial layer is broken. Collagen activates the platelets by inducing the membrane phospholipase A_2, which releases arachidonic acid; this, in turn, is converted to prostaglandin H_2 and, finally, to thromboxane A_2 (TXA_2). This process may take several seconds or minutes to occur, depending on the collagen concentration. Thromboxane A_2 is a very potent inducer of platelet aggregation and is also a vasoconstrictor agent. Thrombin, formed during the clotting process, can also trigger platelet aggregation by a mechanism independent of the ADP or the collagen/TXA_2 pathway. In addition to aggregating and forming a plug that seals the broken endothelium, platelets may also discharge their granules during this

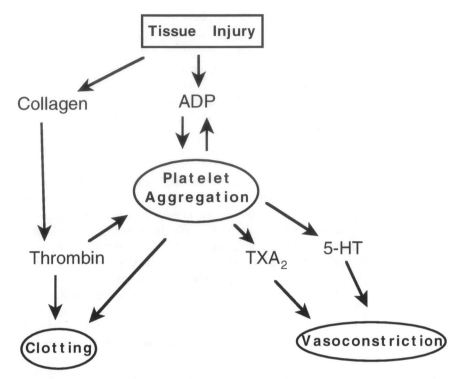

Figure 3.3. Diagram representing the main components and intermediates of the verte-brate hemostatic response. ADP, adenosine diphosphate; 5-HT, 5-hydroxytryptamine (serotonin); TXA_2, thromboxane A_2.

process. These granules contain ADP (whose release recruits more platelets to the platelet plug) and serotonin (a vasoconstrictor). Figure 3.3 summarizes the hemostatic process. The relative importance of each pathway to the hemostatic process in vertebrates in general is not very clear, although a general agreement exists on the importance of the ADP pathway (Law et al., 1992; Leung and Nachman, 1986; Ribeiro, 1987; Vargaftig et al., 1981). Some mammalian spe-cies, like the rat, lack the collagen pathway. Birds and reptiles have nucleated thrombocytes instead of platelets, a much less efficient system, but they compen-sate by having a better protection of the skin with feathers or scales.

Venules and arterioles (but not capillaries) contain a layer of smooth muscle cells, which contract under the influence of TXA_2 or serotonin released by platelets. This contraction tightens the vessel against the platelet plug, preventing further blood loss. Vascular smooth muscle cells additionally respond to many substances that induce either contraction or relaxation.

The endothelial cell, under appropriate stimuli, produces both vasoconstrictor substances (such as the peptide endothelin) and vasodilatory substances (such as nitric oxide and prostacyclin). Nitric oxide (NO), produced by endothelial cells following stimulation with acetylcholine or other peptidic (bradykinin, tachykinins) transmitters, is a potent smooth muscle relaxant (Moncada et al., 1991). NO stimulates the muscle production of cyclic guanosine monophosphate (cGMP), which activates phosphorylating events leading to relaxation.

The variety of mediators of relaxation and constriction of vascular smooth muscle reflects the needs for controlling local blood flow in tissues according to the organism's metabolic demands. These complex demands on the vascular muscle also create a large target for manipulation by blood-sucking arthropods, as will be seen later in this chapter.

Blood clotting is also a redundant phenomenon; it can be triggered by two independent mechanisms, the intrinsic and the extrinsic pathways. The intrinsic pathway starts with factor XII, which is activated by certain solid surfaces, such as subendothelial connective tissue fibrils or, artificially, with glass or kaolin. Activated factor XII is a selective protease, which acts on factor XI and starts a cascade of proteolytic events leading to activation of factor X. Factor X converts prothrombin to thrombin. Thrombin converts fibrinogen to fibrin, which is responsible for the blood clot. Because plasma alone in a glass tube clots through this pathway (without any additional substances), it is called the intrinsic pathway. However, if tissue factors (thromboplastin, a complex mixture of phospholipids and protein) are added to plasma, activation of factor X occurs faster and through a different set of enzymes that constitute the extrinsic pathway. Both pathways converge to the activation of factor X, which leads to thrombin generation and the formation of the fibrin clot. The role of the fibrin clot in hemostasis is to consolidate the platelet plug, preventing it from being washed away by the flowing blood during the hemostatic process.

3.4.1.3.2. Role of Saliva in Blood Feeding

When exposed to a host, blood-feeding insects introduce their mouthparts into their host's skin and then start searching for blood in their probing phase in feeding (see Chapter 6; Friend and Smith, 1971; Gordon and Lumsden, 1939; Lavoipierre, 1964; Lavoipierre et al., 1959). Indeed, blood is present in less than 5% of the skin volume and is not obviously available until some vessel is ruptured. During this probing process, saliva is ejected continuously. When salivary glands are surgically removed from the blood-sucking bug *Rhodnius prolixus,* or the salivary ducts of the mosquito *Aedes aegypti* are cut, these insects display a prolonged probing behavior. Additionally, skin sites probed by salivating *Rhodnius prolixus* are marked by a visible hemorrhagic spot and such spots are not seen at sites probed by salivarectomized insects. However, when these insects are exposed to an artificial feeding apparatus, where the meal is offered homogeneously under a warm membrane, no effect of the surgery is seen, indicating a role for saliva in the blood-searching process in a living host (Ribeiro, 1987, 1988a).

The interpretation of these results is that saliva, injected continuously during probing, acts as an antihemostatic and prevents the arrest of small hemorrhages that form during eventual laceration of the blood vessels. These small hemorrhages collect into a small pool of blood (a hematoma). The hematoma considerably enlarges the percentage of blood in the solid volume that is being probed by the insect, thereby increasing the insect's probability of finding blood. Because salivation and feeding use different channels during probing (the small salivary canals are built in the blades that constitute the walls of the larger feeding canal; see Smith, 1985), the insect is salivating and tasting the meal at the same time. Because of the larger size of the hematomas, in comparison to the blood vessels, the insect will have a larger probability of initially tasting the blood inside a hematoma rather than inside a vessel (unless the insect is very luckily situated on top of a larger superficial vein). When the insect senses the phagostimulants, it terminates the probing behavior and starts the feeding process (see Chapter 6) by pumping blood from the hematoma. If the rate of pumping is faster than the rate of hematoma replenishment from the injured vessel, the walls of the hematoma will collapse toward the ruptured vessel, automatically carrying the insect mouthparts toward the vessel, where feeding may continue in a vessel-feeding mode. If the insect empties the hematoma more slowly than the vessel can supply it with blood, the insect will derive the whole meal from the pool of blood and will be in a pool-feeding mode. Thus, saliva is important in this process by allowing pools of blood to form during probing via its antihemostatic activity.

Blood-sucking insects with piercing mouthparts may feed from vessels (e.g., triatomines) or from pools of blood (e.g., blackflies and sandflies), or from both (e.g., mosquitoes). Other insects, such as the tabanids (horseflies and deerflies), have very coarse mouthparts which are larger than the average skin blood vessel. These mouthparts are adapted to cutting through tough skin, and the insect will feed from the superficial hemorrhages that form from these relatively large skin lacerations.

ANTIPLATELETS

To counteract platelet aggregation, blood-sucking insects and ticks have almost universally in common large amounts of a salivary enzyme that hydrolyzes both adenosine triphosphate (ATP) and adenosine diphosphate (ADP) to orthophosphate and adenosine monophosphate (AMP). This enzyme (ATP-diphosphohydrolase or apyrase) destroys ADP released by injured cells, abolishing one of the important signals for platelet aggregation [for a review see Law et al. (1992) and Table 3.1]. Other substances, such as prostaglandin E_2 (PGE_2) or prostacyclin found in ticks, are also inhibitors of platelet aggregation, in addition to being vasodilators. A platelet inhibitor was found in the salivary glands of deerflies that interferes with fibrinogen binding to platelets (Grevelink et al., 1993), in a manner similar to peptides found in snake venoms (Dennis et al., 1990). It is

Table 3.1. Antihemostatic compounds employed by various blood-sucking arthropods

Activity	Arthropods	Species	Mode of Action	Reference[a]
Anticoagulant[b]	Ticks	*Ornithodorus moubata*	Anti-Xa	Waxman et al. (1990)
		Dermacentor andersoni	Anti-V, Anti-VII	Gordon and Allen (1991)
	Bugs	*Rhodnius prolixus*	Anti-VIII, anti-thrombin	Hellmann and Hawkins (1964, 1965), Friedrich et al. (1993)
	Blackfly	*Eutriatoma maculatus*	Anti-thrombin	Hellmann and Hawkins (1966)
		Simulium vittatum	Anti-X, anti-thrombin	Jacobs et al. (1990)
	Tsetse	*Glossina morsitans*	Anti-thrombin	Parker and Mant (1979)
Antiplatelet	Ticks	*Ornithodorus moubata*	Apyrase	Ribeiro et al. (1991)
		Ixodes dammini	Apyrase, PGI$_2$(?)	Ribeiro et al. (1985a, 1988)
	Bug	*Rhodnius prolixus*	Apyrase, NO	Sarkis et al. (1986), Ribeiro et al. (1990a)
	Fleas	*Oropsylla bacchi*	Apyrase	Ribeiro et al. (1990b)
		Orchopea howardii	Apyrase	Ribeiro et al. (1990b)
		Xenopsylla cheopis	Apyrase	Ribeiro et al. (1990b)
	Mosquitoes	*Anopheles freeborni*	Apyrase	Ribeiro et al. (1985b)
		Anopheles stephensi	Apyrase	Ribeiro et al. (1985b)
		Aedes aegypti	Apyrase	Ribeiro et al. (1984)
	Deerfly	*Chrysops* spp.	Disintegrin	Grevelink et al. (1993)
	Blackfly	*Simulium vittatum*	Apyrase	Cupp et al. (1993)
	Sandflies	*Lutzomyia longipalpis*	Apyrase	Ribeiro et al. (1986)

	Species	Activity	Reference[a]
Vasodilator	*Phlebotomus papatasi*	Apyrase	Ribeiro et al. (1989a)
	Phlebotomus argentipes	Apyrase	Ribeiro et al. (1989a)
	Phlebotomus perniciosus	Apyrase	Ribeiro et al. (1989a)
Tsetse	*Glossina morsitans*	Apyrase	Mant and Paker (1981)
Ticks	*Ixodes dammini*	PGE_2	Ribeiro et al. (1985a)
	Boophilus microplus	PGE_2	Dickinson et al. (1976) Higgs et al. (1976)
	Amblyomma americanum	PGE_2, $PGF_{2\alpha}$	Ribeiro et al. (1992)
Bug	*Rhodnius prolixus*	NO compound	Ribeiro et al. (1990a), Ribeiro et al. (1993)
Mosquitoes	*Aedes aegypti*	Tachykinin	Ribeiro (1992), Champagne and Ribeiro (1994)
	Anopheles albimanus	Peroxidase	Ribeiro and Nussenzveig (1993)
Blackfly	*Simulium vittatum*	Marydilan	Cupp et al. (1994)
Sandfly	*Lutzomyia longipalpis*	Maxidilan	Ribeiro et al. (1989b); Lerner et al. (1991); Lerner and Shoemaker (1992)

[a] References given are not necessarily the original description of the activity, but rather the most recent or comprehensive.

[b] Only anticoagulants which characterized the site of action on the clotting cascade were added. For an extensive review of species with mostly uncharacterized anticlotting activities, see Gooding (1972). Some of the activities reported were also extracted from whole bodies and could derive from organs other than the salivary glands.

possible that this class of platelet inhibitors is widespread in addition to apyrase activity.

VASODILATORS

It is of obvious advantage to blood-feeding arthropods to increase the blood flow of their host's skin, or to counteract vasoconstriction induced by the hemostatic process. In contrast to the common presence of salivary apyrases, insects have found very varied solutions to this problem. For example, the triatomine bug *Rhodnius prolixus* has a salivary hemeprotein that stabilizes, binds, and releases the labile vasodilator NO into the host tissues. Ticks produce vasodilatory prostaglandins E_2 and $F_{2\alpha}$ and prostacyclin. A decapeptide of the tachykinin family is found in the mosquito, *Aedes aegypti*. A larger peptide of novel structure, the most potent vasodilator known, was found in the sandfly, *Lutzomyia longipalpis*, and an oxidase/peroxidase enzyme that destroys catecholamines and serotonin in the mosquito, *Anopheles albimanus*. These insects have very different salivary vasodilators, but all have large amounts of salivary apyrase activity in common (Table 3.1). It is expected that many more novel vasodilators will be found in the saliva of these arthropods.

ANTICOAGULANTS

Since the beginning of the century, anticoagulants have been known to exist in the salivary glands of blood-sucking insects (Cornwall and Patton, 1914). Anticoagulants active against several of the clotting cascade factors have been described. Recently, the health industry has developed a particular interest in those antagonizing active factor X, which would be of potential value in treating thromboembolic diseases. Accordingly, salivary peptides from ticks, biting flies, and leeches have been isolated and are currently under laboratory study or are already in clinical trials (Law et al., 1992; Table 3.1).

3.4.2. Noningestive Functions of Saliva

3.4.2.1. ROLE OF VECTOR SALIVA IN PATHOGEN TRANSMISSION

Many of the pharmacological mediators that influence hemostasis can also modulate immune mechanisms, thereby enhancing pathogen transmission and the infection process. Indeed, sandfly salivary gland homogenates, co-injected with *Leishmania* parasites, enormously increase the virulence of the inoculum. When low numbers of parasites (10–100), such as those delivered in nature, are co-injected with 0.1 pair of salivary glands, infection is readily established; however, infection is aborted in the absence of saliva. Injection of the parasites at the sites probed by the flies also promotes infection (Titus and Ribeiro, 1990). Similarly, experimental transmission of Thogoto virus by ticks is also modified by the presence of ticks, saliva, or salivary gland homogenates at the site of inoculation (Jones et al., 1990, 1992).

Salivary gland infection by *Plasmodium* species or viruses may also influence transmission by decreasing the amount of the salivary compounds that neutralize the host's defense reaction (Grimstad et al., 1980; Rossignol et al., 1984, 1986). Thus, *Plasmodium*-infected mosquitoes exhibit a longer probing time. This change in behavior could increase the efficiency of parasite transmission, because, although salivary gland apyrase levels were decreased, volume of the saliva output was not.

Plant viruses transmitted by aphids have also been reported to be influenced by vector saliva making plant cells either more or less susceptible to infection (Sylvester, 1962; Miles, 1968). The effect of vector saliva in disease transmission should be considered in all cases where needle transmission of the pathogen to the host is inefficient (Ribeiro, 1989).

3.4.2.2. WATER BALANCE

In some insects and in ticks, the salivary glands also function as an important water-regulating organ. During blood feeding in ixodid ticks, blood water is absorbed to the hemocoel from the gut and is injected back into the host as saliva. The volume of saliva produced by a feeding tick vastly exceeds the volume of water lost by the rectum, indicating the major contribution of the salivary glands to water regulation (Kaufman and Phillips 1973a; Kaufman et al., 1980). When off the host, resting ticks produce a hygroscopic saliva that spreads over the mouthparts and collects atmospheric moisture. This saliva is ingested and greatly contributes to water balance (Needham and Teel, 1986). Similarly, in some sap-feeding Homoptera, the anterior midgut does not join the hindgut. The food is absorbed from the midgut to the hemocoel, and most fluid is excreted by the salivary glands (Miles 1972). Insects that do not feed on water-rich food may also use their salivary glands for osmoregulation: when challenged with a hyposmotic liquid load into their hemocoel, cockroaches will produce a hyposmotic saliva that can be reingested and presumably stored in the gut for later uptake (Veenstra, *personal communication*).

3.4.2.3. PHEROMONE PRODUCTION

The German cockroach, *Blattella germanica,* produces a dispersion-inducing contact pheromone in its salivary glands that is presently under study (Faulde et al., 1989, 1990). It is of proteinaceous nature and is produced by nymphs and adults, and it may stay active for a period of at least 15 days. Saliva from nymphs has the highest activity. Saliva collected from crowded females had higher dispersion-inducing ability than did saliva from control gravid females. This activity may act as (a) a defense pheromone of early nymphal stages preventing cannibalism by adult cockroaches and (b) a dispersion pheromone under stress-inducing conditions acting as a regulator of space in order to diminish population density.

3.4.2.4. ANTIMICROBIAL ACTIVITY

Adult male and female *Aedes aegypti* mosquitoes contain a salivary lysozyme activity, which is ingested with the sugar meal and appears in the insect's crop (Pimentel and Rossignol, 1990). This activity may be involved in the control of microbial growth in the crop and gut, and it is analogous to salivary and lacrimal lysozymes of vertebrates. It is likely that other antimicrobial activities will be found in insect saliva, just as they are found in vertebrates (Hart and Powell, 1990).

3.4.2.5. DEFENSE MECHANISMS

Some predacious bugs, such as *Platymeris rhadamanthus,* also use their salivary secretions as a defense mechanism. When disturbed, these insects point their mouthparts at the aggressor and squirt saliva up to a distance of 30 cm. This secretion is very irritating to mucosal membranes, and it deters attacks by rodents if the saliva hits their eyes (Edwards, 1961). This behavior, however, is not widespread among predacious bugs.

3.5. Regulation of Saliva Secretion

Both endocrine and neural regulation of salivary secretion are found in insects (House and Ginsborg, 1985). The role of second messengers of cellular secretion has also been investigated. The classic work by Berridge and colleagues led to the identification of the phosphoinositide pathway in the salivary glands of the blowfly, *Calliphora,* an important intracellular second messenger pathway later found in vertebrates (Berridge, 1987). This and related pathways have been studied in varying detail in other arthropods, namely, the blowfly, *Calliphora* (Bay, 1978; Hansen-Bay, 1978; Trimmer, 1985), the cockroach, *Nauphoeta cinerea* (House, 1980; House and Ginsborg, 1979; House and Smith, 1978), the migratory locust, *Locusta migratoria* (Baines and Tyrer, 1989; Baines et al., 1989; Lafon-Cazal and Bockaert, 1984), and *Amblyomma* ticks (Kaufman, 1978; Kaufman and Phillips, 1973a,b; Kaufman et al., 1980; Pannabecker and Needham, 1985; Sauer et al., 1986).

Isolated salivary glands of many insect and tick species have been used to test the effect of secretagogues (substances that elicit glandular secretion), by measuring the volume and amount of enzyme secreted at the end of the isolated salivary duct kept either in the air, outside the bathing solution, or under mineral oil (Hansen-Bay, 1978; Kaufman, 1978). Serotonin stimulates salivary secretion in blowflies (Hansen-Bay, 1978; Trimmer, 1985) and dopamine stimulates in ixodid ticks (Kaufman, 1978). Electrophysiological techniques, once popular to study electrotonic junctions in larval chironomid salivary glands (Loewenstein et al., 1978; Suzuki, 1989), have not been used widely to investigate the invertebrate cellular response to secretagogues. This technique has attained maturity in vertebrate pancreas and salivary glands (Petersen and Gallacher, 1988) and could be useful in invertebrate studies, as demonstrated by Wuttke and Berry (1991) in the leech salivary gland.

Detection of nerves serving the salivary glands has been accomplished using vital methylene blue staining or, more recently, with cobalt filling techniques, and these observations are scattered in the anatomical literature. Most insect salivary glands are innervated, but the mediators involved in secretion have been identified in very few insects. Serotonin, dopamine, noradrenaline, proctolin, and YGGFMRFamide were found by high-performance liquid chromatography (HPLC) and by immunocytochemical techniques in the salivary glands or salivary gland nerves of *Locusta migratoria* (Ali et al., 1993; Baines and Tyrer, 1989; Baines et al., 1989). Pilocarpine, a cholinomimetic drug, induces ticks to salivate by acting on the nervous ganglia that innervate the salivary glands, but has no effect on the isolated glands (Kaufman, 1978).

Some insects, such as the blowfly *Calliphora,* do not have innervated salivary glands and rely on hormones for stimulation of secretion (Bay, 1978; Hansen-Bay, 1978; Trimmer, 1985). It is interesting that serotonin, long thought of as a mediator acting at the synapse level, reaches pharmacological concentrations in the hemolymph of feeding *Rhodnius prolixus* (Lange et al., 1989), following release from neurohemal organs (Orchard, 1989). Serotonin is also a potent stimulator of water transport in epithelia, such as the midgut (Farmer et al., 1981), epidermal cells (for cuticle plasticization), and malpighian tubules in this bug (Maddrell et al., 1991). Although the salivary gland in this insect is innervated, and hemolymph serotonin may be involved in other functions, other insect species lacking salivary gland nerves may have their salivary glands stimulated, at least in part, by hemolymph serotonin. Indeed, *Calliphora* salivary glands lack innervation, and they are stimulated by serotonin (Trimmer, 1985). Other mediators, possibly of peptidic nature, may also be involved.

Regulation of secretion will likely turn out to be a complex phenomenon, involving multiple neuromediators that can be individually responsible for short bursts of watery saliva, sustained salivation, predominantly proteinaceous (enzymatic) saliva, and so on. Synergism among these mediators should be expected. Some examples of the complexity of this process are given below.

Ticks have morphologically distinct acini implicated in the secretion of hygroscopic substances for atmosphere moisture capture (Binnington, 1978; Needham and Teel, 1986), for producing cement attachment, for water secretion, and for blood feeding (Kemp et al., 1982). These functions are differentiated in time, and the corresponding acini are active depending on the demand. The role of different mediators in each type of secretion remains to be investigated.

Adult female mosquitoes feed on either sugar or blood, and they have specialized regions of the salivary glands to deal with each food type (Orr et al., 1961; Janzen and Wright, 1971; Jensen and Jones, 1957; Wright, 1969). Accordingly, α-glucosidase activity is located in the anterior part of the adult female gland and in male salivary glands, and the enzyme level is reduced following a sugar meal. Apyrase is limited to the posterior, female-specific regions of adult female mosquitoes; it is not reduced following a sugar meal, but is depleted following a blood meal. These observations indicate a dual control of the salivary glands according to the mode of feeding (Marinotti et al., 1990).

Heteroptera have, in addition to the main salivary glands, accessory salivary glands (Baptist, 1941). Dual innervation of the glands has been reported (Miles, 1972). The main glands are rich in protein, while the accessory glands have thin epithelia and are mainly involved in dilute fluid secretion (Goodchild, 1966). The accessory glands could either work alone, providing a fluid-rich, enzyme-poor secretion, or work together with the main secretion, by diluting it. Similarly, the salivary ducts of mosquitoes may play a role similar to that of accessory glands of Hemiptera, because they were shown to be able to contribute to salivary output (Rossignol and Spielman, 1982). Control of each structure involves a complex neural or neuroendocrine effort that depends on the feeding mode of the insect.

References

Agrawal, O. P., and Bahadur, J. (1978). Role of salivary glands in the maintenance of midgut amylase activity in *Periplaneta americana*. *Experientia* 34, 1552.

Agrawal, O. P., and Bahadur, J. (1981). Control of midgut invertase activity in *Periplaneta americana*. The possible non-exocrine role of the salivary glands. *J. Insect Physiol.* 27, 293–295.

Ali, D. W., Orchard, I., and Lange, A. B. (1993). The aminergic control of locust (*Locusta migratoria*) salivary glands: evidence for a dopaminergic and serotonergic innervation. *J. Insect Physiol.* 39, 623–632.

Ammar, E. D. (1984). Muscle cells in the salivary glands of a planthopper *Peregrinus maidis* (Ashmed) and a leafhopper, *Macrosteles fascifrons* (Stal) (Homoptera: Auchenorrhyncha). *Int. J. Insect Morphol. Embryol.* 13, 425–428.

Anderson, O. D. (1982). Enzyme activities in the larval salivary secretion of *Calliphora erythrocephala*. *Comp. Biochem. Physiol.* 72B, 569–576.

Baines, R. A., and Tyrer, N. M. (1989). The innervation of locust salivary glands. II. Physiology of excitation and modulation. *J. Comp. Physiol.* 165A, 407–413.

Baines, R. A., Tyrer, N. M., and Mason, J. C. (1989). The innervation of locust salivary glands. I. Innervation and analysis of transmitters. *J. Comp. Physiol.* 165A, 395–405.

Baptist, B. A. (1941). The morphology and physiology of the salivary glands of Hemiptera–Heteroptera. *Q. J. Microsc. Sci.* 82, 91–139.

Bay, C. M. (1978). The control of enzyme secretion from fly salivary glands. *J. Physiol. (London)* 274, 421.

Berridge, M. J. (1987). Inositol triphosphate and diacylglycerol: two interacting second messengers. *Annu. Rev. Biochem.* 56, 159–193.

Binnington, K. C. (1978). Sequential changes in salivary gland structure during attachment and feeding of cattle tick, *Boophilus microplus. Int. J. Parasitol.* 8, 97–115.

Bowser-Riley, F. (1978). The salivary glands of the cockroach *Nauphoeta cinerea* (Olivier). A study of its innervation by light and scanning electron microscopy. *Cell Tissue Res.* 187, 525–534.

Champagne, D., and Ribeiro, J. M. C. (1984). Sialokinins I and II: Two salivary tachykinins from the yellow fever mosquito, *Aedes aegypti. Proc. Natl. Acad. Sci. USA* 91, 138–142.

Cohen, A. C. (1993). Organization of digestion and preliminary characterization of salivary trypsin-like enzymes in a predaceous Heteropteran, *Zelus renardii. J. Insect Physiol.* 39, 823–829.

Cornwall, J. W., and Patton, W. S. (1914). Some observations on the salivary secretion of the common blood-sucking insects and ticks. *Indian J. Med. Res.* 2, 569–593.

Cupp, M. S., Cupp, E. W., and Ramberg, F. B. (1993). Salivary gland apyrase in black flies (*Simulium vittatum*). *J. Insect Physiol.* 39, 817–821.

Cupp, M. S., Ribeiro, J. M. C., and Cupp, E. W. (1994). Vasodilative activity in black fly salivary glands. *Am. J. Trop. Med. Hyg.* 50, 241–246.

Dennis, M. S., Henzel, W. J., Piti, R. M., Lipari, M. T., Napier, M. A., Deisher, T. A., Bunting, S., and Lazarus, R. A. (1990). Platelet glycoprotein IIb–IIIa protein antagonists from snake venoms: evidence for a family of platelet-aggregation inhibitors. *Proc. Natl. Acad. Sci. USA* 87, 2471–2475.

Dickinson, R. G., O'Hagan, J. E., Shotz, M., Binnington, K. C., and Hegarty, M. P. (1976). Prostaglandin in the saliva of the cattle tick *Boophilus microplus. Aust. J. Exp. Biol. Med. Sci.* 54, 475–486.

Edwards, J. S. (1961). The action and composition of the saliva of an assassin bug *Platymeris rhadamanthus. J. Exp. Biol.* 38, 61–77.

Eliason, P. A. (1963). Feeding adult mosquitoes on solid sugars. *Nature* 200, 289.

Farmer, J., Maddrell, S. H. P., and Spring, J. H. (1981). Absorption of fluid by the midgut of *Rhodnius. J. Exp. Biol.* 94, 301–316.

Faulde, M., Fuchs, M., and Nagl, W. (1989). Dispersion inducing proteins in the saliva of several cockroach species (Blattodea: Blattellidae, Blattidae, Blaberidae). *Entomol. Generalis* 14, 203–210.

Faulde, M., Fuchs, M., and Nagl, W. (1990). Further characterization of a dispersion-inducing contact pheromone in the saliva of the German cockroach, *Blattella germanica* L. (Blattodea: Blattellidae). *J. Insect Physiol.* 36, 353–359.

Freymuller, E., Milder, R., Jankevicius, J. V., Jankevicius, S. I., and Plessman Camargo, E. (1990). Ultrastructural studies on the trypanosomatid *Phytomonas serpens* in the salivary glands of a phytophagous hemipteran. *J. Protozool.* 37, 225–229.

Friedrich, R., Kroger, B., Biajolan, S., Lemaire, H. G., Hoffken, H. W., Reuschenbach, P., Otte, M., and Dodt, J. (1993). A Kazal-type inhibitor with thrombin specificity from *Rhodnius prolixus*. *J. Biol. Chem.* 268, 16216–16222.

Friend, W. G., and Smith, J. J. B. (1971). Feeding in *Rhodnius prolixus:* mouthpart activity and salivation, and their correlation with changes of electrical resistance. *J. Insect Physiol.* 17, 233–243.

Goodchild, A. J. P. (1966). Evolution of the alimentary canal in the Hemiptera. *Biol. Rev.* 41, 97–140.

Gooding, R. H. (1972). Digestive process of haematophagous insects. I. A literature review. *Quaest. Entomol.* 8, 5–60.

Gordon, J. R., and Allen, J. R. (1991). Factors V and VII anticoagulant activities in the salivary glands of feeding *Dermacentor andersoni* ticks. *J. Parasitol.* 77, 167–170.

Gordon, R. M., and Lumsden, W. H. R. (1939). A study of the behaviour of the mouthparts of mosquitoes when taking up blood from living tissue; together with some observations on the ingestion of microfilariae. *Ann. Trop. Med. Parasitol.* 33, 259–278.

Grevelink, S. A., Youssef, D. E., Loscalzo, J., and Lerner, E. A. (1993). Salivary-gland extracts from the deerfly contain a potent inhibitor of platelet aggregation. *Proc. Natl. Acad. Sci.* 90, 9155–9158.

Grimstad, P. R., Ross, Q. E., and Craig, G. B., Jr. (1980). *Aedes triseriatus* (Diptera: Culicidae) and La Cross virus. II. Modification of mosquito feeding behavior by virus infection. *J. Med. Entomol.* 17, 1–7.

Hansen-Bay, C. M. (1978). Control of salivation in the blow fly *Calliphora*. *J. Exp. Biol.* 75, 189–201.

Hart, B. L., and Powell, K. L. (1990). Antibacterial properties of saliva: role in maternal periparturient grooming and in licking wounds. *Physiol. Behav.* 48, 383–386.

Hellmann, K., and Hawkins, R. I. (1964). Anticoagulant and fibrinolytic activities from *Rhodnius prolixus* Stahl. *Nature* 201, 1008–1009.

Hellmann, K., and Hawkins, R. I. (1965). Prolixin-S and Prolixin-G: two anticoagulants from *Rhodnius prolixus* Stahl. *Nature* 207, 265–267.

Hellmann, K., and Hawkins, R. I. (1966). An antithrombin (Maculatin) and a plasminogen activator extractable from the blood-sucking Hemipteran, *Eutriatoma maculatus*. *Br. J. Haematol.* 12, 376–384.

Higgs, G. A., Vane, J. R., Hart, R. J., Porter, C., and Wilson, R. G. (1976). Prostaglandins in the saliva of the cattle tick, *Boophilus microplus* (Canestrini) (Acarina, Ixodidae). *Bull. Entomol. Res.* 66, 665–670.

House, C. R. (1980). Physiology of invertebrate salivary glands. *Biol. Rev.* 55, 417–473.

House, C. R., and Ginsborg, B. L. (1979). Pharmacology of cockroach salivary secretion. *Comp. Biochem. Physiol.* 63C, 1–6.

House, C. R., and Ginsborg, B. L. (1985). Salivary gland. In: Kerkut, G. A. and Gilbert,

L. I. (eds.), *Comprehensive Insect Physiology, Biochemistry and Pharmacology*, vol. 11. Pergamon Press, Oxford, pp. 196–224.

House, C. R., and Smith, R. K. (1978). On the receptors involved in the nervous control of salivary secretion by *Nauphoeta cinerea* Olivier. *J. Physiol. (London)* 279, 457–471.

Jacobs, J. W., Cupp, E. W., Sardana, M., and Friedman, P. A. (1990). Isolation and characterization of a coagulation factor Xa inhibitor from black fly salivary glands. *Thromb. Haemost.* 64, 235–238.

James, A. A., Blackmer, K., and Racioppi, J. V. (1989). A salivary gland-specific, maltase-like gene of the vector mosquito, *Aedes aegypti. Gene* 75, 73–83.

James, A. A., Blackmer, K., Marinotti, O., Ghosn, C. R., and Racioppi, J. V. (1991). Isolation and characterization of the gene expressing the major salivary gland protein of the female mosquito, *Aedes aegypti. Mol. Biochem. Parasitol.* 44, 245–254.

Janzen, H. G., and Wright, K. A. (1971). The salivary glands of *Aedes aegypti* (L.): an electron microscope study. *Can. J. Zool.* 49, 1343–1345.

Jensen, D. V., and Jones, J. C. (1957). The development of the salivary glands in *Anopheles albimanus* Widemann (Diptera, Culicidae). *Ann. Entomol. Soc. Am.* 50, 464–469.

Jones, L. D., Davies, C. R., Williams, T., Cory, J., and Nuttall, P. A. (1990). Non-viraemic transmission of Thogoto virus: vector efficiency of *Rhipicephalus appendiculatus* and *Amblyomma variegatum. Tr. R. Soc. Trop. Med. Hyg.* 84, 846–948.

Jones, L. D., Kaufman, W. R., and Nuttall, P. A. (1992). Modification of the skin feeding site by tick saliva mediates virus transmission. *Experientia* 48, 779–782.

Kaufman, W. R. (1978). Actions of some transmitters and their antagonists on salivary secretion in a tick. *Am. J. Physiol.* 235, R76–81.

Kaufman, W. R., and Phillips, J. E. (1973a). Ion and water balance in the ixodid tick *Dermacentor andersoni.* I. Routes of ion and water excretion. *J. Exp. Biol.* 58, 523–536.

Kaufman, W. R., and Phillips, J. E. (1973b). Ion and water balance in the ixodid tick *Dermacentor andersoni.* II. Mechanism and control of salivary secretion. *J. Exp. Biol.* 58, 537–547.

Kaufman, W. R., Aeschimann, A. A., and Diehl, P. A. (1980). Regulation of body volume by salivation in a tick challenged with fluid loads. *Am. J. Physiol.* 238, R102–R112.

Kemp, D. H., Stone, B. F., and Binnington, K. C. (1982). Tick attachment and feeding: role of mouthparts, feeding apparatus, salivary gland secretions and the host response. In: Obenchain, F. D., and Galun R. (eds.). *Physiology of Ticks.* Pergamon Press, pp. 119–168.

Lafon-Cazal, M., and Bockaert, J. (1984). Pharmacological characterization of dopamine-sensitive adenylate cyclase in the salivary glands of *Locusta migratoria* L. *Insect Biochem.* 14, 541–545.

Lange, A. B., Orchard, I., and Barrett, F. M. (1989). Changes in the haemolymph serotonin levels associated with feeding in the blood-sucking bug, *Rhodnius prolixus*. *J. Insect Physiol.* 35, 393–399.

Laurema, S., Varis, A.-L., and Miettinen, H. (1985). Studies on enzymes in the salivary glands of *Lygus regulipennis* (Hemiptera, Miridae). *Insect Biochem.* 15, 211–224.

Lavoipierre, M. M. J. (1964). Feeding mechanisms on blood-sucking arthropods. *Nature* 208, 302–303.

Lavoipierre, M. M. J., Dickerson, G., and Gordon, R. M. (1959). Studies on the methods of feeding of blood sucking arthropods. I. The manner in which triatomine bugs obtain their blood meal, as observed in the tissues of the living rodent, with some remarks on the effects of the bite on human volunteers. *Ann. Trop. Med. Parasitol.* 53, 235–250.

Law, J., Ribeiro, J. M. C., and Wells, M. (1992). Biochemical insights derived from diversity in insects. *Annu. Rev. Biochem.* 61, 87–112.

Lerner, E. A., and Shoemaker, C. B. (1992). Maxadilan: cloning and functional expression of the gene encoding this potent vasodilator peptide. *J. Biol. Chem.* 267, 1062–1066.

Lerner, E. A., Ribeiro, J. M. C., Nelson, R. J., and Lerner, M. R. (1991). Isolation of maxadilan, a potent vasodilatory peptide from the salivary glands of the sand fly *Lutzomyia longipalpis*. *J. Biol. Chem.* 266, 11234–11236.

Leung, L., and Nachman, R. (1986). Molecular mechanisms of platelet aggregation. *Annu. Rev. Med.* 37, 179–186.

Loewenstein, W. R., Kanno, Y., and Socolar, S. J. (1978). The cell to cell channel. *Fed. Proc.* 37, 2645–2650.

Ma, R., Reese, J. C., Black, W. C., IV, and Bramel-Cox, P. (1990). Detection of pectinesterase and polygalacturonase from salivary secretions of living greenbugs, *Schizaphis graminum* (Homoptera: Aphididae). *J. Insect Physiol.* 36, 507–512.

Maddrell, S. H. P., Herman, W. S., Mooney, R. L., and Overton, J. A. (1991). 5-Hydroxytryptamine: a second diuretic hormone in *Rhodnius prolixus*. *J. Exp. Biol.* 156, 557–566.

Mant, M. J., and Parker, K. R. (1981). Two platelet aggregation inhibitors in tsetse (*Glossina*) saliva with studies of roles of thrombin and citrate *in vitro* platelet aggregation. *Br. J. Haematol.* 48, 601–608.

Marinotti, O., James, A., and Ribeiro, J. M. C. (1990). Diet and salivation in female *Aedes aegypti* mosquitoes. *J. Insect Physiol.* 36, 545–548.

McAllan, J. W., and Adams, J. B. (1961). The significance of pectinase in plant penetration by aphids. *Can. J. Zool.* 39, 305–310.

Miles, P. W. (1968). Insect secretions in plants. *Annu. Rev. Phytopathol.* 6, 137–164.

Miles, P. W. (1972). The saliva of hemiptera. *Adv. Insect Physiol.* 9, 183–255.

Miles, P. W., and Peng, Z. (1989). Studies on the salivary physiology of plant bugs: detoxification of phytochemicals by the salivary peroxidase of aphids. *J. Insect Physiol.* 35, 865–872.

Moncada, S., Palmer, R. M. J., and Higgs, E. A. (1991). Nitric oxide: physiology, pathophysiology, and pharmacology. *Pharmacol. Rev.* 43, 109–142.

Needham, G. R., and Teel, P. D. (1986). Water balance by ticks between blood meals. In: Sauer, J. R., and Hair, J. A. (eds.), *Morphology, Physiology, and Behavioral Biology of Ticks*. John Wiley & Sons, New York, pp. 100–151.

Orchard, I. (1989). Serotonergic neurohemal tissue in *Rhodnius prolixus:* synthesis, release and uptake of serotonin. *J. Insect Physiol.* 35, 943–947.

Orr, C. W. M., Hudson, A., and West, A. S. (1961). The salivary glands of *Aedes aegypti*. Histological–histochemical studies. *Can. J. Zool.* 39, 265–272.

Pannabecker, T., and Needham, G. R. (1985). Effects of octopamine on fluid secretion by isolated salivary glands of a feeding tick. *Arch. Insect. Biochem. Physiol.* 2, 217–226.

Parker, K. R., and Mant, M. J. (1979). Effects of tsetse salivary gland homogenate on coagulation and fibrinolysis. *Thromb. Haemost. (Stuttg)* 42, 743–751.

Peng, Z., and Miles, P. W. (1988). Studies on the salivary physiology of plant bugs: function of the catechol oxidase of the rose aphid. *J. Insect Physiol.* 34, 1027–1033.

Petersen, O. H., and Gallacher, D. V. (1988). Electrophysiology of pancreatic and salivary acinar cells. *Annu. Rev. Physiol.* 50, 65–80.

Pimentel, G. E., and Rossignol, P. A. (1990). Age dependence of salivary bacteriolytic activity in adult mosquitoes. *Comp. Biochem. Physiol.* 96B, 549–551.

Rees, A. R., and Offord, R. E. (1969). Studies on the protease and other enzymes from the venom of *Lethocerus cordofanus*. *Nature* 2212, 675–677.

Ribeiro, J. M. C. (1987). Role of arthropod saliva in blood feeding. *Annu. Rev. Entomol.* 32, 463–478.

Ribeiro, J. M. C. (1988a). How mosquitoes find blood. *Misc. Publ. Entomol. Soc. Am.* 68, 18–24.

Ribeiro, J. M. C. (1988b). Role of saliva in tick–host interactions. *Exp. Appl. Acarol.* 7, 15–20.

Ribeiro, J. M. C. (1989). Vector saliva and its role on pathogen transmission. *Exp. Parasitol.* 69, 104–106.

Ribeiro, J. M. C. (1992). Characterization of vasodilator from the salivary glands of the yellow fever mosquito, *Aedes aegypti*. *J. Exp. Biol.* 165, 61–71.

Ribeiro, J. M. C., and Nussenzveig, R. H. (1993). The salivary catechol oxidase/peroxidase activities of the mosquito, *Anopheles albimanus*. *J. Exp. Biol.* 179, 273–287.

Ribeiro, J. M. V., Sarkis, J. J. F., Rossignol, P. A., and Spielman, A. (1984). Salivary apyrase of *Aedes aegypti:* characterization and secretory fate. *Comp. Biochem. Physiol.* 79B, 81–86.

Ribeiro, J. M. C., Makoul, G., Levine, J., Robinson, D., and Spielman, A. (1985a). Antihemostatic, antiinflammatory and immunosuppressive properties of the saliva of a tick, *Ixodes dammini*. *J. Exp. Med.* 161, 332–344.

Ribeiro, J. M. C., Rossignol, P. A., and Spielman, A. (1985b). Salivary gland apyrase determines probing time in anopheline mosquitoes. *J. Insect Physiol.* 31, 689–692.

Ribeiro, J. M. C., Rossignol, P. A., and Spielman, A. (1986). Blood feeding strategy of a capillary feeding sandfly, *Lutzomyia longipalpis. Comp. Biochem. Physiol.* 83A, 683–686.

Ribeiro, J. M. C., Makoul, G., and Robinson, D. (1988). *Ixodes dammini:* evidence for salivary prostacyclin secretion. *J. Parasitol.* 74, 1068–1069.

Ribeiro, J. M. C., Modi, G. B., and Tesh, R. B. (1989a). Salivary apyrase activity of some old world phlebotomine sand flies. *Insect Biochem.* 19, 409–412.

Ribeiro, J. M. C., Vachereau, A., Modi, G. B., and Tesh, R. B. (1989b). A novel vasodilator peptide from the salivary glands of the sand fly *Lutzomyia longipaplpis. Science* 243, 212–214.

Ribeiro, J. M. C., Gonzales, R., and Marinotti. O. (1990a). Characterization of a salivary vasodilator in the blood-sucking bug, *Rhodnius prolixus. Br. J. Pharmacol.* 101, 932–936.

Ribeiro, J. M. C., Vaughan, J. A., and Azad, A. F. (1990b). Characterization of the salivary apyrase activity of three rodent flea species. *Comp. Biochem. Physiol.* 95B, 215–218.

Ribeiro, J. M. C., Endris, T. M., and Endris, R. (1991). Saliva of the tick *Ornithodorus moubata* contains anti-platelet and apyrase activities. *Comp. Biochem. Physiol.* 100B, 109–112.

Ribeiro, J. M. C., Evans, P. M., MacSwain, J. L., and Sauer, J. (1992). *Amblyomma americana:* characterization of salivary prostaglandins E_2 and $E_{2\alpha}$ by RP-HPLC/bioassay and gas chromatography–mass spectrometry. *Exp. Parasitol.* 74, 112–116.

Ribeiro, J. M. C., Hazzard, J. M. H., Nussenzveig, R. H., Champagne, D., and Walker, F. A. (1993). Reversible binding of nitric oxide by a salivary nitrosylhemeprotein from the blood sucking bug, *Rhodnius prolixus. Science* 260, 539–541.

Ross, D. H., and Craig, D. A. (1980). Mechanisms of fine particle capture by larval black flies (Diptera: Simuliidae). *Can. J. Zool.* 58, 1186–1192.

Rossignol, P. A., and Spielman, A. (1982). Fluid transport across the ducts of the salivary glands of a mosquito (*Aedes aegypti*). *J. Insect Physiol.* 28, 579–583.

Rossignol, P. A., Ribeiro, J. M. C., and Spielman, A. (1984). Increased intradermal probing time in sporozoite-infected mosquitoes. *Am. J. Trop. Med. Hyg.* 33, 17–20.

Rossignol, P. A., Ribeiro, J. M. C., and Spielman, A. (1986). Increased biting rate and reduced fertility in sporozoite-infected mosquitoes. *Am. J. Trop. Med. Hyg.* 35, 277–279.

Sarkis, J. J. F., Guimaraes, J. A., and Ribeiro, J. M. C. (1986). The salivary apyrase of *Rhodnius prolixus:* kinetics and purification. *Biochem. J.* 233, 885–891.

Sauer, J. R., Mane, S. D., Schmidt, S. P., and Essenberg, R. C. (1986). Molecular basis for salivary fluid secretion in ixodid ticks. In: Sauer, J. R., and Hair, J. A. (eds.), *Morphology, Physiology, and Behavioral Biology of Ticks.* John Wiley & Sons, New York, pp. 55–74.

Smith, J. J. B. (1985). Feeding mechanisms. In: Kerkut, G. A., and Gilbert, L. I. (eds.), *Comprehensive Insect Physiology, Biochemistry and Pharmacology,* vol. 4. Pergamon, Oxford, pp. 33–86.

Suzuki, K. (1989). Hydrocortisone and dexamethasone increase intercellular communication in salivary glands of *Chironomus thummi* larvae. *Tokai J. Exp. Clin. Med.* 14, 321–328.

Sylvester, E. S. (1962). Mechanisms of plant virus transmission by aphids. In: Maramorosch, K. (ed.), *Biological Transmission of Disease Agents*. Academic Press, New York, pp. 11–31.

Titus, R. G., and Ribeiro, J. M. C. (1990). The role of vector saliva in transmission of arthropod-borne diseases. *Parasitol. Today* 6, 157–160.

Trimmer, B. A. (1985). Serotonin and the control of salivation in the blow fly *Calliphora*. *J. Exp. Biol.* 114, 307–328.

Vargaftig, B. B., Chignard, M., and Benveniste, J. (1981). Present concepts on the mechanism of platelet aggregation. *Biochem. Pharmacol.* 30, 263–271.

Veivers, P. C., Musca, A. M., O'Brien, R. W., and Slayton, M. (1982). Digestive enzymes of the salivary glands and gut of *Mastotermes darwiniensis* (Termite). *Insect Biochem.* 12, 35–40.

Waxman, L., Smith, D. E., Arcuri, K. E., and Vlasuk, G. P. (1990). Tick anticoagulant peptide (TAP) is a novel inhibitor of blood coagulation factor Xa. *Science* 248, 593–596.

Wright, K. A. (1969). The anatomy of salivary glands of *Anopheles stephensi* Liston. *Can. J. Zool.* 47, 579–587.

Wuttke, W. A., and Berry, M. S. (1991). Initiation and modulation of action potentials in salivary gland cells of *Haementeria ghilianii* by putative transmitters and cyclic nucleotides. *J. Exp. Biol.* 157, 101–122.

Regulation of a Meal

4

Chemosensory Regulation of Feeding

R. F. Chapman

4.1. Introduction

In this chapter, I discuss the chemosensory regulation of feeding once the insect has made contact with a potential food source. The central questions are: what initiates feeding, and what maintains feeding? The factors leading to the end of feeding are discussed extensively in the following chapters.

Most of the discussion will focus on contact chemoreception (gustation or "taste"). This does not imply that I consider olfaction to be unimportant. Many insects have olfactory sensilla on the mouthparts (see below), and odors may be in high concentration within the boundary layer of air above any object. Because of their small size, insects may normally operate within, or at least partly in, this boundary layer. For these reasons, it seems very likely that olfaction plays a significant role in regulating feeding behavior close to the food source. However, this possible role has been largely ignored because of the obvious importance of odor in host-finding from a distance (see Visser, 1986), and also because of the difficulty of working with odor in this context.

In fact, there is no clear distinction between olfaction and gustation. For example, sensilla which structurally and electrophysiologically are considered to be gustatory have also been shown to respond to odors (Städler and Hanson, 1975). Dethier (1993) suggests that the most obvious objective separation of gustation and olfaction rests on the fact that the latter has a clearly defined neural center with a glomerular structure, while the former does not. The neural projections of insect peripheral chemoreceptors within the central nervous system are considered in Section 4.2.4.

The structure and physiology of insect chemoreceptors is described by Zacharuk (1985), Morita and Shiraishi (1985), and Frazier (1992). Städler (1984) gives a general account of contact chemoreception. The sensilla of immature insects, including those associated with feeding, are described by Zacharuk and Shields (1991).

4.2. The Sensory Array

Chemoreceptors associated with ingestion are present on the mouthparts and, in many insects, also on the tarsi. In addition to these, many insects appear to employ the antennae in monitoring their food, by vibrating the antennae (antennating) on or close to the surface of the food. In many species there are contact chemoreceptors on the tips of the antennae, but olfactory receptors may also be employed at this time. In caterpillars, where the antennae are small and closely associated with the mouthparts, they are usually assumed to be directly involved in food selection (e.g., de Boer, 1993).

The antennae are not included in the following account.

4.2.1. Chemoreceptors on the Mouthparts of Biting and Chewing Insects

Amongst the mandibulate insects, chemoreceptors are often present on the labrum and labium, and they are probably universally present on parts of the maxilla. In addition, a small number of sensilla occur in the walls of the cibarium, but chemoreceptors never occur on the mandibles (Zacharuk, 1985). The Thysanura and orthopteroid orders appear to differ from mandibulate holometabolous insects in having much larger numbers of chemosensitive sensilla on the mouthparts (Figs. 4.1 and 4.2). The general difference in numbers is illustrated in Table 4.1 for the maxillary palp and galea of some species. Although most of the endopterygotes listed in the table are small relative to the orthopteroids, it should be noted that first-stage grasshoppers—which are the same size as, or even smaller than, the endopterygotes—nevertheless have greater numbers of sensilla than most endopterygote larvae. The large number of sensilla on the palps of *Leptinotarsa decemlineata* appears to be an exception to this generalization, but they are mainly multiporous and, presumably, have an olfactory function (see below). Considering only the uniporous sensilla, the number of sensilla on the palps of *L. decemlineata* is comparable with that in other beetles (Table 4.1). Further data may be found in Chapman (1982).

Most of the sensilla on the mouthparts are uniporous and thus, presumably, are contact chemoreceptors, but multiporous receptors, presumed to be olfactory, are also present. These are generally uncommon. On the tip of a maxillary palp of *Locusta migratoria* they comprise about 5% (Blaney, 1977), and in *Gryllus bimaculatus* about 7% (Klein, 1981), of the total numbers of sensilla. Larger numbers do sometimes occur, however, and in the gryllacridid *Bothriogryllacris pinguipes,* multiporous sensilla constitute almost 40% of the total (Bland and Rentz, 1991). Amongst the mandibulate holometabolous insects, caterpillars usually have three or four olfactory receptors amongst the small number of

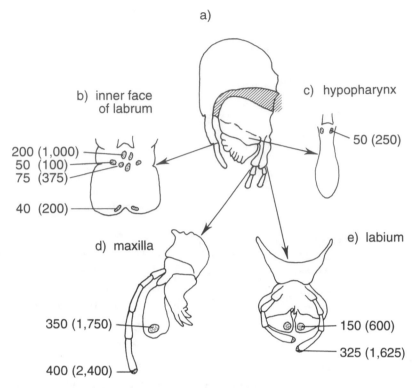

a)

b) inner face
of labrum

c) hypopharynx

50 (250)

200 (1,000)
50 (100)
75 (375)

40 (200)

d) maxilla

e) labium

350 (1,750)

150 (600)

325 (1,625)

400 (2,400)

Figure 4.1. Diagram of the head of a grasshopper showing the positions of the groups of chemoreceptors. (**a**) Sagittal section of the head showing the positions of the mouthparts. (**b–e**) Individual mouthparts showing the surfaces that come into contact with the food as it passes toward the mouth. The position of each of the major sensillum groups is indicated by a shaded area. This distribution is found in all grasshoppers, but the numbers in each group vary. The number adjacent to each group is the approximate number of sensilla in that group, and the number in parentheses is the approximate number of chemosensory neurons in the group in an adult of *Locusta migratoria*. The total exceeds 16,000 neurons.

sensilla on the maxillary palp and, possibly, two others on the galea. Multiporous sensilla are not present on the palps of the few beetles that have been examined except for *Leptinotarsa decemlineata*. This species has about 370 chemoreceptors on the maxillary palps and 130 on the labial palps. Of these, approximately 88% and 54%, respectively, are multiporous (Sen, 1988). This large number of putative olfactory sensilla on the palps is exceptional amongst the phytophagous insects so far examined. It may be a feature of chrysomelid beetles.

There is some evidence that the number of contact chemoreceptors on the mouthparts varies according to feeding habits in the orthopteroid orders, but not in endopterygotes. The detritivorous and scavenging orthopteroids listed in Table

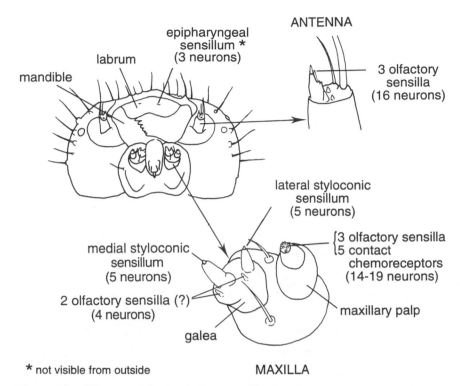

ANTENNA

epipharyngeal
sensillum *
(3 neurons)

labrum

mandible

3 olfactory
sensilla
(16 neurons)

lateral styloconic
sensillum
(5 neurons)

medial styloconic
sensillum
(5 neurons)

⎰3 olfactory sensilla
⎱5 contact
  chemoreceptors
  (14-19 neurons)

2 olfactory sensilla (?)
(4 neurons)

maxillary palp

galea

* not visible from outside   MAXILLA

Figure 4.2. Diagram of the head of a caterpillar seen from below with enlargements
of an antenna and a maxilla. Chemosensory sensilla are labeled, and the number of
neurons associated with each group of sensilla is given. (Partly after Schoonhoven, 1987.)

4.1 usually have many more sensilla than do the phytophagous grasshoppers.
By contrast, there is no pattern related to feeding habit in the small number of
beetles examined. Amongst the grasshoppers (Acridoidea), there is a tendency
for species with more restricted feeding habits to have fewer sensilla in some of
groups on the mouthparts (Fig. 4.3). Amongst the caterpillars (the only other
group of insects in which a number of species have been examined), however,
the number is the same irrespective of the feeding habits of the insects. This is
apparent from the data of Grimes and Neunzig (1986a,b), who examined 41
species from 24 families of Lepidoptera. These species range from monophagous,
such as *Bombyx mori,* through oligophagous species, like *Danaus plexippus* and
Pieris rapae, to the highly polyphagous *Trichoplusia ni.* All the species examined
have eight sensilla on the tip of the maxillary palp and two styloconic and up
to three basiconic (presumed olfactory) sensilla on the galea irrespective of their
feeding habits or taxonomic position. Grimes and Neunzig (1986a) do suggest,
however, that caterpillars living inside plant tissue have more multiporous sensilla
amongst those at the tip of the palp than do larvae living on the outside of plants.

Table 4.1. *Numbers of chemosensitive sensilla on the tip of a maxillary palp and on the galea of various mandibulate insects*

Order	Species	Food habit	Numbers of sensilla on Palp[a]	Numbers of sensilla on Galea	Source
Blattodea	*Periplaneta americana*	Detritivore	2650[b]	125	Wieczorek (1978), Petryszak (1975)
Orthoptera	*Bothriogryllus pinguipes*	Seed eater	820+518[c]	ND[d]	Bland and Rentz (1991)
	Pareremus sp.	Scavenger?	1735+24[c]	ND	Bland and Rentz (1991)
	Cnemotettix bifasciatus	Scavenger?	1698+55[c]	ND	Bland and Rentz (1991)
	Penalva sp.	Scavenger?	3038+0[c]	ND	Bland and Rentz (1991)
	Gryllus bimaculatus	Detritivore	3250+250	ND	Klein (1981)
	Conocephalus upoluensis	Seed eater?	636+21[c]	ND	Bland and Rentz (1991)
	Idiostatus inermoides	Scavenger?	800+160[c]	ND	Bland and Rentz (1991)
	Torbia viridissima	Scavenger?	60+25[c]	ND	Bland and Rentz (1991)
	Many species of Acridoidea	Phytophagous	35–500	30–350	Bland and Rentz (1991), Chapman and Thomas (1978)
Coleoptera	*Coccinella septempunctata* (A)[e]	Predaceous	23+0[f]	ND	Fu-shun et al. (1982)
	Hypera postica (L)[e]	Phytophagous	12+0	11	Bland (1983)
	Hypera postica (A)	Phytophagous	22+0	ND	Bland (1984)
	Leptinotarsa decemlineata (A)	Phytophagous	45+325	ND	Sen (1988)
	Speophyes luciulus (L)	Scavenger	10+1	2	Corbière-Tichané (1971)
	Dendroctonus ponderosai (A)	Wood	14+6?	2?	Whitehead (1981)
Lepidoptera	Many species (L)	Phytophagous	8+3	2+2[g]	Grimes and Neunzig (1986a,b)

[a] In this column, $x+y$ indicates numbers of uniporous + multiporous sensilla.

[b] Includes some multiporous sensilla.

[c] Includes all sensilla on the terminal segment of the palp, not just those at the tip.

[d] ND, no data.

[e] A, adult; L, larva.

[f] Labial palp.

[g] Probably multiporous.

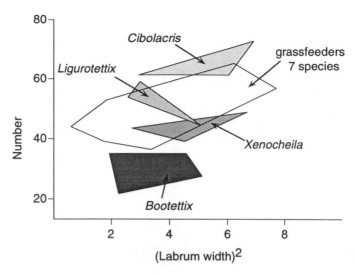

Figure 4.3. Numbers of sensilla in the most distal group of the labrum of different grasshoppers (see Fig. 4.1) in relation to the size of the insect. The square of the labrum width is used as a measure of size. Each polygon shows the results from one or a group of species. *Cibolacris* is polyphagous, *Ligurotettix* is oligophagous, and *Bootettix* and *Xenocheila* are monophagous. All these grasshoppers belong to one subfamily, Gomphocerinae, most of which are oligophagous on grasses. (After Chapman and Fraser, 1989.)

The insect groups also differ in the changes in numbers of sensilla that occur during development. In the blattids and acridids, the numbers of sensilla increase with the size of the insect because more sensilla are added at each molt (Fig. 4.4) (Chapman, 1982). By contrast, amongst caterpillars, first-stage larvae have the same number of sensilla as do final-stage larvae even though the surface area of the mouthparts may increase more than 100-fold (Baker et al., 1986). More fragmentary evidence suggests that these different patterns of development are general for orthopteroid insects versus endopterygote larvae.

4.2.2. Chemoreceptors on the Mouthparts of Fluid-Feeding Insects

The situation in fluid-feeding insects is quite different. The Hemiptera, with their piercing and sucking mouthparts, have no chemoreceptors on the stylets that enter the tissues of the host. The only sensilla associated with the food canal are those in the cibarium. Consequently, the insects do not taste the food they are about to feed on until they begin to ingest it. A few chemoreceptors may be present on the labrum, but most are on the labium, and even here there are only small numbers (Backus, 1988). These sensilla are in a position only to monitor chemicals on the outside of the host. Aphids have no chemoreceptors on the tip of the proboscis.

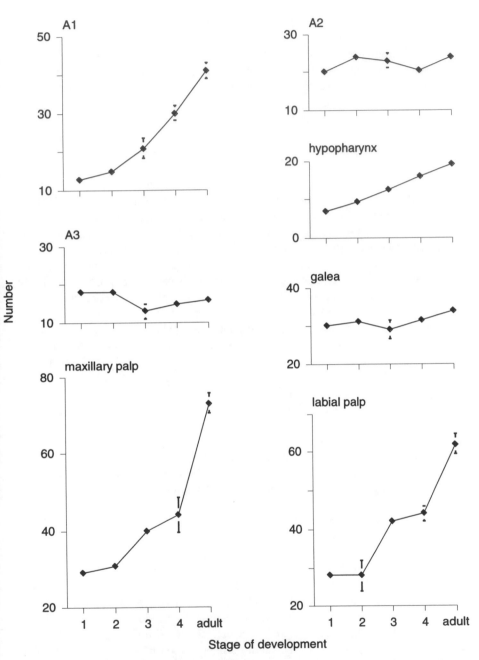

Figure 4.4. Changes in numbers of sensilla in groups on the mouthparts during the development of males of the grasshopper, *Bootettix argentatus*. See Fig. 4.1 for the positions of the groups. (After Chapman and Fraser, 1989.)

Amongst the small number of sensilla on the tip of the labium of the brown planthopper, *Nilaparvata lugens,* is one multiporous sensillum. Another multiporous sensillum occurs low down on the shaft of the labium (Foster et al., 1983).

Female mosquitoes use a piercing and sucking mechanism to obtain blood. Like the Hemiptera, they have a small number of chemoreceptors on the labellar lobes and in the cibarial cavity. In addition, they have two pairs of contact chemoreceptors near the tip of the labrum. Because the labrum enters the tissues of the host, these sensilla are in a position to monitor the quality of the host's blood as it enters the food canal. A number of olfactory sensilla are present on the maxillary palps (McIver, 1982).

The Lepidoptera and cyclorrhaphous Diptera are also fluid feeders, but, in general, do not pierce tissues. In Lepidoptera, there are some sensilla in the food canal in addition to those in the cibarium (Faucheax and Chauvin, 1980). The cyclorrhaphous Diptera do not have sensilla in the food canal apart from the cibarial receptors. In both groups, however, there are chemoreceptors on the outside of the sucking tube, formed by the galeae in the Lepidoptera, and on the labellum in the Diptera (see, for example, Faucheux and Chauvin, 1980; Dethier, 1976). These sensilla do usually come into direct contact with the food because the tip of the proboscis is immersed in the fluid to be ingested. Chemoreceptors are also present on the maxillary palps in Diptera and the labial palps in Lepidoptera. Some of them, at least, have an olfactory function. In some species of Lepidoptera, a terminal depression in the palp encloses a large number of sensilla, as many as 1750 in *Manduca* (Kent et al., 1986). These are CO_2 receptors (Stange, 1992).

The honeybee, *Apis mellifera,* has a few contact chemoreceptors on the tip of the glossa and about 70 on the area proximal to the tip. A few are also present on the galeae and labial palps which surround the glossal tongue to form the food canal (Whitehead and Larsen, 1976). The sensilla on the galeae and labial palps are probably in a position to monitor the food prior to ingestion, while those on the glossa taste it as it passes along the food canal.

4.2.3. Chemoreceptors on the Legs

Uniporous sensilla that appear to be contact chemoreceptors have been demonstrated on the tarsi of many insects: Collembola (Crouau et al., 1987), Orthoptera (e.g., Chapman and Fraser, 1989; Fudalewicz-Niemczyk et al., 1980), adults of numerous Diptera [see Colwell and Berry (1993) for references] and Lepidoptera (see Kent and Griffin, 1990), and *Apis* (Hymenoptera) (Whitehead and Larsen, 1976). The behavioral and electrophysiological studies of Dethier (1976) and subsequently of many other authors have demonstrated unequivocally that these sensilla do function as contact chemoreceptors in these orders (e.g., Chapman et al., 1991) and also in the Coleoptera (Rees, 1969). They are not necessarily always connected with feeding, and in many cases a role in oviposition (e.g.,

Roessingh et al., 1992) or the detection of pheromones (e.g., Städler et al., 1994) has been established. A role in feeding is, nevertheless, obvious in fluid-feeding insects where tarsal stimulation results in proboscis extension (Blaney and Simmonds, 1990; Dethier, 1976), and they are probably also important in mandibulate insects (White and Chapman, 1990). In several species, neurophysiological responses to food-related stimuli, such as inorganic salts and sugars, have been demonstrated.

Tarsal sensilla are usually present in considerable numbers. Large grasshoppers may have over 200 on the tarsi of each of the fore and middle legs and over 100 on the hind legs. Similar numbers are present in larger flies, such as *Musca domestica*. Lepidoptera and mosquitoes have smaller numbers (Chapman, 1982). Contact chemoreceptors have not been recorded on the legs of holometabolous larvae.

A putative olfactory sensillum is present on the foreleg of the proturan *Acerentomon majus* (Dallai and Nosek, 1981), which lacks antennae and extends the forelegs in front of the body. Olfactory sensilla are not known to occur on the legs of other insects.

4.2.4. Organization of Connections Within the Central Nervous System

In all animals, the central processing of olfactory information appears to be quite different from the processing of gustatory information. Olfactory information is integrated in areas of discrete neuropil, forming glomeruli, and the glomeruli are grouped together to form an olfactory bulb or lobe. In contrast, there is no comparable organization of gustatory input (Dethier, 1993).

In insects, the olfactory lobes lie in the deutocerebrum. Their structure and functioning is reviewed by Homberg et al. (1989) and by Boeckh and Ernst (1987). Most of the sensory input to the olfactory lobes comes from sensilla on the antennae; but where multiporous sensilla or proven olfactory receptors are present on the mouthparts, they also appear to connect with the olfactory lobes, although this has only been investigated in a few species.

Caterpillars have a relatively poorly developed antennal center, but a majority of fibers from the antennae of larval *Manduca sexta* run directly to this center. In addition, some fibers extend from the maxilla and the labium to, or close to, the antennal center (Kent and Hildebrand, 1987). While it has not been proved, it seems likely that these fibers originate in the multiporous sensilla.

The sensory axons from multiporous sensilla on the palps of Diptera, Lepidoptera, and orthopteroid insects all appear to project to the olfactory lobe. In *Drosophila melanogaster*, the projections end in at least five different glomeruli (Singh and Nayak, 1985). In the adults of *M. sexta* and other moths, the fibers from the labial pit sensilla, which are sensitive to CO_2, extend to a single glomerulus in each antennal lobe (Kent et al., 1986). Some fibers from the maxillary palps of *L. migratoria* and *P. americana* are known to reach a structure called the *lobus glomeratus,* an area immediately behind the antennal lobe (Ernst et al., 1977). It is probable, though not proved, that these fibers come from the multiporous sensilla on the palp tips.

For uniporous, gustatory sensilla, however, the picture is quite different. The axons from the neurons associated with uniporous sensilla on the legs of grasshoppers terminate in the ganglion of the segment of which the leg is an appendage (White and Chapman, *unpublished data*). Similarly, most of the axons from the sensilla on the tips of the palps of grasshoppers (95% uniporous) terminate in the subesophageal ganglion. At least a majority of the axons from gustatory sensilla on the maxillae of the caterpillar of *Manduca sexta* also end in the subesophageal ganglion (Kent and Hildebrand, 1987). Sensilla on the labrum and in the cibarial cavity connect with the brain. In adult flies, the contact chemoreceptors on the labellum project to the subesophageal ganglion (Nayar and Singh, 1985; Yetman and Pollack, 1986). In none of these cases is there evidence of any organization comparable with the olfactory glomeruli. Nayar and Singh (1985) suggest that receptors coding for different modalities on the labellum of *D. melanogaster* do end in discrete areas of the subesophageal ganglion, but Yetman and Pollack (1986) found no such separation in *Phormia regina.*

These differences in the central organization of chemosensory inputs, converging directly to the olfactory lobe on the one hand and ending diffusely in various ganglia on the other, are important because they will affect the way in which information is processed. Consequently, even though a uniporous sensillum may be stimulated by an odor, the information it conveys to the central nervous system is not likely to be processed in the olfactory lobe; it will be treated as a gustatory stimulus.

4.3. Behavioral Hierarchies

In a normally feeding insect, the sensilla on the legs and mouthparts may be used in a sequence that initiates feeding. This is most completely studied in the blowfly, *P. regina.* In this species, contact of tarsal receptors with potential food, such as a drop of sugar solution, leads to extension of the proboscis. This brings sensilla on the aboral surface of the labellum into contact with the food. Stimulation of these sensilla causes spreading of the labellar lobes on the surface of the sugar drop. This, in turn, brings small sensilla between the pseudotracheae into contact with the sugar, leading to ingestion (see Dethier, 1976; Edgecomb et al., 1987). Presumably the cibarial sensilla monitor the food as it finally passes into the gut.

A very similar sequence of responses occurs in a mosquito feeding on sugar. When the tarsi are stimulated with sugar, the insect turns towards the side stimulated and probes with its proboscis until the sensilla on the labellum contact the drop. Sucking is initiated when the sensilla on the labella are stimulated at the same time as those on the labrum. Stimulation of only one of these groups of sensilla is not sufficient (Pappas and Larsen, 1978; but see Friend, 1985).

In mosquitoes and blowflies, proboscis extension requires stimulation of the tarsal sensilla by a higher concentration of sugar than is required for the subsequent labellar response (see Clements, 1992). The threshold concentration at which probing occurs in response to tarsal stimulation also varies with time since the previous feed. Immediately after a sugar meal, the threshold is high, but it declines with time, returning to the prefeeding level after several days.

Similar behavioral sequences are known in phytophagous insects. In a grasshopper, contact of the tarsi or antennae with a leaf may lead the insect to lower its head and bring the sensilla on the tips of the palps into a rapid sequence of brief contacts. This behavior is known as *palpation*. On the basis of the chemosensory information received, the insect may reject the plant or bite it (Chapman and Sword, 1993). Biting releases the fluids within the leaf; these fluids flow over the mouthparts, probably following tracts of wettable hairs on the epipharyngeal face of the labrum (Cook, 1977). These hair tracts direct the flow toward two main groups of chemoreceptors, the A1 and A2 groups on the epipharynx (see Fix. 4.1). At this point, again, the insect may reject the food, or it may start to ingest it. Essentially similar behavior occurs at the start of feeding by the caterpillar of *Spodoptera exempta* (Blaney and Simmonds, 1987).

These behavioral sequences are not fixed. For example, a grasshopper will palpate a potential food source although it is standing on a totally different object. Insects deprived of their palps will, after a brief interval, continue to feed, although experiments on caterpillars (Blom, 1978; Hanson and Dethier, 1973) and locusts (Mordue, 1979) show that the insects have a reduced ability to discriminate between different foods at the start of feeding.

Information derived from receptors early in the behavioral sequence may be overridden by information from others higher in the hierarchy. In an extreme example, White and Chapman (1990) describe a nymph of *Schistocerca americana* holding its forelegs high in the air to avoid contact with a maize leaf coated with the deterrent compound nicotine hydrogen tartrate while the insect was eating the same leaf. Clearly, the strong negative signal from the tarsal receptors was overridden by positive signals from the cibarial sensilla which were stimulated by fluids released from within the leaf when the insect bit it. Biting, in this case, occurred when the insect encountered an irregularity on the leaf surface. Locusts and grasshoppers, lacking a suitable food, soon start to bite objects with suitable narrow edges provided the objects are not chemically deterrent.

The importance of different elements in the hierarchy may also vary with different food qualities, presumably reflecting differences in the sensitivities of the different receptors. In a behavioral study using larvae of *M. sexta,* de Boer (1991, 1993) showed that insects reared on tomato rejected *Vigna* (cowpea). This behavior is mediated primarily by the lateral styloconic sensilla of the galea

(Fig. 4.2) because the larvae continued to reject cowpea with all the other sensilla on the mouthparts ablated; but if only these sensilla were removed, the larvae no longer discriminated against the plant. The lateral styloconic sensilla contain a deterrent cell, but so does the medial sensillum (Peterson et al., 1993). In contrast to these results, insects reared on cowpea exhibited a preference for leaf discs of this plant over wet filter-paper discs. If all the sensilla on the head were ablated, the insects exhibited no preference; but if only the antennae remained, the preference was maintained. This was also true with only the maxillary palps. Both of these organs carry olfactory receptors, although the palps also have contact chemoreceptors; and it seems probable that in this case, odor played a critical role in final selection of the plant. Insects with only the epipharyngeal, galeal, and labial sensilla remaining preferred filter paper. All of these structures have only contact chemoreceptors. In an earlier study on the same species, Hanson and Dethier (1973) concluded that larvae without olfactory sensilla exhibited a reduced ability to discriminate between the leaves of tomato and Jerusalem cherry.

4.4. Sensitivity of Chemoreceptors to Different Chemicals

The contact chemoreceptors of many insects possess four chemosensitive cells, although more neurons are present in the palp tip sensilla of grasshoppers and some of the sensilla in the cricket, *Gryllus bimaculatus* (Klein, 1981). The sensitivity ranges of these cells have been most fully examined in cyclorrhaphous flies, notably by Dethier and his associates (Dethier, 1976; Hanson, 1987), and in caterpillars, again by Dethier (see Dethier, 1973; Dethier and Kuch, 1971) but also by Schoonhoven (1973, 1987) and his colleagues (Blom, 1978; Ma, 1972; van Drongelen, 1979) and by Blaney and Simmonds (1988). In these insects it is common for one cell to respond primarily to inorganic salts and another to sugars. Even amongst these cells, the sensitivity may vary from sensillum to sensillum. For example, the lateral styloconic sensillum on the galea of larval *Pieris brassicae* (see Fig. 4.2) is sensitive only to sucrose and glucose, while the medial sensillum responds to a wide range of sugars including pentose sugars, and, weakly, trisaccharides and the polyhydric alcohols inositol and sorbitol (Ma, 1972). Comparable "salt" and "sugar" cells have been recorded in a wide range of insects.

The other neurons within a sensillum may also be relatively specific, although the compounds to which they respond vary from insect species to species and from sensillum to sensillum within a species. Amongst the flies, one cell may be sensitive to water and, in *Phormia,* to very dilute salt solutions. It is called the "water" cell, to distinguish it from the "salt" cell. The "water" and "salt" cells differ in the ranges of NaCl to which they are sensitive (Fig. 4.5).

Cells responding to amino acids are present in some sensilla of caterpillars

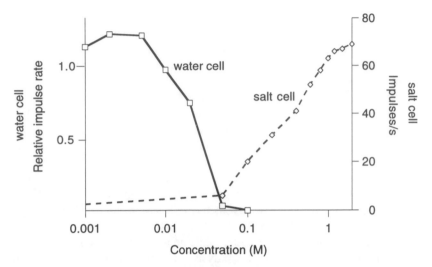

Figure 4.5. Responses of the "water" cell and the "salt" cell in the labellar hairs of *Phormia* species to stimulation with sodium chloride solutions. Notice that values for the "water" cell are expressed relative to the unstimulated firing rate; "salt" cell responses are action potentials per second. (Data from Rees, 1968, 1970b.)

and chrysomelid beetles, and probably in many other insects. Fourteen amino acids stimulated a specific cell in the larva of *Pieris brassicae,* the most effective being histidine, phenylalanine, and 4-hydroxyproline. On the other hand, the first two of these were amongst the least stimulating for *Pieris rapae.* Eight amino acids produced no response in the sensilla of either species (van Loon and van Eeuwijz, 1989). Mitchell and Schoonhoven (1974) examined the sensitivity of sensilla on the mouthparts of *L. decemlineata* to 13 amino acids. Six of the acids produced a response at 0.1 M, with γ-aminobutyric acid and L-alanine being most effective.

In a few species, cells responding to specific chemical components of the food are known to be present. The best known example of these is the possession by brassica-feeding insects of a cell that responds to glucosinolates. Glucosinolates are characteristic compounds in Brassicaceae. Such cells have been identified in a number of pierid caterpillars, in the larva of *Mamestra brassicae* (Wieczorek, 1976), and in the larva of the turnip beetle, *Entomoscelis americana* (Mitchell and Gregory, 1981; also see Städler, 1992). Another example is the possession of a neuron sensitive to hypericin in the tarsal sensilla of the beetle, *Chrysolina brunsvicensis.* Hypericin is characteristic of the host plant, *Hypericum* (Rees, 1969). Other examples do occur, but may not be as widespread as is often presumed. There are insufficient data to draw any firm conclusion.

Many species that feed on vertebrate blood have cells that respond to adenosine nucleotides (Friend, 1989; also see Chapter 6). The normal role of these receptors in insects such as *Rhodnius prolixus* and *Glossina* species is not at all clear because only the cibarial sensilla normally come into contact with blood. In culicine mosquitoes, however, sensilla on the tip of the labrum may be sensitive to nucleotides. These insects are often sensitive to very low concentrations of nucleotides. Behavioral responses are detectable at ATP concentrations ranging from 2×10^{-4} M in some mosquitoes to as low as 4×10^{-6} M in the stable fly, *Stomoxys calcitrans* (Ascoli-Christensen et al., 1990b), and 1.3×10^{-8} M in female *Glossina tachinoides* (Galun, 1988).

In caterpillars, one of the cells in the styloconic sensilla responds to chemicals from a number of different chemical classes which produce aversive behavior (see Peterson et al., 1993). This has been called the "deterrent" cell. There is also evidence for a deterrent cell in grasshoppers (Chapman et al., 1991), but no cell with an obvious link to aversive behavior has been described in Coleoptera or Diptera.

The specificity ranges of these cells, while generally clear, in most cases have not been rigorously tested. Critical analysis sometimes reveals a broader range of sensitivities than is commonly assumed. For example, in *Protophormia terraenovae* the activity of the "salt" receptors on the labellum is enhanced by sugars. It appears that these cells do possess sugar-specific binding sites on the membrane of the dendrite (Schnuch and Hansen, 1990) Similarly, the "sugar" cells in sensilla of the beetle, *Entomoscelis americana*, and in blowflies are known to respond to some amino acids (Mitchell and Gregory, 1979). Other examples of cells responding to apparently unusual stimuli are known; and it is probable that this is a common, if not a usual, phenomenon. Nevertheless, the characterizations that are usually used are probably generally valid, although they clearly should not be regarded as implying absolute specificity.

Amongst the orthopteroid insects, the physiology of contact chemoreceptors has been examined extensively only in grasshoppers. Here the picture is different. At least some of the cells are less specific than is commonly the case in endopterygote insects. There is evidence from studies on the maxillary palp sensilla of *L. migratoria* (Blaney, 1975) and the tarsal sensilla of *Schistocerca americana* (White and Chapman, 1990) that the responses to inorganic salts and sugars are mediated by the same two cells.

Only in a few cases has the physiology of multiporous sensilla associated with the mouthparts been studied. One study involves the antennal sensilla of larval *Manduca sexta* (Dethier and Schoonhoven, 1969). They observed that each cell from which they recorded responded to a range of odors. For example, in one preparation one cell responded to linalool, allyl isothiocyanate, and allyl sulfide, as well as to the odors of tomato (a host plant) and geranium (which is not eaten). It was inhibited by ionone and the odor of cabbage (which is eaten, although not a normal host). A second cell in the same sensillum exhibited a similar response pattern except that it did respond to ionone.

The sensilla in the labial pit organ of adult Lepidoptera were shown to respond to CO_2 (Stange, 1992), but their behavioral role is not known.

4.5. Variability of Sensory Response

The responses of peripheral chemoreceptors are not constant even with a constant stimulus. Bernays et al. (1972) first showed that the sensitivity of contact chemo-receptors on the maxillary palps of *Locusta migratoria* varied in relation to the time since feeding. Immediately after a meal, the sensilla were relatively unresponsive. The responsiveness of the sensilla increased until, about 2 hr after feeding, a majority of the sensilla responded when stimulated. Bernays et al. attributed these changes to the closing and opening of the terminal pore of each sensillum and subsequently showed (Bernays and Chapman, 1972) that the decrease in sensitivity was a consequence of crop filling, the effect of which was to cause the release of a hormone from the corpora cardiaca. Homogenates of the corpora cardiaca and blood from newly replete insects acted on the sensilla (Bernays and Mordue, 1973). Although there is some controversy, sensitivity of labellar sensilla of *Phormia regina* also appears to be reduced by feeding. Rachman (1979) argues that her data provide no evidence of a statistically significant effect, but the examples in her Fig. 2 do show a marked depression of activity within 30 min of feeding. By 1 hr post feeding, the firing rate had recovered to the prefeeding level. Her earliest comparison was made at 1 hr. Her flies received meals of 10 µl of fructose, which would only have half filled the crop. Angioy et al. (1979) measured the electrical resistance through tarsal and labellar chemoreceptors. Flies were fed on sucrose solution. On average, they consumed 21 µl, filling the crop. Immediately after feeding, the resistance through the sensilla of fed flies was much higher than through the sensilla of control flies, suggesting that the pores had closed. This effect persisted for at least an hour. It is known that the pore at the tip of contact chemoreceptors of *Calliphora vicina* may be opened or closed in response to different stimuli (van der Wolk et al., 1984). Whether it also changes in relation to feeding is not known, although the resistance measurements of Angioy et al. (1979) suggest that it does. In longer-term experiments, Omand and Zabara (1981) showed markedly higher firing rates in the sensilla of flies maintained without food compared with those with continuous access to sucrose. It should be noted, however, that factors other than these changes in receptor sensitivity also contribute significantly to the behavioral responses of flies to tarsal stimulation (Edgecomb et al., 1987).

In addition to these general effects on sensitivity of all the cells in a sensillum, Simpson et al. (1991) have shown that, in *L. migratoria,* specific feedbacks have differential effects on the sensory cells within a sensillum. The responsiveness of neurons in the terminal sensilla of the palps to amino acids in an insect fed a high-protein low-carbohydrate diet is greatly reduced, but the responsiveness of sucrose-sensitive cells is unaffected. In insects fed a low-protein, high-carbohydrate diet, the converse is true; that is, the activity of the sugar-sensitive cell is reduced. It was demonstrated that these effects correlated with titers of the relevant compounds in the hemolymph. In the course of a meal on a high-protein diet, the titer of amino acids in the hemolymph increased. The associated decrease in the response of the neurons to amino acids could be mimicked by injecting the appropriate mixture (Fig. 4.6; also see Chapter 9). Not all the amino acids were equally effective. The effect depended on a mixture of eight amino acids: leucine, glutamine, serine, methionine, phenylalanine, lysine, valine, and alanine. The evidence suggests that the amino acids acted directly on the sensilla (Simpson and Simpson, 1992). A similar variation in sensitivity in relation to the nature of the food previously eaten is known to occur in the caterpillar of *Spodoptera littoralis* (Simmonds et al., 1992).

Variation in responsiveness also occurs over longer time scales. These changes may be related to the previous experience of the insect (Chapter 10) or to its ontogeny (Chapter 11).

Changes in sensitivity as a consequence of experience have generally been demonstrated with respect to aversive responses. For example, the sensitivity to salicin of the deterrent cell in the caterpillar of *Manduca sexta* is reduced if the larva is forced to feed on a diet containing salicin (Schoonhoven, 1969). The response in the sensilla of larval *Spodoptera* species is similarly reduced after previous exposure to azadirachtin (Simmonds and Blaney, 1984). These changes

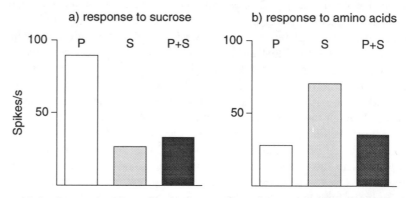

Figure 4.6. Responses of sensilla on the maxillary palp of *Locusta migratoria* after feeding for 4 hr on artificial diets containing protein but no carbohydrate (P), containing digestible carbohydrates but no protein (S), and containing both (P + S). The sensilla were stimulated with (**a**) 0.01M sucrose or (**b**) a 0.01M mixture of amino acids. (After Simpson et al., 1991.)

may be critical features of the process of habituation to food that is initially unpalatable (see Chapter 10). Interestingly, the activity of a cell responding to inositol is also reduced after feeding on a diet containing inositol (Schoonhoven, 1969), yet inositol is a weak phagostimulant (Städler and Hanson, 1978).

Ontogenetic changes in the sensitivity of contact chemoreceptors have been demonstrated in a number of insects. Simmonds et al. (1991) showed that in the larva of *Spodoptera littoralis* the response of the lateral styloconic sensillum to sucrose was lowest on the day of ecdysis, while in the medial sensillum there was an upward trend in sensitivity through the stadium. The responses to inositol were similar in both sensilla, with a markedly higher level after day 2, while the response to sinigrin peaked in mid-stadium and then declined to a low level on the day of ecdysis (Schoonhoven et al., 1991). Simpson et al. (1990) showed that in adult *L. migratoria,* both the proportion of sensilla responding and the sensitivity of those that did respond had declined considerably by the end of the period of somatic growth. Similar changes occur in the sensilla of adult *Phormia terraenovae* (Rees, 1970a; and see Blaney et al., 1986).

Changes over the course of a day have also been recorded. The responsiveness of the lateral styloconic sensillum of *Spodoptera littoralis* to sucrose was greater in the mornings than in the afternoons for most of the final stadium (Fig. 4.7). On the first day of the stadium, however, the converse was true, while on day 2 there was little difference with time of day. Parallel changes occurred in the responses of the medial sensillum (Simmonds et al., 1991).

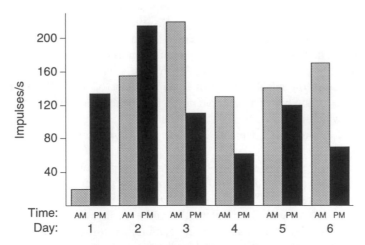

Figure 4.7. Temporal variation in the response of the lateral sensillum on the galea of the larva of *Spodoptera littoralis* (see Fig. 4.2) to stimulation with 0.05 M sucrose. AM, morning; PM, afternoon. The numbers 1–6 represent days in the final larval stadium. (After Simmonds et al., 1991.)

4.6. Peripheral Interactions Resulting from Mixtures of Chemicals

The response to a normal food, usually containing a variety of different chemicals, is, however, not simply the sum of the inputs to be expected when the various components are tested singly. It has been demonstrated in flies, beetles, caterpillars, and grasshoppers that interactions occur between the chemicals or between the peripheral sensory neurons. The signals transmitted to the central nervous system are a reflection of these interactions.

The interactions of behaviorally deterrent compounds with sugars have been particularly studied. Alkaloids produce an aversive response from many insects, yet in *Phormia regina* and *L. decemlineata* there is no evidence indicating the presence of a deterrent cell in the peripheral receptors. High concentrations of the alkaloids commonly produce irregular bursting activity in many of the neurons, suggesting a damage response (Mitchell and Harrison, 1985). However, low concentrations may not have this effect, although they do significantly reduce food intake. They apparently do this by reducing the input from the cells responding to phagostimulatory compounds present in the mixture. A similar effect is observed in insects which do have deterrent cells, namely, the caterpillars and grasshoppers. Here, as the concentration of alkaloid is increased, the responses of the cells responding to sugar are decreased until the cells are finally silenced completely (Fig. 4.8b) (Chapman et al., 1991). The leaf beetle, *E. americana,* apparently has no deterrent cell, but several alkaloids stimulate the "glucosinolate cell." Here, too, Mitchell and Sutcliffe (1984) showed that 10 mM sparteine markedly reduced the response to 10 mM sucrose in a sensillum on the galea.

This effect is not restricted to alkaloids. In the grasshopper, *Schistocerca americana,* a similar effect is produced by salicin, a phenolic glycoside; in some caterpillars, this effect is produced by tannic acid (Dethier, 1982). Mitchell and Sutcliffe (1984) cite other examples. Moreover, Blaney and Simmonds (1990) showed that in the tarsal sensilla of several species of noctuid moth there is a mutual interaction between azadirachtin (which inhibits proboscis extension) and sucrose (which produces proboscis extension). In a mixture, not only was the activity of the sugar cell depressed, so was the activity of the deterrent cell (Fig. 4.8a).

There is also evidence showing that similar types of interactions occur between chemicals that are not deterrents. This has been particularly well studied by Mitchell and Gregory (1979) in *E. americana.* The sugar-sensitive cell in a galeal sensillum of this beetle is stimulated by amino acids as well as sugars. When stimulated by a mixture of sucrose and alanine, the firing rate of the sugar cell is intermediate between the rates for each compound at the same concentration when tested alone. In the stable fly, *Stomoxys calcitrans,* ATP decreases the activity of the salt-sensitive cell (Ascoli-Christensen et al., 1991), but increases the activity of the sugar-sensitive cell, although ATP alone has no effect on this cell (Ascoli-Christensen et al., 1990a). Similar effects of ATP on the activity

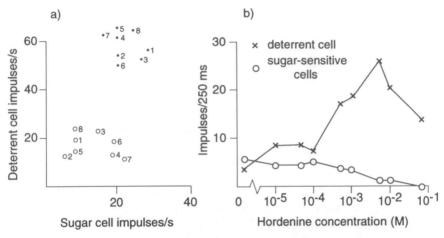

Figure 4.8. Interactions between neurons within a sensillum. (**a**) Responses of tarsal sensilla of *Helicoverpa armigera* when stimulated with sucrose or azadirachtin. Each point represents the response of one sensillum (1–8). Sucrose stimulated the sugar cell, but not the deterrent cell; azadirachtin stimulated the deterrent cell, but not the sucrose cell. Dots show the responses of these cells when each sensillum was stimulated with each of the two chemicals independently. When the chemicals are applied at the same concentrations in a mixture (open circles), the firing rates of both cells are reduced in most insects. (After Blaney and Simmonds, 1990). (**b**) Responses of sensilla on the tibia of a grasshopper, *Schistocerca americana,* to stimulation with a mixture of sucrose and hordenine. As the hordenine concentration was increased, the firing rate of the deterrent cell increased and remained high, but the response to sucrose decreased and finally disappeared altogether. The sucrose concentration was constant throughout the experiment. (After Chapman et al., 1991.)

of salt- and sugar-sensitive cells occur in *Phormia regina* (Liscia et al., 1987). Dethier (1976) summarizes some of the other interactions between stimulating chemicals that occur in this species.

It is evident that there are two types of interaction occurring. In one, only a single cell is involved; in the other, two separate cells participate. The mechanisms by which these interactions occur are not well understood. Where a single cell is involved, as in the interaction of sugars and amino acids in *E. americana,* it is very likely that there are acceptor sites for the two classes of compound on the membrane of the one cell. The interaction may then be due to competitive inhibition as Mitchell and Gregory (1979) suggest. Mitchell (1987) and Wieczorek (1976) also suggest that in *L. decemlineata* and the caterpillar of *Mamestra brassicae,* alkaloids compete with amino acids and sugars, respectively, for acceptor sites on the cell membranes of the same neurons. However, Dethier and Bowdan (1989) demonstrated that competitive interactions were not responsible for the reduction in the activity of the sugar cells of *P. regina* produced by a range of alkaloids.

In the grasshopper, *S. americana,* White et al. (1990) demonstrated a direct interaction between the neurons within a sensillum occurring after the cells became active. Stimulation of tarsal sensilla with a solution of nicotine hydrogen tartrate with sodium chloride sometimes caused the deterrent cell to fire continuously. More frequently, however, the activity of the deterrent cell was interrupted by the activity of one of the salt-sensitive cells. It appeared that the firing of this cell inhibited the activity of the deterrent cell, and sometimes the input from the deterrent cell was repeatedly interrupted (Fig. 4.9). Similar interactions have subsequently been observed in other species of grasshopper and in other sensilla (Ascoli-Christensen and Chapman, *unpublished data*). Some of the data of Mitchell and Sutcliffe (1984) on *E. americana* can also, perhaps, be interpreted in this way. For example, in their Fig. 3, which shows the interaction of nicotine and sucrose, the firing of the cell responding to nicotine seems to be interrupted when the sugar-sensitive cell fires.

The mechanism by which this interaction might occur is not known. Although contacts between neurons in contact chemoreceptors are well known (Moulins and Noirot, 1972), no structure is known which might mediate a negative electrical

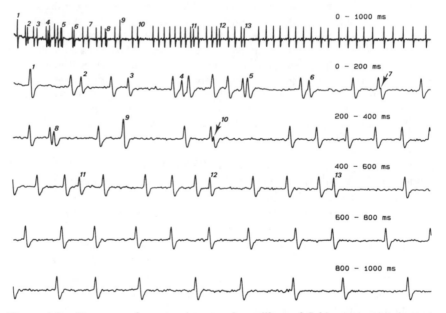

Figure 4.9. Response of neurons in a tarsal sensillum of *Schistocerca americana* to stimulation by 10 mM nicotine hydrogen tartrate in 50 mM NaCl. The trace is expanded in the lower parts of the figure for clarification. Numbers indicate action potentials of a salt cell. Spikes 1 and 9 are believed to be an action potential of the salt cell superimposed on an action potential of the deterrent cell (the larger spike). Whenever the salt cell fired, there was a delay in the firing of the deterrent cell, except for spike 4. (After White et al., 1990.)

interaction. A similar interaction has been described between mechanosensory cells in the tympanal organs of moths (Coro and Pérez, 1983). It is analogous to the lateral inhibition which is well known in compound eyes.

4.7. Interpreting the Sensory Code

The taste system of insects has no identifiable center in the central nervous system that is comparable with the antennal lobe in the brain where all olfactory information is processed. Largely because of this lack of an identifiable center, there is no physiological information on how information from the peripheral contact chemoreceptors is processed when it reaches the central nervous system. Our ideas stem from how individual experimenters *believe* the system may work.

The functions of the neurons in the peripheral receptors is inferred from parallel studies of behavior. The cells which respond to compounds eliciting feeding are assumed to provide some positive information with respect to feeding, while those that respond to compounds inhibiting feeding provide negative information. The sugar-sensitive cells are amongst the former, and the deterrent-sensitive cells are amongst the latter. Higher concentrations of salt inhibit insect feeding, and presumably the input of the "salt" cell has a negative effect. On the other hand, stimulation by water promotes proboscis extension in flies with a water deficit (Dethier, 1976), so this cell probably has a positive input.

There is evidence from studies on a variety of insects that indicates a strong correlation between the amount of any test substrate eaten and the concentration of phagostimulatory chemicals. There is also ample evidence showing that the amount eaten decreases when the concentration of deterrent chemicals is increased. In experiments using single compounds with different species of caterpillar, Schoonhoven and Blom (1988), Ma (1972), Blaney and Simmonds (1988), and Simmonds et al. (1991) showed that the amount eaten was correlated with the firing rate of certain neurons in the sensilla on the mouthparts. With phagostimulatory compounds, the rate of firing was positively correlated with the amount eaten; with deterrent compounds, the correlation was negative (Fig. 4.10).

It appears that each sensory axon carries unambiguous information: feed or do not feed. These axons are called *labeled lines*. The interpretation of these signals does not require complex central processing, simply the integration of the various inputs.

There are no qualitative differences in the information provided to the central nervous system by the cells transmitting "acceptance" and those transmitting "rejection" information; both types produce action potentials of similar form, and, in general, the responses of the cells are dependent on the concentration of the stimulating compound. The interpretation of these signals must therefore occur centrally, and it may be presumed that the differences depend on the

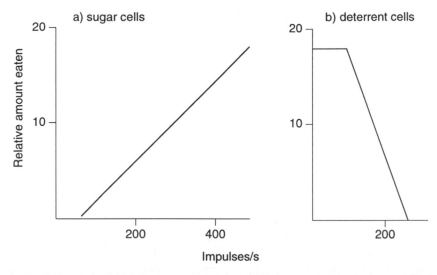

Figure 4.10. Relationship between sensory input and amount eaten by the larva of *Pieris brassicae*. (**a**) Amount eaten in relation to the combined activities of the sugar-sensitive cells in the lateral and medial galeal sensilla and the epipharyngeal sensillum when stimulated by sucrose. (**b**) Amount eaten in relation to the combined activities of two deterrent cells when stimulated by strychnine in the presence of 0.02 M sucrose. Amount eaten was based on the production of feces. (Based on data in Schoonhoven and Blom, 1988.)

excitatory or inhibitory qualities of synapses at some point in the neural pathway. Information from cells signaling acceptance presumably involves excitatory synapses; the circuits from cells signaling rejection possibly include an inhibitory synapse. There is, however, no experimental data to support this idea.

On the basis of the behavioral correlations, Schoonhoven (Schoonhoven, 1987; Schoonhoven and Blom, 1988) developed a simple model illustrating the manner in which inputs from different peripheral receptors might be integrated in the central nervous system (Fig. 4.11). This model, which is based on experimental data with *Pieris brassicae,* indicates that even the inputs from cells signaling acceptance were not all equally effective. This may indicate the occurrence of synaptic interactions of different strengths along the input pathways.

Peripheral interactions between sensory neurons, as described above, will have the effect of modifying quantitatively the different inputs, but they will not alter them qualitatively. A simple central summation of the inputs can still lead to an appropriate response.

In *L. decemlineata,* which is not known to possess deterrent cells, Mitchell and McCashin (1994) suggest that peripheral interaction is critically important in distinguishing host from nonhost plants. They suggest that the activity of one cell (cell 1) in a galeal sensillum is necessary for feeding to occur. It is stimulated

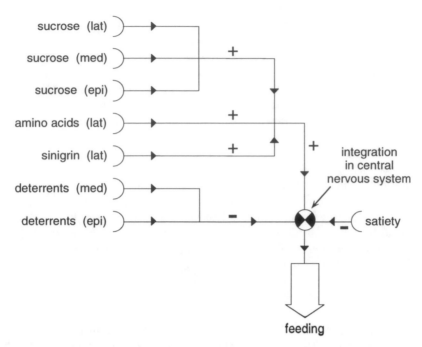

feeding

Figure 4.11. Schematic representation of how the inputs from different mouthpart receptors might be integrated within the central nervous system to regulate feeding in the caterpillar of *Pieris brassicae*. Inputs from the sucrose, amino acid, and sinigrin cells in the lateral (lat) and medial (med) sensilla on the galea and those on the epipharynx (epi) would have positive effects (+) tending to induce feeding. Inputs from the deterrent cells would have negative effects (−) tending to inhibit feeding. The degree of satiety of the insect would also be important, tending to inhibit feeding when the gut was already full. "Feeding" or "not-feeding" results from the balance of these positive and negative inputs. (After Schoonhoven, 1987.)

by a range of different compounds including solutions of alcohols normally present in leaves. The activity of this cell alone, however, is not sufficient to elicit feeding, and it must be accompanied by one or more other cells. These cells, like cell 1, have an excitatory effect, but they are ineffective without cell 1. The activity of cell 1 is suppressed by a number of alkaloids. Perhaps it is also suppressed by chemicals in nonhost plants, and this accounts for the host plant specificity of these insects [but see also comments below on a paper by Sperling and Mitchell (1991)]. In such a system all the inputs are excitatory, but one line is ineffective without another.

It seems unlikely, however, that such simple systems operate in all insects and for all stimuli. In locusts and grasshoppers, sugars are phagostimulatory, but high concentrations of sodium chloride inhibit feeding. Yet both types of compound stimulate the same cells (see above). How does the insect differentiate between them? One possibility is that the totality of input from all the receptors is, in some way, analyzed within the central nervous system. This is called *across-fiber patterning*. It requires that, although different cells respond to the same compounds, they exhibit differences in their relative sensitivities to these compounds. For example, the sensilla on the tips of the maxillary palps of *L. migratoria* vary in their relative sensitivities to single concentrations of sodium chloride and fructose (Blaney, 1975). The total input from some sensilla was consistently higher to fructose, that from others was higher to sodium chloride, and others were equally responsive to the two compounds. Using analysis of variance, Blaney was able to use the data from 20 sensilla to distinguish fructose from sodium chloride. While it is not to be supposed that the insect uses analysis of variance, it demonstrates that the compounds can be separated on the basis of the neural inputs from a number of sensilla. In this analysis, Blaney did not distinguish between the different cells within a sensillum, having concluded that the sensillum, not the individual neurons, was the functional unit. An analysis that included differences between cells would be even more powerful. Maes (Maes and Harms, 1986; Maes and Ruifrok, 1986) discusses the use of across-fiber patterning by *Calliphora vicina* in distinguishing between different inorganic salts.

Across-fiber patterning does not exclude the possibility that a labeled line system also exists. The deterrent cells of grasshoppers, for example, may be labeled lines, while other distinctions require an across-fiber system. Even in caterpillars, Dethier and Crnjar (1982) argue that the two may coexist. From their studies of *Manduca sexta* they conclude that a labeled line system would be appropriate for signaling acceptance or rejection. They do not, however, believe that such a system would enable the insect to discriminate between different acceptable plants. They argue that the plants are not discriminated from each other because they elicit different intensities of a single stimulus, but because they elicit qualitatively different perceptions. Certainly, the patterns of input from the cells in the galeal contact chemoreceptors are different when stimulated by the saps of different acceptable plants, and Dethier and Crnjar suggest that the use of across-fiber patterns—or ensemble coding as they refer to it—is one way in which the caterpillar might discriminate between different acceptable plants.

For across-fiber patterning to provide such information, there must be a minimal number of cells responding. This number will, of course, depend on the extent to which cells differ in their relative sensitivities to different compounds. The more extreme this difference, the smaller the number of inputs needed to make a decision. In this respect, it is interesting that caterpillars, apparently using a labeled line system, have very few sensilla, while grasshoppers, which perhaps use across-fiber patterning, have many sensilla. It is clear that not all of these sensilla are necessary to make a decision. During palpation by *L. migratoria,* only about 40 sensilla on each of the maxillary palps may come into contact with the food (Blaney and Chapman, 1970). If two neurons in each sensillum respond to salt and sugar, some 80 neurons will contribute to the input from each palp. Given two palps, and perhaps the labial palps as well, the amount of information conveyed to the central nervous system is clearly very large. Chapman (1988), in discussing this issue, examined the numbers of sensilla in each of the sensillum groups shown in Fig. 4.1. He found that, amongst polyphagous species, 16 was the smallest number of sensilla in any group. In species with restricted diet breadth, the lowest number was eight. Subsequently, a minimum of seven was found in the hypopharyngeal sensillum groups of first-stage nymphs of *Bootettix argentatus,* a monophagous grasshopper (Chapman and Fraser, 1989). Chapman (1988) suggests that species with a restricted range of host plants can have a more restrictive filtering system within the central nervous system because decision-making does not need to be so flexible as in polyphagous species.

Thysanura and orthopteroid insects in general have large numbers of sensilla associated with the mouthparts. Phylogenetically, they are less advanced than the hemipteroid and endopterygote insects, which generally have very few contact chemoreceptors on the mouthparts. Perhaps this reflects an evolutionary change in the method by which gustatory information was processed in the central nervous system, with the more "primitive" insects depending to a great extent on across-fiber patterning, and the more derived using labeled lines.

It is important to bear in mind that both the labeled line system and across-fiber patterning are hypothetical. There is no direct evidence that insects employ either of these systems. Mitchell et al. (1990) point out that the use of across-fiber patterning is dependent on relatively limited variation in the response of a cell to any given stimulus. But it is known that the response of a cell to repeated stimulation may be highly variable. In studies with the fly, *Sarcophaga bullata,* they found that the variability of input from one cell was much less when stimulated with two foods that were highly palatable, but greater with a less palatable food. They suggested that the degree of variability of the input might itself provide at least a part of the code; a high level of variability would indicate lack of acceptability. This idea is extended in later studies. Sperling and Mitchell (1991) suggest that variability of input may provide a "coarse-grained" filter that permits species of *Leptinotarsa* to distinguish solanaceous plants from others, while across-fiber patterning or a labeled line system, perhaps both, may operate in separating more closely related plants. Dethier and Crnjar (1982) also demonstrate that plants with different acceptabilities for the larva of *M. sexta* could be separated on the basis of the variabilities in the responses of different cells.

From their studies on the olfactory responses of the larva of *Manduca sexta,* Dethier and Schoonhoven (1969) concluded that, considering the input from all the receptors on the antenna, the insect would be able to distinguish suitable from unsuitable plants from the across-fiber pattern. However, they suggest that an alternative possibility would be for the insects to make use of the temporal patterns produced, because the timing of increasing or decreasing firing rate was consistent with one neuron but differed across neurons.

4.8. Chemosensory Input and the Regulation of Feeding

4.8.1. Regulation of a Meal

It seems to be true of most insects that the start of ingestion is triggered by chemosensory input. There are many examples to illustrate the importance of phagostimulation at the start of feeding. A fully hydrated fly does not extend its proboscis when its tarsi are stimulated with water, but stimulation with a sucrose solution results in proboscis extension and feeding (Dethier, 1976). Higher concentrations of sugars or amino acids, which increase the firing rate of neurons in the tarsal sensilla, elicit proboscis extension in a greater proportion of adult *Spodoptera littoralis* (Blaney and Simmonds, 1990). The monophagous grasshopper, *Bootettix argentatus*, rejects all nonhosts without attempting to feed on them, usually on the basis of information from the sensilla on the palps. However, if the nonhost is dipped in a solution of nordihydroguaiaretic acid, a chemical which is characteristic of the host plant, the grasshopper attempts to eat it (Chapman et al., 1988).

It is important to recognize, however, that such effects are only produced in the context of other conditions favoring feeding (see Chapters 5, 6, 8, 9, and 11). The stimuli are ineffective in a freshly replete insect. It may be that, in some or all of these cases, reduction in the sensitivity of receptors effectively prevents the insect from perceiving the stimulus, or that internal inhibitory mechanisms negate the phagostimulatory effects.

In some situations, as when an insect has been deprived of food for a long period, it may begin to feed in the absence of suitable chemosensory input. For example, there are many examples of locusts and grasshoppers eating plants that are rejected by well-fed insects (e.g., Bernays et al., 1976). In some cases this may be the result of a shortage of water, and perhaps water in the unusual host acts as a phagostimulant. There are also examples in the literature of locusts eating inert materials, such as nylon screen or polystyrene. Perhaps in these situations, the palps provide no chemosensory information because potentially stimulating chemicals are absent from the material. This situation is analogous to palpectomy experiments.

The question arises whether sustained chemosensory input is necessary to sustain feeding. There is no published information dealing directly with this topic. There are numerous examples of insects eating larger amounts of food when the concentration of a phagostimulant is higher. For example, Ascoli-Christensen et al. (1990b) show that the volume of fluid ingested by the stable fly, *Stomoxys calcitrans,* increases as the concentration of adenine nucleotides in solution is increased. The amount of wheat eaten in a single meal is increased by almost 50% in nondeprived nymphs of *L. migratoria* if the wheat had been dipped in a sucrose solution (Simpson et al., 1988). However, these effects could simply result from the heightened level of excitation induced in the central nervous system at the beginning of feeding by the high concentration of phagostimulant as was demonstrated in *L. migratoria* (Bernays and Chapman, 1974; and see Chapter 5). The elegant experiments of Blaney and Duckett (1975) on *L. migratoria* can also be interpreted in this way. They covered the tips of the maxillary palps with short capillary tubes so that the terminal sensilla could not make contact with the food. When these tubes were filled with 0.1 M sucrose, the insects ate longer meals on an artificial substrate. When they were filled with a solution of 0.1 M sodium chloride, the insects ate for a shorter time. In both cases, meal durations were in excess of 5 min. Sensory adaptation would probably be complete in much less than 1 min, so these changes in the duration of feeding probably reflect the initial level of excitation produced by the stimulating compounds. All the other sensilla on the mouthparts were functioning normally in these experiments.

The necessity of sustained stimulation during feeding remains an open question.

4.8.2. Long-Term Regulation of Feeding

The temporal variation in the activity of chemoreceptors has, in some insects, been clearly correlated with feeding behavior. For example, the daily variations in sensory response of larval *Spodoptera littoralis* shown in Fig. 4.9 correlate moderately well with the amounts eaten in the mornings and afternoons, and from day to day (Fig. 4.12a) (Simmonds et al., 1991). However, the correlation is less good with some other compounds, such as inositol (Fig. 4.12b) (Schoonhoven et al., 1991). Here the correlation is good over the first 4 days of the stadium, but subsequently there is no obvious relationship between the amounts eaten and the activity of the chemoreceptors. It is tempting to suggest that the changes in feeding are caused by the changes in receptor sensitivity, but perhaps it is more likely that the two factors are each governed by some separate system.

4.9. Conclusions

Although much information has been accumulated, there is no complete understanding of the precise mechanisms by which peripheral chemoreceptors regulate feeding by insects. It is clear that peripheral interactions between chemicals and neurons play a major part in determining the nature of the input to the central nervous system, but there is little understanding of the nature of the mechanisms that result in the interactions. With respect to taste, the central nervous system remains a tightly sealed black box. We shall have no convincing understanding of the central mechanisms involved in integrating peripheral inputs until the box is breached. Only then shall we able to determine if Dethier (Dethier and Crnjar,

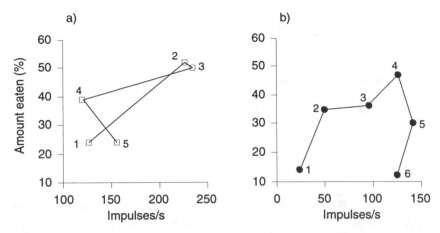

Figure 4.12. Relationship between the electrophysiological response and the amount eaten on successive days of the final larval stadium of *Spodoptera littoralis*. Numbers indicate days. Electrophysiological response is the average of the lateral and medial styloconic sensilla of the galea. (**a**) Response of 0.05 M sucrose. (Data from Simmonds et al., 1991.) (**b**) Response to 0.1 M inositol. (Data from Schoonhoven et al., 1991.)

1982) is right in believing that an insect may have "qualitatively different perceptions" of the foods that it encounters, or if this perception is in the mind of the observer.

Acknowledgments

I would like to thank Drs. E. A. Bernays, J. G. Glendinning, and S. J. Simpson for reviewing the manuscript.

References

Angioy, A. M., Liscia, A., and Pietra, P. (1979). Influence of feeding conditions on wing, labellar and tarsal hair resistance in *Phormia regina* (Meig.). *Experientia* 35, 60–61.

Ascoli-Christensen, A., Sutcliffe, J. F., and Albert, P. J. (1990a). Effect of adenine nucleotides on labellar chemoreceptive cells of the stable fly, *Stomoxys calcitrans*. *J. Insect Physiol*. 36, 339–344.

Ascoli-Christensen, A., Sutcliffe, J. F., and Straton, C. J. (1990b). Feeding responses of the stable fly, *Stomoxys calcitrans* L., to blood fractions and adenine nucleotides. *Physiol. Entomol*. 15, 249–259.

Ascoli-Christensen, A., Sutcliffe, J. F., and Albert, P. J. (1991). Purinoreceptors in blood feeding behaviour in the stable fly, *Stomoxys calcitrans. Physiol. Entomol.* 16, 145–152.

Backus, E. A. (1988). Sensory systems and behaviours which mediate hemipteran plant-feeding: a taxonomic overview. *J. Insect Physiol.* 34, 151–165.

Baker, G. T., Parrott, W. L., and Jenkins, J. N. (1986). Sensory receptors on the larval maxillae and labia of *Heliothis zea* (Boddie) and *Heliothis virescens* (F.) (Lepidoptera: Noctuidae). *Int. J. Insect Morphol. Embryol.* 15, 227–232.

Bernays, E. A., and Chapman, R. F. (1972). The control of changes in peripheral sensilla associated with feeding in *Locusta migratoria* (L.). *J. Exp. Biol.* 57, 755–763.

Bernays, E. A., and Chapman, R. F. (1974). The regulation of food intake by acridids. In: Barton Browne, L. (ed.), *Experimental Analysis of Insect Behaviour*. Springer-Verlag, Berlin, pp. 48–59.

Bernays, E. A., and Mordue (Luntz), A. J. (1973). Changes in the palp tip sensilla of *Locusta migratoria* in relation to feeding: the effects of different levels of hormone. *Comp. Biochem. Physiol.* 45A, 451–454.

Bernays, E. A., Blaney, W. M., and Chapman, R. F. (1972). Changes in chemoreceptor sensilla on the maxillary palps of *Locusta migratoria* in relation to feeding. *J. Exp. Biol.* 57, 745–753.

Bernays, E. A., Chapman, R. F., Macdonald, J., and Salter, J. E. R. (1976). The degree of oligophagy in *Locusta migratoria* (L.). *Ecol. Entomol.* 1, 223–230.

Bland, R. G. (1983). Sensilla on the antennae, mouthparts, and body of the larva of the alfalfa weevil, *Hypera postica* (Gyllenhal) (Coleoptera: Curculionidae). *Int. J. Insect Morphol. Embryol.* 1, 261–272.

Bland, R. G. (1984). Mouthpart sensilla and mandibles of the adult alfalfa weevil *Hypera postica* and the Egyptian alfalfa weevil *H. brunneipennis* (Coleoptera: Curculionidae). *Ann. Entomol. Soc. Am.* 77, 720–724.

Bland, R. G., and Rentz, D. C. F. (1991). External morphology and abundance of mouthpart sensilla in Australian Gryllacrididae, Stenopelmatidae, and Tettigoniidae. *J. Morphol.* 207, 315–325.

Blaney, W. M. (1975). Behavioural and electrophysiological studies of taste discrimination by the maxillary palps of larvae of *Locusta migratoria* (L.). *J. Exp. Biol.* 62, 555–569.

Blaney, W. M. (1977). The ultrastructure of an olfactory sensillum on the maxillary palps of *Locusta migratoria. Cell Tissue Res.* 184, 397–409.

Blaney, W. M., and Chapman, R. F. (1970). The functions of the maxillary palps of Acrididae (Orthoptera). *Entomol. Exp. Appl.* 13, 363–376.

Blaney, W. M., and Duckett, A. M. (1975). The significance of palpation by the maxillary palps of *Locusta migratoria* (L.): an electrophysiological and behavioural study. *J. Exp. Biol.* 63, 701–712.

Blaney, W. M., and Simmonds, M. S. J. (1987). Experience: a modifier of neural and behavioural sensitivity. In: Labeyrie, V., Fabres, G., and Lachaise, D. (eds.), *Insects–Plants: Proceedings of the 6th International Symposium on Plant–Insect Relationships*. Junk, Dordrecht, pp. 237–241.

Blaney, W. M., and Simmonds, M. S. J. (1988). Food selection in adults and larvae of three species of Lepidoptera: a behavioural and electrophysiological study. *Entomol. Exp. Appl.* 49, 111–121.

Blaney, W. M., and Simmonds, M. S. J. (1990). A behavioural and electrophysiological study of the role of tarsal chemoreceptors in feeding by adults of *Spodoptera, Heliothis virescens* and *Helicoverpa armigera. J. Insect Physiol.* 36, 743–756.

Blaney, W. M., Schoonhoven, L. M., and Simmonds, M. S. J. (1986). Sensitivity variations in insect chemoreceptors; a review. *Experientia* 42, 13–19.

Blom, F. (1978). Sensory activity and food intake: a study of the input–output relationships in two phytophagous insects. *Neth. J. Zool.* 28, 277–340.

Boeckh, J., and Ernst, K.-D. (1987). Contribution of single unit analysis in insects to an understanding of olfactory function. *J. Comp. Physiol.* A161, 549–565.

Chapman, R. F. (1982). Chemoreception: the significance of receptor numbers. *Adv. Insect Physiol.* 16, 247–356.

Chapman, R. F. (1988). Sensory aspects of host-plant recognition by Acridoidea: questions associated with the multiplicity of receptors and variability of response. *J. Insect Physiol.* 34, 167–174.

Chapman, R. F., and Fraser, J. (1989). The chemosensory system of the monophagous grasshopper, *Bootettix argentatus* Bruner (Orthoptera: Acrididae). *Int. J. Insect Morphol. Embryol.* 18, 111–118.

Chapman, R. F., and Sword, G. (1993). The importance of palpation in food selection by a polyphagous grasshopper (Orthoptera: Acrididae). *J. Insect Behav.* 6, 79–91.

Chapman, R. F., and Thomas, J. G. (1978). The numbers and distribution of sensilla on the mouthparts of Acridoidea. *Acrida* 7, 115–148.

Chapman, R. F., Bernays, E. A., and Wyatt, T. (1988). Chemical aspects of host-plant specificity in three *Larrea*-feeding grasshoppers. *J. Chem. Ecol.* 14, 557–575.

Chapman, R. F., Ascoli-Christensen, A., and White, P. R. (1991). Sensory coding for feeding deterrence in the grasshopper *Schistocerca americana. J. Exp. Biol.* 158, 241–259.

Clements, A. N. (1992). *The Biology of Mosquitoes,* vol. 1. Chapman & Hall, London.

Colwell, D. D., and Berry, N. M. (1993). Tarsal sensilla of the warble flies *Hypoderma bovis* and *H. lineatum* (Diptera: Oestridae). *Ann. Entomol. Soc. Am.* 86, 756–765.

Cook, A. G. (1977). The anatomy of the clypeo-labrum of *Locusta migratoria* (L.). (Orthoptera: Acrididae). *Acrida* 6, 287–306.

Corbière-Tichané, G. (1971). Ultrastructure du système sensoriel de la maxille chez la larve du coléoptère cavernicole *Speophyes lucidulus* Delar. (Bathysciinae). *J. Ultrastruct. Res.* 36, 318–341.

Coro, F., and Pérez, M. (1983). Peripheral interaction in the tympanic organ of a moth. *Naturwissenschaften* 70, 99–100.

Crouau, Y., Bauby, A., and Deharveng, L. (1987). Fine structure of the tibiotarsal and pretarsal sensory organs in *Monobella grassei banyulensis* Deharveng (Collembola: Neanuridae). *Int. J. Insect Morphol. Embryol.* 16, 245–261.

Dallai, R., and Nosek, J. (1981). Ultrastructure of sensillum t_1 on the foretarsus of *Acerentomon majus* Berlese (Protura: Acerentomidae). *Int J. Insect Morphol. Embryol.* 10, 321–330.

de Boer, G. (1991). Effect of diet experience on the ability of different larval chemosensory organs to mediate food discrimination by the tobacco hornworm, *Manduca sexta. J. Insect Physiol.* 37, 763–769.

de Boer, G. (1993). Plasticity in food preference and diet-induced differential weighting of chemosensory information in larval *Manduca sexta. J. Insect Physiol.* 39, 17–24.

Dethier, V. G. (1973). Electrophysiological studies of gustation in lepidopterous larvae. II. Taste spectra in relation to food-plant discrimination. *J. Comp. Physiol.* 82, 103–134.

Dethier, V. G. (1976). *The Hungry Fly.* Harvard University Press, Cambridge.

Dethier, V. G. (1982). Mechanism of host-plant recognition. *Entomol. Exp. Appl.* 31, 49–56.

Dethier, V. G. (1993). The role of taste in food intake: a comparative view. In: Simon, S. A., and Roper, S. D. (eds.), *Mechanisms of Taste Transduction.* CRC Press, Boca Raton, FL, pp. 3–25.

Dethier, V. G., and Bowdan, E. (1989). The effect of alkaloids on sugar receptors and the feeding behaviour of the blowfly. *Physiol. Entomol.* 14, 127–136.

Dethier, V. G., and Crnjar, R. M. (1982). Candidate codes in the gustatory system of caterpillars. *J. Gen. Physiol.* 79, 549–569.

Dethier, V. G., and Kuch, J. H. (1971). Electrophysiological studies of gustation in lepidopterous larvae. I. Comparative sensitivity to sugars, amino acids, and glycosides. *Z. Vergl. Physiol.* 72, 343–363.

Dethier, V. G., and Schoonhoven, L. M. (1969). Olfactory coding by lepidopterous larvae. *Entomol. Exp. Appl.* 12, 535–543.

Edgecomb, R. S., Murdock, L. L., Smith, A. B., and Stephen, M. D. (1987). Regulation of tarsal taste threshold in the blowfly, *Phormia regina. J. Exp. Biol.* 127, 79–94.

Ernst, K.-D., Boeckh, J., and Boeckh, V. (1977). A neuroanatomical study on the organization of the central antennal pathways in insects. II. Deutocerebral connections in *Locusta migratoria* and *Periplaneta americana. Cell Tissue Res.* 176, 285–308.

Faucheux, M. J., and Chauvin, G. (1980). Les pièces buccales des adultes de cinq lépidoptères tinéydes kératophages communs dans l'ouest de la France. III. Les récepteurs sensoriels des maxilles. *Bull. Soc. Sci. Natl. Ouest France* 2, 16–25.

Foster, S., Goodman, L. J., and Duckett, J. G. (1983). Ultrastructure of sensory receptors on the labium of the rice brown planthopper. *Cell Tissue Res.* 230, 353–366.

Frazier, J. L. (1992). How animals perceive secondary plant compounds. In: Rosenthal, G. A., and Berenbaum, M. R. (eds.), *Herbivores. Their Interactions with Secondary Plant Metabolites,* vol. 2. Academic Press, San Diego, pp. 89–134.

Friend, W. G. (1985). Effects of mating, nutritional condition, and mouthpart separation on ingestion and destination of sugar and water by the mosquito *Culiseta inornata. Physiol. Entomol.* 10, 137–144.

Friend, W. G. (1989). Phagostimulation in vector insect: comparative purinergic responses. In: Borovsky, D., and Spielman, A. (eds.), *Host Regulated Developmental Mechanisms in Vector Arthropods*. IFAS, Vero Beach, FL, pp. 149–158.

Fudalewicz-Niemczyk, W., Oleksy, M., and Rościszewska, M. (1980). The peripheral nervous system of the larva of *Gryllus domesticus* L. (Orthoptera), part III. Legs. *Acta Biol. Cracov Ser. Zool.* 22, 51–63.

Fu-shun, Y., Junde, Q., and Xiu-fen, X. (1982). The fine structure of the chemoreceptors on the labial palps of *Coccinella septempunctata*. *Acta Entomol. Sinica* 25, 135–140.

Galun, R. (1988). Recognition of very low concentrations of ATP by *Glossina tachinoides* Westwood. *Experientia* 44, 800.

Grimes, L. R., and Neunzig, H. H. (1986a). Morphological survey of the maxillae in last-stage larvae of the suborder Ditrysia (Lepidoptera): palpi. *Ann. Entomol. Soc. Am.* 79, 491–509.

Grimes, L. R., and Neunzig, H. H. (1986b). Morphological survey of the maxillae in last-stage larvae of the suborder Ditrysia (Lepidoptera): mesal lobes (laciniogaleae). *Ann. Entomol. Soc. Am.* 79, 510–526.

Hanson, F. E. (1987). Chemoreception in the fly: the search for the liverwurst receptor. In: Chapman, R. F., Bernays, E. A., and Stoffolano, J. G. (eds.), *Perspectives in Chemoreception and Behavior*. Springer-Verlag, New York, pp. 99–122.

Hanson, F. E., and Dethier, V. G. (1973). Role of gustation and olfaction in food plant discrimination in the tobacco hornworm, *Manduca sexta*. *J. Insect Physiol.* 19, 1019–1034.

Homberg, U., Christensen, T., and Hildebrand, J. G. (1989). Structure and function of the deutocerebrum in insects. *Annu. Rev. Entomol.* 34, 477–501.

Kent, K. S., and Griffin, L. M. (1990). Sensory organs of the thoracic legs of the moth *Manduca sexta*. *Cell Tissue Res.* 259, 209–223.

Kent, K. S., and Hildebrand, J. G. (1987). Cephalic sensory pathways in the central nervous system of larval *Manduca sexta* (Lepidoptera: Sphingidae). *Philos. Trans. R. Soc. London Ser B* 315, 1–36.

Kent, K. S., Harrow, I. D., Quartararo, P., and Hildebrand, J. G. (1986). An accessory olfactory pathway in Lepidoptera: the labial pit organ and its central projections in *Manduca sexta* and certain other sphinx and silk moths. *Cell Tissue Res.* 245, 237–245.

Klein, U. (1981). Sensilla on the cricket palp. *Cell Tissue Res.* 219, 229–252.

Liscia, A., Angioy, A. M., Crnjar, R., and Barbarossa, I. T. (1987). Effects of ATP and cAMP on salt and sugar responses of chemosensilla in the blowfly. *Comp. Biochem. Physiol.* 88A, 455–459.

Ma, W. C. (1972). Dynamics of feeding responses in *Pieris brassicae* Linn. as a function of chemosensory input: a behavioural, ultrastructural and electrophysiological study. *Meded. Landbouwhogesch. Wageningen* 72, 11.

Maes, F. W., and Harms, G. (1986). Neural coding of salt taste quality in the blowfly *Calliphora vicina*. I. Temporal coding. *J. Comp. Physiol.* 159A, 75–88.

Maes, F. W., and Ruifrok, A. C. C. (1986). Neural coding of salt taste quality in the blowfly *Calliphora vicina*. II. Ensemble coding. *J. Comp. Physiol.* 159A, 89–96.

McIver, S. B. (1982). Sensilla of mosquitoes (Diptera: Culicidae). *J. Med. Entomol.* 19, 489–535.

Mitchell, B. K. (1987). Interactions of alkaloids with galeal chemosensory cells of the Colorado potato beetle. *J. Chem. Ecol.* 13, 2009–2022.

Mitchell, B. K., and Gregory, P. (1979). Physiology of the maxillary sugar sensitive cell in the red turnip beetle, *Entomoscelis americana*. *J. Comp. Physiol.* 132, 167–178.

Mitchell, B. K., and Gregory, P. (1981). Physiology of the lateral galeal sensillum in red turnip beetle larvae (*Entomoscelis americana* Brown): responses to NaCl, glucosinolates and other glucosides. *J. Comp. Physiol.* 144A, 495–501.

Mitchell, B. K., and Harrison, G. D. (1985). Effects of *Solanum* glycoalkaloids on chemosensilla in the Colorado potato beetle. A mechanism of feeding deterrence? *J. Chem. Ecol.* 11, 73–83.

Mitchell, B. K., and McCashin, B. G. (1994). Tasting green leaf volatiles by larvae and adults of Colorado potato beetles, *Leptinotarsa decemlineata*. *J. Chem. Ecol.* 20, 753–769.

Mitchell, B. K., and Schoonhoven, L. M. (1974). Taste receptors in Colorado beetle larvae. *J. Insect Physiol.* 20, 1787–1793.

Mitchell, B. K., and Sutcliffe, J. F. (1984). Sensory inhibition as a mechanism of feeding deterrence: affects of three alkaloids on leaf beetle feeding. *Physiol. Entomol.* 9, 57–64.

Mitchell, B. K., Smith, J. J. B., Albert, P. J., and Whitehead, A. T. (1990). Variance: a possible coding mechanism for gustatory sensilla on the labellum of the fleshfly *Sarcophaga bullata*. *J. Exp. Biol.* 150, 19–36.

Mordue (Luntz), A. J. (1979). The role of the maxillary and labial palps in the feeding behaviour of *Schistocerca gregaria*. *Entomol. Exp. Appl.* 25, 279–288.

Morita, H., and Shiraishi, A. (1985). Chemoreception physiology. In: Kerkut, G. A., and Gilbert, L. I. (eds.), *Comprehensive Insect Physiology, Biochemistry and Pharmacology*, vol. 6. Pergamon, Oxford, pp. 133–170.

Moulins, N., and Noirot, C. (1972). Morphological features bearing on transduction and peripheral integration in insect gustatory organs. In: Schneider, D. (ed.), *Olfaction and Taste IV*. Wissenschaftliche Verlagsgesellschaft MBH, Stuttgart, pp. 49–55.

Nayar, S. V., and Singh, R. N. (1985). Primary sensory projections from the labella to the brain of *Drosophila melanogaster* Meigen (Diptera: Drosophilidae). *Int. J. Insect Morphol. Embryol.* 14, 115–129.

Omand, E., and Zabara, J. (1981). Response reduction in dipteran chemoreceptors after sustained feeding or darkness. *Comp. Biochem. Physiol.* 70A, 469–478.

Pappas, L. G., and Larsen, J. R. (1978). Gustatory mechanisms and sugar-feeding in the mosquito, *Culiseta inornata*. *Physiol. Entomol.* 3, 115–119.

Peterson, S. C., Hanson, F. E., and Warthen, J. D., Jr. (1993). Deterrence coding by a larval *Manduca* chemosensory neurone mediating rejection of a non-host plant, *Canna generaalis* L. *Physiol. Entomol.* 18, 285–295.

Petryszak, A. (1975). The sensory peripheric nervous system of the *Periplaneta americana* (L.) (Blattoidea). I. Mouthparts. *Zesz. Nauk. Uniw. Jagiellon.* 21, 41–79.

Rachman, N. J. (1979). The sensitivity of the labellar sugar receptors of *Phormia regina* in relation to feeding. *J. Insect Physiol.* 25, 733–739.

Rees, C. J. C. (1968). The effect of aqueous solutions of some 1:1 electrolytes on the electrical response of the type 1 ('salt') chemoreceptor cell in the labella of *Phormia*. *J. Insect Physiol.* 14, 1331–1364.

Rees, C. J. C. (1969). Chemoreceptor specificity associated with choice of feeding site by the beetle *Chrysolina brunsvicensis* on its food plane, *Hypericum hirsutum*. *Entomol. Exp. Appl.* 12, 565–583.

Rees, C. J. C. (1970a). Age dependency of response in an insect chemoreceptor sensillum. *Nature (London)*, 227, 740–742.

Rees, C. J. C. (1970b). The primary process of reception in the Type 3 ("Water") receptor cell of the fly, *Phormia terranovae*. *Proc. R. Soc. London* 174B, 469–490.

Roessingh, P., Städler, E., Fenwick, G. R., Lewis, J. A., Nielsen, J. K., Hurter, J., and Ramp, T. (1992). Oviposition and tarsal chemoreceptors of the cabbage root fly are stimulated by glucosinolates and host plant extracts. *Entomol. Exp. Appl.* 65, 267–282.

Schnuch, M., and Hansen, K. (1990). Sugar sensitivity of a labellar salt receptor of the blowfly *Protophormia terraenovae*. *J. Insect Physiol.* 36, 409–417.

Schoonhoven, L. M. (1969). Sensitivity changes in some insect chemoreceptors and their effect on food selection behaviour. *K. Ned. Akad. Wet. Amsterdam Ser. C*, 72, 491–498.

Schoonhoven, L. M. (1973). Plant recognition by lepidopterous larvae. *Symp. R. Entomol. Soc. London* 6, 87–99.

Schoonhoven, L. M. (1987). What makes a caterpillar eat? The sensory code underlying feeding behavior. In: R. F. Chapman, E. A. Bernays and J. G. Stoffolano (eds.), *Perspectives in Chemoreception and Behavior*. Springer-Verlag, New York, pp. 69–97.

Schoonhoven, L. M., and Blom, F. (1988). Chemoreception and feeding behaviour in a caterpillar: towards a model of brain functioning in insects. *Entomol. Exp. Appl.* 49, 123–129.

Schoonhoven, L. M., Simmonds, M. S. J., and Blaney, W. M. (1991). Changes in the responsiveness of the maxillary styloconic sensilla of *Spodoptera littoralis* to inositol and sinigrin correlate with feeding behaviour during the final larval stadium. *J. Insect Physiol.* 37, 261–268.

Sen, A. (1988). Ultrastructure of the sensory complex on the maxillary and labial palpi of the Colorado potato beetle, *Leptinotarsa decemlineata*. *J. Morphol.* 195, 159–175.

Simmonds, M. S. J., and Blaney, M. S. J. (1984). Some neurophysiological effects of azadirachtin on lepidopterous larvae and their feeding response. In: Schmutterer, H., and Ascher, K. R. S. (eds.), *Proceedings of the 2nd International Neem Conference* QTZ, Eschborn, pp. 163–180.

Simmonds, M. S. J., Schoonhoven, L. M., and Blaney, W. M. (1991). Daily changes in the responsiveness of taste receptors correlate with feeding behaviour in larvae of *Spodoptera littoralis*. *Entomol. Exp. Appl.* 61, 73–81.

Simmonds, M. S. J., Simpson, S. J., and Blaney, W. M. (1992). Dietary selection behaviour in *Spodoptera littoralis:* the effects of conditioning diet and conditioning period on neural responsiveness and selection behaviour. *J. Exp. Biol.* 162, 73–90.

Simpson, C. L., Chyb, S., and Simpson, S. J. (1990). Changes in chemoreceptor sensitivity in relation to dietary selection by adult *Locusta migratoria*. *Entomol. Exp. Appl.* 56, 259–268.

Simpson, S. J., and Simpson, C. L. (1992). Mechanisms controlling modulation by haemolymph amino acids of gustatory responsiveness in the locust. *J. Exp. Biol.* 168, 269–287.

Simpson, S. J., Simmonds, M. S. J., Wheatley, A. R., and Bernays, E. A. (1988). The control of meal termination in the locust. *Anim. Behav.* 36, 1216–1227.

Simpson, S. J., James, S., Simmonds, M. S. J. and Blaney, W. M. (1991). Variation in chemosensitivity and the control of dietary selection behaviour in the locust. *Appetite* 17, 141–154.

Singh, R. N., and Nayak, S. V. (1985). Fine structure and primary sensory projections of sensilla on the maxillary palp of *Drosophila melanogaster* Meigen (Diptera: Drosophilidae). *Int. J. Insect Morphol. Embryol.* 14, 291–306.

Sperling, J. L. H., and Mitchell, B. K. (1991). A comparative study of host recognition and the sense of taste in *Leptinotarsa*. *J. Exp. Biol.* 157, 439–459.

Städler, E. (1984). Contact chemoreception. In: Bell, W. J., and Cardé, R. T. (eds.) *Chemical Ecology of Insects*. Chapman & Hall, New York, pp. 3–35.

Städler, E. (1992). Behavioral responses of insects to plant secondary compounds. In: Rosenthal, G. A., and Berenbaum, M. R. (eds.), *Herbivores. Their Interactions with Secondary Plant Metabolites,* vol. 2. Pergamon Press, San Diego, pp. 45–88.

Städler, E., and Hanson, F. E. (1975). Olfactory capabilities of the "gustatory" chemoreceptors of the tobacco hornworm larvae. *J. Comp. Physiol.* 104, 97–102.

Städler, E., and Hanson, F. E. (1978). Food discrimination and induction of preference for artificial diets in the tobacco hornworm, *Manduca sexta*. *Physiol. Entomol.* 3, 121–133.

Städler, E., Ernst, B., Herter, J., and Boller, E. (1994). Tarsal contact chemoreceptor for the host marking pheromone of the cherry fruit fly, *Rhagoletis cerasi:* response to natural and synthetic compounds. *Physiol. Entomol.* 19, 139–151.

Stange, G. (1992). High resolution measurement of atmospheric carbon dioxide concentration changes by the labial palp organ of the moth *Heliothis armigera* (Lepidoptera: Noctuidae). *J. Comp. Physiol.* A171, 317–324.

van Drongelen, W. (1979). Contact chemoreception of host plant specific chemicals in larvae of various *Yponomeuta* species (Lepidoptera). *J. Comp. Physiol.* A134, 265–279.

van der Wolk, F. M., Koerten, H. K., and van der Starre, H. (1984). The external morphology of contact-chemoreceptive hairs of flies and the motility of the tips of these hairs. *J. Morphol.* 180, 37–54.

van Loon, J. J. A., and van Eeuwijz, F. A. (1989). Chemoreception of amino acids in larvae of two species of *Pieris. Physiol. Entomol.* 14, 459–469.

Visser, J. H. (1986). Host odor perception in phytophagous insects *Annu. Rev. Entomol.* 31, 121–144.

White, P. R., and Chapman, R. F. (1990). Tarsal chemoreception in the polyphagous grasshopper *Schistocerca americana:* behavioural assays, sensilla distributions and electrophysiology. *Physiol. Entomol.* 15, 105–121.

White, P. R., Chapman, R. F., and Ascoli-Christensen, A. (1990). Interactions between two neurons in contact chemosensilla of the grasshopper, *Schistocerca americana. J. Comp. Physiol.* 167, 431–436.

Whitehead, A. T., and Larsen, J. R. (1976). Ultrastructure of the contact chemoreceptors of *Apis mellifera* L. (Hymenoptera: Apidae). *Int. J. Insect Morphol. Embryol.* 5, 301–315.

Whitehead, A. T. (1981). Ultrastructure of sensilla of the female mountain pine beetle, *Dendroctonus ponderosae* Hopkins (Coleoptera: Scolytidae). *Int. J. Insect Morphol. & Embryol.* 10, 19–28.

Wieczorek, H. (1976). The glycoside receptor of the larvae of *Mamestra brassicae* L. (Lepidoptera, Noctuidae). *J. Comp. Physiol.* A106, 153–176.

Wieczorek, H. (1978). Biochemical and behavioural studies of sugar reception in the cockroach. *J. Comp. Physiol.* A124, 353–356.

Yetman, S., and Pollack, G. S. (1986). Central projections of labellar taste hairs in the blowfly, *Phormia regina* Meigen. *Cell Tissue Res.* 245, 555–561.

Zacharuk, R. Y. (1985). Antennae and sensilla. In: Kerkut, G. A., and Gilbert, L. I. (eds.), *Comprehensive Insect Physiology, Biochemistry and Pharmacology,* vol. 6. Pergamon, Oxford, pp. 1–69.

Zacharuk, R. Y., and Shields, V. D. (1991). Sensilla of immature insects. *Annu. Rev. Entomol.* 36, 331–354.

5

Regulation of a Meal: Chewing Insects

S. J. Simpson

5.1. Introduction

The importance of understanding the regulation of meals when investigating food intake was elegantly stated 40 years ago by John Brobeck (Brobeck, 1955):

> . . . the total amount of food eaten is always the product of two factors, the number of meals multiplied by the intake of the average meal. . . . Any procedure altering food intake does so through some change in one or both of these, while any constancy observed in intake from time to time is a result either of a uniformity of frequency and intake or of some proportional and opposed change in them. Of these factors, the number of meals (i.e., the frequency of feeding) is determined by just where in the interfeeding interval a new feeding activity begins, but the average intake depends upon how long the eating persists and how rapidly it proceeds.

Accordingly, there are two questions to be answered in considering the regulation of feeding. Why do meals start, and why do they stop? There is now a considerable body of data addressing both of these questions in two groups of chewing insects, acridids and caterpillars. Gaining such data has involved analyzing feeding patterns, manipulating available food, and experimentally perturbing putative control systems.

This chapter will be structured around the two basic questions of why meals start and stop. Throughout, it will be assumed that insects have ample, nutritious food available to them. First, however, the phenomenon at issue must be defined: what, then, is a meal?

5.2. The Definition of a Meal

Meals are bouts of ingestion, separated from each other by periods during which sustained feeding does not occur. The difficulty lies in deciding (a) what period without feeding is deemed to be a gap between two meals, rather than a pause within a meal, and (b) what period constitutes "sustained" feeding—that is, distinguishes meals from sampling events? Objective answers to these questions have been obtained by analyzing the pattern of feeding in a range of chewing insects with *ad libitum* access to food. The most widely used technique has been log-survivorship analysis (see Simpson, 1990), although more recently log-frequency analysis has been preferred (Sibly et al., 1990; Berdoy, 1993). The task is made easier when there is a clear behavioural correlate of meal termination, such as leaving the feeding site and perching, as usually occurs in acridids (Simpson, 1982, 1990; Chapman and Beerling, 1990).

By investigating the frequency distribution of gaps between feeding events, it is possible to ascertain whether such events are grouped into bouts and to define the "bout criterion." Gaps shorter than the bout criterion are intrabout pauses, whereas longer ones are interbout intervals. Bout criteria from a number of studies are shown in Table 5.1.

By using the bout criterion, it is possible to break an individual insect's feeding pattern into a series of feeding episodes. The next question is, Which of these episodes are meals and which are isolated sampling events? Investigating the frequency distribution of feeding bout lengths shows whether there is a "meal criterion," separating meals from "nibbles." In nymphs of *Locusta migratoria* the meal criterion is ~50 sec (Simpson, 1994).

Recasting the definition in these terms yields the following: a meal is a period of ingestion longer than the meal criterion, often containing intrameal pauses and separated from other meals by periods longer than the bout criterion.

5.3. Why Does a Meal Begin?

The timing of meal initiation is determined by the developmental and nutritional state of the insect, prevailing environmental conditions, and the nature and availability of food. While total intake of nutrients over a given period of days often shows little variation between individuals (Raubenheimer and Simpson, 1993; Simpson and Raubenheimer, 1993a; Chambers et al., 1995), the detailed feeding patterns of individual insects generally vary considerably, even under controlled environmental conditions when ample, nutritionally appropriate food is available (Blaney et al., 1973; Simpson, 1990). Much of the complexity and variability of feeding patterns results from the interplay of factors which determine meal initiation. While some of these are clearly homeostatic in function, others are not and introduce a probabilistic element to meal initiation. The difficulty lies in untangling and identifying the various causal relationships.

Another complication is the fact that intermeal intervals are composed of several behavioral subcomponents. For instance, a locust usually leaves the food and spends some time locomoting until a resting site is reached. Then the animal

Table 5.1. *Bout criteria distinguishing intrameal pauses from intermeal intervals, as derived from ad libitum feeding patterns for 10 species (all chewers, except the last two)*

Species	Stage	Food	Lab/field	Bout criterion (min)	Reference[b]
Acridids					
Locusta migratoria	Fifth instar	Wheat seedling	Lab	4 (2–6)	1
		Artificial diets	Lab	4	2
	Adult (m + f)	Artificial diets	Lab	4	3
Schistocerca gregaria	Fifth instar	Artificial diets	Lab	2–3	4
Schistocerca americana	First instar	Wheat seedling	Lab	2–9	5
Hypochlora alba	Adult (f)	Artemesia leaves	Lab	4	6
Melanoplus sanguinipes	Adult (f)	Lolium seedling	Lab	4	6
Taeniopoda eques	Adult (f)	Various plants	Field	8	7
Lepidoptera					
Manduca sexta	Fifth instar	Tobacco leaves	Lab	1.6 (1.5–2.5)	8
		Artificial diet	Lab	1.9 (1.5–2.5)	8
		Artificial diet	Lab	2	9
		Tomato leaves	Lab	2	10
		Artificial diets	Lab	3.5 (2.3–4.4)	11
Coleoptera					
Leptinotarsa decemlineata	Adult (f)	Potato leaves	Lab	4.8 (3–10.5)	12
Diptera					
Lucilia cuprina	Adult (m)	Sugar solutions	Lab	2 (1.5–2.5)	13
Homoptera					
Schizaphis graminum	Adult (m, f)	Sorghum plants	Lab	15	14

[a]Values shown are means and/or ranges. f, females; m, males.

[b]References:, 1, Simpson (1982); 2, Simpson and Abisgold (1985); 3, Chyb and Simpson (1990); 4, Raubenheimer and Simpson (1990); 5, Chapman and Beerling (1990); 6, Blust and Hopkins (1990); 7, Raubenheimer and Bernays (1993); 8, Reynolds et al. (1986); 9, Bowdan (1988a); 10, Bowdan (1988b); 11, Timmins et al. (1988); 12, Mitchell and Low (1994); 13, Simpson et al. (1989); 14, Montllor (1991).

becomes quiescent. Following postprandial quiescence the insect commences locomotion, which may be undirected with respect to food, or else directed toward food stimuli (Simpson and Ludlow, 1986; Simpson, 1990). If appropriate food is not located, then the animal resettles, only to recommence locomotion soon afterwards. Eventually food is contacted and feeding initiated. There is some evidence that the thresholds for the initiation of locomotion following postprandial quiescence and that for the initiation of feeding are dissociated, at least in locusts, although this distinction has received relatively little attention [see Simpson (1990) for discussion on this point].

First, let us consider homeostatic feedbacks controlling intermeal intervals, in particular the influence of the previous meal.

5.3.1. The Inhibitory Effect of the Previous Meal

The likelihood that an insect (or any animal) provided with food will commence ingestion increases as a function of time since the last meal ended. Quantifying this function provides a basis for investigating the control of meal initiation. One way of doing this is to analyze the frequency distribution of intermeal intervals for insects feeding *ad libitum*. This was done for fifth-instar nymphs of *L. migratoria* by fitting proportional hazards models using generalized linear interactive modeling (GLIM) (Simpson and Ludlow, 1986).

Figure 5.1A shows the average change in the tendency for a fifth-instar *Locusta migratoria* nymph to commence feeding on seedling wheat with time since the last meal ended. Initially the probability of feeding is low, but it rises sharply as time progresses and reaches an asymptote after about 2 hr. The shape of this function is influenced by the size of the previous meal, with large meals inhibiting feeding more and for longer than small ones (Simpson and Ludlow, 1986). Additionally, the nutritional quality of the previous meal affects the probability of feeding, with foods containing a high density of nutrients delaying the initiation of the next meal [Simpson and Abisgold, 1985; Simpson, 1990; Chyb and Simpson, 1990; see Timmins et al., (1988) for similar results in caterpillars; also see Chapter 9].

The relationship between meal size and the duration of the following intermeal interval is also demonstrated in locusts in a much less statistically effective manner by correlating meal size with following intermeal length (Simpson, 1982). Other studies on acridids (Chapman and Beerling, 1990; Raubenheimer and Bernays, 1993), caterpillars (Reynolds et al., 1986; Bowdan, 1988a) and beetles (Mitchell and Low, 1994) have failed to demonstrate a positive correlation between meal size and duration of the following intermeal interval. However, given the limited power of correlations based on small data sets, it would be unwise to deduce that no such relationship exists. It is more likely that other causal factors obscure the effect, requiring the use of more statistically powerful

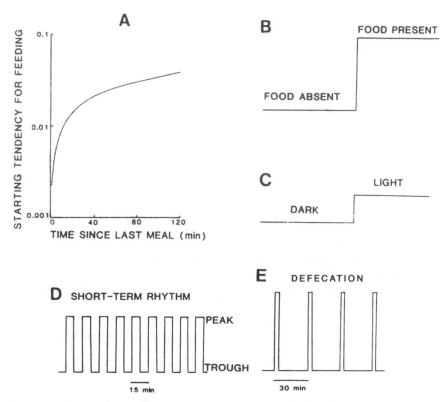

Figure 5.1. The influence of several factors on the tendency of a nymphal *L. migratoria* to commence a meal, as determined by deriving proportional hazards from feeding patterns under various experimentally manipulated conditions. (**A**) The starting tendency for feeding plotted on a log scale against time since an average meal. (**B**) The effect of the presence of food stimuli nearby on the tendency to commence locomotion (leading to feeding if food is present). (**C–E**) The effect of other factors on the tendency to feed. Factors B–E are plotted on the same log scale as A. They multiply the probability of feeding independently of time since the last meal, which means that their interactive effects can be derived from the figure by simply summing them with each other and the curve in A. (After Simpson and Ludlow, 1986).

techniques, such as fitting proportional hazards models, to expose the relationship.

Having established that taking a meal inhibits further feeding as a particular function of the time since the last meal, its size and the nutrient composition, the next task is to discover the component mechanisms underlying these effects. This involves looking for correlated physiological changes and then manipulating such variables to establish causality. The only detailed information that exists for a chewing insect comes from work on acridids, notably *L. migratoria*.

As food enters the gut during feeding, it stimulates stretch receptors located on the anterior wall of the crop (Bernays and Chapman, 1973). It may also enter the anterior part of the midgut and shunt food already there backwards to distend the ileum, wherein lie other stretch receptors (Simpson, 1983). Distension of these gut regions inhibits further feeding (Bernays and Chapman, 1973; Simpson, 1983; Roessingh and Simpson, 1984; see Section 5.4.2). Additionally, stretching of the crop elicits release of hormones from the storage lobes of the corpora cardiaca. Such hormones have a range of effects, including stimulating diuresis (Mordue, 1969), increasing gut motility (Cazal, 1969), reducing locomotor activity (Bernays and Chapman, 1974a; Bernays, 1980), and increasing the electrical resistance of the palp tips, probably by shutting the terminal pores of taste sensilla and thereby reducing the flow of chemosensory information to the central nervous system (Bernays et al., 1972; Bernays and Chapman, 1972a; Bernays and Mordue, 1973).

The identity of these hormonal factors is not known. In vertebrates a large number of hormonal agents have been implicated in signaling satiety, including insulin, glucagon, adrenalin, cholecystokinin, bombesin, gastrin, calcitonin, satietin, vasopressin, and adipocyte satiety factor (Le Magnen, 1985; Forbes, 1992). Immunoreactivity to many of these substances and to other biogenic amines and peptides, including FMRF-amides, has been found in the central nervous system and gut of locusts and other chewing insects (Jenkins et al., 1989; Montuenga et al., 1989; Robb and Evans, 1990; Schoofs et al., 1993; Peef et al., 1993; Zitnan et al., 1993), although their functions in the control of feeding behavior are not known.

During and after feeding, the composition of the hemolymph changes. Solute concentrations rise as water passes into the gut, both directly down an osmotic gradient and indirectly via ingested saliva, and as nutrients from digesta are absorbed from the gut. In *L. migratoria* these changes in osmolality and nutrient titers influence locomotion and meal initiation in a variety of ways which are discussed in detail in Chapter 9 (Simpson and Raubenheimer, 1993b).

The role of volumetric feedback from the gut of caterpillars in the regulation of either intermeal interval or meal size is equivocal (Bowdan, 1988a; Timmins and Reynolds, 1992; see Section 5.4.2), but there is evidence for nutrient feedbacks being involved in meal initiation (Friedman et al., 1991; Simmonds et al., 1992; Timmins and Reynolds, 1992; see Chapter 9). In larval *Manduca sexta* there is tonic inhibition of chewing from peripheral structures innervated by the paired nerves of the thoracic ganglia (Griss et al., 1991; Rowell and Simpson, 1992). The nature of these structures is unknown, but they could either be stretch receptors associated with the body wall or chemoreceptive organs monitoring blood composition. It was hypothesized that this tonic inhibition sets a threshold which must be exceeded by inputs from chemo- and mechanoreceptors before feeding will begin (Rowell and Simpson, 1992).

5.3.2. *Other Causal Factors*

So far, the inhibitory consequences of previous meals have been considered in the control of meal initiation, with food-finding behavior and feeding commencing when these inhibitory influences have declined. In addition, there are also endogenous and exogenous factors which enhance or inhibit the probability of locomoting and feeding. These include circadian rhythms (Brady, 1974), short-term rhythms (Simpson, 1981), defecation (Simpson and Ludlow, 1986), and light, visual, and chemosensory stimuli (Kennedy and Moorhouse, 1969; Blaney et al., 1973; Bernays and Simpson, 1982; Simpson and Ludlow, 1986; Simpson, 1990).

In locusts, excitatory factors including a short-term endogenous rhythm, defecation, light intensity, and stimuli provided by nearby food all appear to act independently of time since the last meal, simply multiplying the probability that feeding will commence at any moment (Simpson and Ludlow, 1986; Fig. 5.1). The effect of such factors on the probability of feeding may be substantial: the act of defecation temporarily raises the tendency to feed sevenfold in a locust, while gently flicking the outside of a pot housing a larval *Manduca sexta* sitting on food almost invariably provokes ingestion (Simpson, *personal observations*).

The interaction between these factors and the inhibitory consequences of previous meals results in complex and variable feeding patterns (see Section 5.5). To date there is no evidence that such excitatory factors are specific to feeding. Rather they seem to act by generally arousing the animal (Simpson, 1990). The possibility exists, however, that as yet unknown systemic signals are involved in specifically exciting feeding behavior, as seems to be the case in vertebrates (Campfield and Smith, 1990).

Another class of feeding-specific causal factors—learned cues—is considered in Chapters 9 and 10.

5.4. Why Does a Meal End?

Once initiated, three parameters define a meal: the amount of food eaten (meal size), the duration of the meal, and the rate of ingestion during the meal. Between them these constitute two degrees of freedom: when two are known, the third can be derived. When recording feeding behavior it is usual to measure meal size and duration directly and to derive ingestion rate subsequently. Although this is how we do it as ethologists, it does not necessarily reflect what happens inside the animal. In fact, the weight of evidence suggests that meal size and ingestion rate are the physiologically regulated variables, while meal duration arises passively (Simpson et al., 1988; Simpson, 1990). (This does not imply, of course, that meal duration is not "functionally regulated" through natural selection.) The following three statements would appear to apply to animals as diverse as rats and locusts:

1. Meals are terminated when negative feedbacks accruing during the meal counterbalance (a) net levels of excitation present as the meal begins and (b) excitation generated during the meal. Hence meal size is a positive function of excitation (e.g., Dethier and Gelperin, 1967; Barton Browne, 1975).

2. The overall rate of ingestion during a meal is also some positive function of feeding excitation (Simpson et al., 1988), as well as being influenced by the physical properties of the food (e.g., leaf toughness).

3. Meal duration is the result of the interplay between the functions defining the relationship between excitation and meal size and between excitation and overall ingestion rate (Simpson et al., 1988).

The nature of the relationship between meal size, overall ingestion rate, and meal duration is shown in Fig. 5.2 for three species of chewing insect fed *ad libitum* on standardized food plants: *Locusta migratoria*, *Schistocerca americana*, and *Leptinotarsa decemlineata*. In the two acridids, ingestion rate increases proportionally with meal size up to a limit beyond which it either reaches a plateau (*L. migratoria*) or declines (*S. americana*). In *L. decemlineata*, ingestion rate at first remains constant with increasing meal size, and then increases for larger meals. As a result, in the two grasshoppers, meal duration is conserved over a wide range of meal sizes, but increases for very large meals, while in *Leptinotarsa* the inverse pattern is found.

5.4.1. The Net Level of Feeding Excitation as a Meal Begins

The net level of feeding excitation present at the time feeding begins is a function of several variables: (a) the amount of feeding inhibition remaining from earlier meals (see Section 5.3.1); (b) excitatory input generated as a result of the presence of food nearby, or as a result of approaching, contacting, and ingesting food (Bernays and Chapman, 1974a; Barton Browne et al., 1975a; Bernays and Simpson, 1982; Simpson and Ludlow, 1986; Simpson et al., 1988); (c) a number of other factors, including, in the case of a locust, whether the animal has just defecated, the phase of its internal rhythm, light intensity, and whether other insects are nearby (Simpson and Ludlow, 1986; Simpson et al., 1988).

As already discussed in Section 5.3, these same factors determine whether or not a meal will commence: provided that net levels of feeding excitation exceed the feeding threshold, ingestion will begin. Once the feeding threshold has been crossed, the extent to which the combined total of excitatory influences exceeds this threshold then translates into meal size and rate of ingestion. Hence, meal size increases with time since the last meal, as does the rate of ingestion. These effects are evident both during *ad libitum* feeding, where gaps between meals

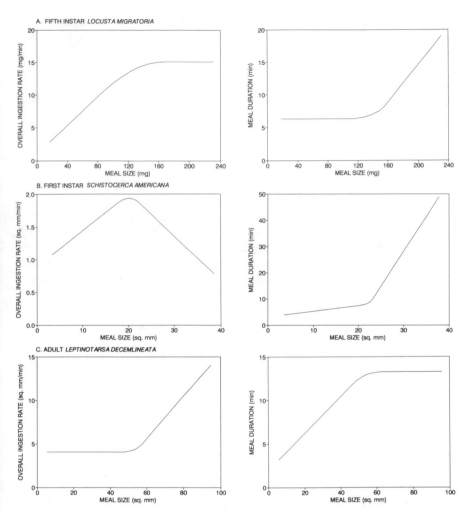

Figure 5.2. Stylized plots of the relationships between meal size, meal duration, and overall rate of ingestion during a meal in three species. Data are from Simpson et al., 1988 (**A**); Chapman and Beerling, 1990 (**B**), and Mitchell and Low, 1994 (**C**).

are "self-imposed" (Reynolds et al, 1986; Bowdan, 1988b; Chapman and Beerling, 1990; Mitchell and Low, 1994; Simpson and Simpson, 1990), and after periods of experimenter-imposed deprivation (Sinoir, 1968; Bernays and Chapman, 1972b; Roessingh and Simpson, 1984; Bowdan, 1988a, Simpson et al., 1988).

The excitatory effect of contacting and ingesting food comes in large part from input from chemoreceptors responding to phagostimulants (Bernays and Simpson, 1982). This input serves substantially and rapidly to raise the level of central excitation, an effect which has been documented most comprehensively by Dethier et al. (1965) for the black blowfly, but is also seen in chewing insects [e.g., Bernays and Chapman (1974a) and Barton Browne et al. (1975a) for acridids and Nakamuta (1985) for *Coccinella septempunctata*]. The function of such a positive feedback during the early stages of a meal is probably to prevent the animal from "dithering" about the feeding threshold (Houston and Sumida, 1985; Simpson, 1990). It also serves to ensure that "area-concentrated searching" occurs following loss of contact with food during a meal (Dethier, 1957; Bernays and Chapman, 1974a; Nakamuta, 1985).

The amplitude and duration of the excitatory effect of chemostimulation is a function of the intensity of phagostimulatory input (the product of the number of receptors stimulated and their firing rate), the time since the last meal, and presumably also the size and nutritional quality of previous meals (Dethier et al., 1965; Barton Browne et al., 1975a; Nakamuta, 1985; see Section 5.3). The intensity of phagostimulatory input is in turn a function of the chemical properties of the food and of internal feedbacks affecting chemoreceptor responsiveness (Bernays et al., 1972; Bernays and Chapman, 1972a; Bernays and Mordue, 1973; Abisgold and Simpson, 1988; Simpson et al., 1991; Simpson and Simpson, 1992; see Chapter 9). The presence of deterrent compounds in the food will negate phagostimulatory excitation (Bernays and Chapman, 1972b, 1974a, 1977; Chapman et al., 1991; Schoonhoven and Blom, 1988).

The generation of enhanced central excitation is unable to sustain feeding for long, however, if phagostimulatory input ceases. Continued phagostimulation is required throughout the meal, presumably acting to maintain the levels of excitation generated by the initial contact with food (Barton Browne et al., 1975a; Bernays and Chapman, 1974a).

5.4.2. Sources of Inhibition Accruing Throughout a Meal

Ingestion has short-term inhibitory consequences which eventually negate levels of excitation present in excess of the feeding threshold. When the net level of excitation drops below the feeding threshold, the meal ends. This is a classic negative feedback loop, which provides an exponentially falling level of feeding excitation as the meal progresses. In *L. migratoria* this exponential decline in excitation is seen in a change in instantaneous rate of ingestion throughout the meal (Simpson et al., 1988). On average, meals of seedling wheat commence at the same starting rate (20 mg/min), which then declines exponentially at a rate of decay which is a negative function of meal size (and hence feeding excitation) (Simpson et al., 1988; Simpson, 1990). The basis for the falling rate of ingestion is both (a) an increase in time spent pausing rather than ingesting and (b) a fall in the rate of chewing (Simpson et al., 1988). Declining ingestion rate throughout a meal has also been reported in first instar nymphs of *S. ameri-*

cana, largely as a result of more short pauses occurring in later parts of the meal (Chapman and Beerling, 1990).

What then are these short-term inhibitory feedbacks which contribute to meal termination? Once again, these are best known in locusts.

Of particular importance is volumetric feedback provided by stretch receptors located on the anterior wall of the crop, innervated by the posterior pharyngeal nerves (Bernays and Chapman, 1973; Roessingh and Simpson, 1984), and the ileum, innervated from branches of the rectal nerves (Simpson, 1983; Roessingh and Simpson, 1984). The balance of inputs from these two sources depends on the physical nature of the food and the amount of food already in the alimentary tract. Unlike meals of tough, mature grass, meals of soft, seedling grass are not restricted to the crop and thus enter the midgut, pushing any food there backwards to straighten the ileal fold (Simpson, 1983).

Other sources of volumetric feedback related to hemolymph or body volume, body mass, or blood pressure may exist, as was suggested for the locust *Chortoicetes terminifera* by Barton Browne et al. (1976). Additional evidence for such inputs came from experiments on *S. gregaria* (Roessingh and Simpson, 1984), although it seems that they might play a more important role in the control of the initiation of drinking rather than of meal termination (Bernays, 1977). It may also be that information regarding blood volume might be integrated in the central nervous system with inputs carrying information about blood nutrient concentrations to produce an overall measure of the total quantity in the hemolymph of key nutrients (and hence the animal's nutritional state; see Chapter 9). This would be particularly useful under circumstances where, for instance, high concentrations of nutrients in the hemolymph reflect dehydration rather than nutritional repletion.

In addition to volumetric feedbacks, it is likely that rapid changes in blood osmolality (Bernays and Chapman, 1974b,c; Abisgold and Simpson, 1987) and nutrient composition (Abisgold and Simpson, 1987, 1988; Simpson and Simpson, 1992) inhibit further feeding during a meal. As is discussed fully in Chapter 9, changes in nutrient concentrations may progressively reduce phagostimulatory input during a meal by desensitizing mouthpart gustatory receptors (Simpson and Simpson, 1992).

Phagostimulatory input during feeding may also decline as a result of adaptation of gustatory receptors (Barton Browne et al., 1975b; Blaney and Duckett, 1975), although the current consensus is that such changes are only likely to be of real significance when feeding on chemically simple substrates (Bernays and Simpson, 1982).

The role of volumetric feedbacks in the control of meal termination in caterpillars is unclear. After analyzing feeding patterns and the distribution of food in the gut, Bowdan (1988a) proposed that volumetric feedbacks from the foregut and anterior third of the midgut terminate meals in larval *Manduca sexta*. However, Timmins and Reynolds (1992) could find no evidence of either hyperphagia following sectioning of the recurrent nerve (which innervates these gut regions) or of reduced meal sizes after cannulating paraffin into the foregut, midgut, or hindgut. However, it is likely that cutting the recurrent nerve interfered with salivation and foregut emptying and that the excitatory effects of cannulation may well have overridden any inhibitory effects of cannulation (Miles and Booker, 1994; Timmins and Reynolds, 1992).

What is clear from the experiments of Timmins and Reynolds (1992) is that nutrient feedbacks can terminate feeding in *M. sexta*. Cannulating nutrient solutions into the gut resulted in reduced meal size. Whether this inhibition resulted from changes in hemolymph composition (see above; Chapter 9) or from input from chemoreceptors within the gut is unknown. As yet there is no direct evidence for gut sensory chemoreceptors in insects, although they have been postulated (Champagne and Bernays, 1991; Timmins and Reynolds, 1992). Such receptors are known to occur in mammals (Mei, 1985; Deutsch, 1990).

5.5. Putting It All Together: A Simulation of Locust Feeding

For *Locusta migratoria* there now exists not only a good understanding of the nature of the component elements in the control of feeding, but also quantitative parameter estimates for the relative strength and interaction of many of these elements. This provides the possibility, for the first time in any insect, to produce a full simulation model of feeding, which can be used both to test the internal consistency of the existing logical structure and to predict the effect of changes in the internal and external environment of the simulated animal. An outcome from such a simulation is shown in Fig. 5.3.

The figure shows a 280-min period during the life of a fifth-instar nymph kept at a temperature of 30°C with constant access to seedling wheat. The level of a variable called *feeding excitation* (representing total activity in some central neural network regulating feeding behavior) is plotted on a log scale against time. Two behavioral thresholds are shown, for locomotion and feeding. Others for palpation and biting will lie between these, with that for palpation being lower than the threshold for biting. The assumption made here is that all these behaviors share a causal factor with feeding (although this does not imply that they share *all* their causal factors), and they differ in their threshold with respect to it. The level of feeding excitation is influenced by: (a) a short-term rhythm (15-min period), which runs continuously; (b) declining inhibition from the previous meal; (c) defecation, the effect of which lasts 3 min; (d) central excitation, generated during feeding in proportion to the level of inhibition remaining from the last meal, and to the chemical properties of the food; (e) negative feedbacks (gut stretch, rapid osmotic, and nutrient feedbacks) accruing exponentially throughout a bout of feeding; and (f) inhibitory inputs (continued gut stretch, hormonal effects, the influence of hemolymph osmolality and nutrient

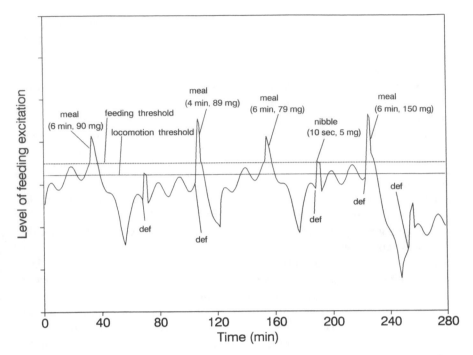

Figure 5.3. Output from a simulation of feeding in *Locusta migratoria*. See text for explanation.

titers) which persist or become apparent after the meal ends, the magnitude and time course of which are a function of meal size.

The parameter values used come mainly from Simpson and Ludlow (1986) and Simpson et al. (1988). Similarly, the decision to use a log-scale for feeding excitation is based on these studies, because it allows multiplicative effects to be added (see Fig. 5.1) and straightens exponential relationships.

As the simulation commences, the animal is resting on its perch between meals. Twenty minutes later the combination of rising excitation due to the removal of inhibition from the last meal and a peak in the internal rhythm leads to the initiation of walking. The locust wanders about randomly and contacts food but does not feed. After 5 min the insect settles, only to recommence locomotion a few minutes later. This time the insect walks randomly until it wanders across the food, whereupon it palpates, bites, and initiates feeding, which in turn result in generation of central excitation. The meal lasts 6 min, during which 90 mg of wheat is eaten. Feeding ceases when inhibitory feedbacks bring excitation below the feeding threshold. There then follows a period of walking and palpating lasting 2 min, after which the insect perches and becomes quiescent. Feeding excitation continues to fall for the next 17 min as inhibitory inputs stimulated by the meal (hormonal effects and changes in blood composition) continue to accrue. After this the balance between acquisition and removal of feeding-induced inhibition changes so that thereafter feeding excitation rises again.

Sixty-five minutes after the simulation began, the locust defecates, which causes a rise in excitation lasting 3 min. However, because it is not long since the previous meal and when there is a trough in the rhythms, only a shuffling of the legs and a brief period of walking results. Forty minutes later the insect defecates again, and this time the thresholds for locomotion and feeding (and hence also palpating and biting) are exceeded simultaneously. The locust commences directed locomotion. It walks straight to the food, palpating and even biting the perch as it goes. The meal lasts only 4 min, during which 89 mg is eaten. The early termination relative to the last meal is caused by the excitatory effect of the recent defecation wearing off in the middle of the meal.

There then follows a similar sequence of events over the next 160 min, with the locust taking two further meals and a nibble. All meals except the second are of the same duration, but they vary nearly twofold in size, the difference being due to the fact that instantaneous ingestion rate during a meal is a function of the level of feeding excitation present at that instant.

The combination of causal factors interact to produce complex and variable feeding patterns which nevertheless yield certain underlying relationships between meal sizes and intermeal durations—just like real locusts. The probabilistic element which is so apparent during *ad libitum* feeding is reproduced. For instance, having the model animal defecate only a minute later can profoundly change the subsequent pattern of feeding. The model also helps explain why even apparently well-fed locusts will often bite inert objects (Blaney and Chapman, 1970), without having to evoke more elaborate explanations (e.g., Blaney et al., 1985).

While the simulation in Fig. 5.3 contains relatively few elements and is based upon an insect with access to a single, homogeneous food plant, it is a simple task to incorporate other factors. The influences of the spatial distribution, temporal availability, phagostimulatory property, and nutritional quality of foods are readily included (e.g., see Fig. 9.7), as are changes in light phase, temperature, and so on.

5.6. Conclusions

Now is an exciting time in the study of the control of feeding behavior in chewing insects. Twenty years of detailed experimental work, much of it stimulated by Reg Chapman, has brought us to the stage where a complete explanation is in sight for at least one species, *Locusta migratoria*.

Additionally, comparative studies, especially of feeding patterns, are beginning to provide the basis for functional interpretation of feeding mechanisms. Evolved differences between species in the weighting of presumably similar control mechanisms has led to fundamental differences in feeding patterns (e.g., Blust and Hopkins, 1990; Bernays et al., 1992). As well as highlighting differences, comparative studies have demonstrated that certain aspects of meal regulation are apparently highly conserved, as exemplified in some of the fundamental similarities seen between the feeding patterns of *L. migratoria* (Simpson, 1982, 1990) in the laboratory and *Taeniopoda eques* in the field (Raubenheimer and Bernays, 1993). Observation of feeding behavior in the field is a welcome and necessary addition to the study of meal regulation, and it is gratifying to note that results obtained so far are readily interpreted in the light of laboratory systems.

A major gap remains in the study of meal regulation, however; this void—the central neural basis of feeding—was identified more than 10 years ago (Bernays and Simpson, 1982) and still remains largely unfilled. Efforts have commenced (Griss et al., 1991; Rowell and Simpson, 1992; Simpson, 1992; Schachtner and Braunig, 1993; Rorbacher, 1994a,b), but there is much to do. The other area which needs study is the role played by the various peptides and biogenic amines found in abundant association with the gut and central nervous system. Initial work (Cohen et al., 1988; Schachtner and Braunig, 1993) looks promising, but care is needed in negotiating the minefield that this area of research presents when attempting to attribute a causal role to any particular compound.

References

Abisgold, J. D., and Simpson, S. J. (1987). The physiology of compensation by locusts for changes in dietary protein. *J. Exp. Biol.* 129, 329–346.

Abisgold, J. D., and Simpson, S. J. (1988). The effect of dietary protein levels and haemolymph composition on the sensitivity of the maxillary palp chemoreceptors of locusts. *J. Exp. Biol.* 135, 215–229.

Barton Browne, L. (1975). Regulatory mechanisms in insect feeding. *Adv. Insect Physiol.* 11, 1–116.

Barton, Browne, L., Moorhouse, J. E., and van Gerwen, A. C. M. (1975a). An excitatory state generated during feeding in the locust, *Chortoicetes terminifera*. *J. Insect Physiol.* 21, 1731–1735.

Barton Browne, L., Moorhouse, J. E., and van Gerwen, A. C. M. (1975b). Sensory adaptation and the regulation of meal size in the Australian plague locust, *Chortoicetes terminifera*. *J. Insect Physiol.* 21, 1633–1639.

Barton Browne, L., Moorhouse, J. E., and van Gerwen, A. C. M. (1976). A relationship between weight loss during a period of food deprivation and the sizes of subsequent meals taken by the Australian plague locust, *Chortoicetes terminifera*. *J. Insect Physiol.* 22, 89–94.

Berdoy, M. (1993). Defining bouts of behaviour: a three process model. *Anim. Behav.* 46, 387–396.

Bernays, E. A. (1977). The physiological control of drinking behaviour in *Locusta migratoria*. *Physiol. Entomol.* 2, 261–273.

Bernays, E. A. (1980). The post-prandial rest in *Locusta migratoria* and its hormonal regulation. *J. Insect Physiol.* 26, 119–123.

Bernays, E. A., and Chapman, R. F. (1972a). The control of changes in peripheral sensilla associated with feeding in *Locusta migratoria* (L.). *J. Exp. Biol.* 57, 755–763.

Bernays, E. A., and Chapman, R. F. (1972b). Meal size in nymphs of *Locusta migratoria*. *Entomol. Exp. Appl.* 15, 399–410.

Bernays, E. A., and Chapman, R. F. (1973). The regulation of feeding in *Locusta migratoria*. Internal inhibitory mechanisms. *Entomol. Exp. Appl.* 16, 329–342.

Bernays, E. A., and Chapman, R. F. (1974a). The regulation of food intake by acridids. In: Barton Browne, L. (ed.), *Experimental Analysis of Insect Behaviour*. Springer-Verlag, Berlin, pp. 48–59.

Bernays, E. A., and Chapman, E. A. (1974b). Changes in haemolymph osmotic pressure in *Locusta migratoria* in relation to feeding. *J. Entomol. Ser. A* 48, 149–155.

Bernays, E. A., and Chapman, R. F. (1974c). The effects of haemolymph osmotic pressure on the meal size of nymphs of *Locusta migratoria* L. *J. Exp. Biol.* 61, 473–480.

Bernays, E. A., and Chapman, R. F. (1977). Deterrent chemicals as a basis of oligophagy in *Locusta migratoria*. *Ecol. Entomol.* 2, 1–18.

Bernays, E. A., and Mordue, A. J. (1973). Changes in palp tip sensilla of *Locusta migratoria* in relation to feeding: the effects of different levels of hormone. *Comp. Biochem. Physiol. A* 45, 451–454.

Bernays, E. A., and Simpson, S. J. (1982). Control of food intake. *Adv. Insect Physiol.* 16, 59–118.

Bernays, E. A., Blaney, W. M., and Chapman, R. F. (1972). Changes in chemoreceptor sensilla on the maxillary palps on *Locusta migratoria* in relation to feeding. *J. Exp. Biol.* 57, 745–753.

Bernays, E. A., Bright, K., Howard, J. J., Raubenheimer, D., and Champagne, D. (1992). Variety is the spice of life: frequent switching between foods in the polyphagous grasshopper *Taeniopoda eques* Burmeister (Orthoptera: Acrididae). *Anim. Behav.* 44, 721–731.

Blaney, W. M., and Chapman, R. F. (1970). The functions of the maxillary palps of Acrididae (Orthoptera). *Entomol. Exp. Appl.* 13, 363–376.

Blaney, W. M., and Duckett, A. M. (1975). The significance of palpation of the maxillary palps of *Locusta migratoria* (L.): an electrophysiological and behavioural study. *J. Exp. Biol.* 63, 701–712.

Blaney, W. M., Chapman, R. F., and Wilson, A. (1973). The pattern of feeding of *Locusta migratoria* (Orthoptera, Acrididae). *Acrida* 2, 119–137.

Blaney, W. M., Winstanley, C., and Simmonds, M. S. J. (1985). Food selection by locusts: an analysis of rejection behaviour. *Entomologia Exp. Appl.* 38, 35–40.

Blust, M. H., and Hopkins, T. L. (1990). Feeding patterns of a specialist and a generalist grasshopper: electronic monitoring on their host plants. *Physiol. Entomol.* 15, 261–267.

Bowdan, E. (1988a). The effect of deprivation on the microstructure of feeding by the tobacco hornworm caterpillar. *J. Insect Behav.* 1, 31–50.

Bowdan, E. (1988b). Microstructure of feeding by tobacco hornworm caterpillars, *Manduca sexta. Entomol. Exp. Appl.* 47, 127–136.

Brobeck, J. R. (1955). Neural regulation of food intake. *Ann. N.Y. Acad. Sci.* 63, 44–55.

Brady, J. (1974). The physiology of insect circadian rhythms. *Adv. Insect Physiol.* 10, 1–116.

Campfield, L. A., and Smith, F. J. (1990). Systemic factors in the control of food intake. Evidence for patterns as signals. In: Stricker, E. M. (ed.), *Handbook of Behavioral Neurobiology,* vol. 10. Plenum, New York, pp. 183–205.

Cazal, M. (1969). Actions d'extraits de corpora cardiaca sur le peristaltisme intestinal de *Locusta migratoria. Arch. Zool. Exp. Gen.* 110, 83–89.

Chambers, P. G., Simpson, S. J., and Raubenheimer, D. (1995). Behavioural mechanisms of nutrient balancing in *Locusta migratoria* nymphs. *Anim. Behav.*

Champagne, D. E., and Bernays, E. A. (1991). Phytosterol suitability as a factor mediating food aversion learning in the grasshopper *Schistocerca americana. Physiol. Entomol.* 16, 391–400.

Chapman, R. F., and Beerling, E. A. M. (1990). The pattern of feeding of first instar nymphs of *Schistocerca americana. Physiol. Entomol.* 15, 1–12.

Chapman, R. F., Ascoli-Christensen, A., and White, P. R. (1991). Sensory coding for feeding deterrence in the grasshopper *Schistocerca americana. J. Exp. Biol.* 158, 241–259.

Chyb, S., and Simpson, S. J. (1990). Dietary selection in adult *Locusta migratoria. Entomol. Exp. Appl.* 56, 47–60.

Cohen, R. W., Friedman, S., and Waldbauer, G. P. (1988). Physiological control of nutrient self-selection in *Heliothis zea* larvae: the role of serotonin. *J. Insect Physiol.* 34, 935–940.

Dethier, V. G. (1957). Communication by insects: physiology of dancing. *Science* 125, 331–336.

Dethier, V. G., and Gelperin, A. (1967). Hyperphagia in the blowfly. *J. Exp. Biol.* 47, 191–200.

Dethier, V. G., Solomon, R. L., and Turner, L. H. (1965). Sensory input and central excitation and inhibition in the blowfly. *J. Comp. Physiol. Psychol.* 60, 303–313.

Deutsch, J. A. (1990). Gastric factors. In: Stricker, E. M. (ed.), *Handbook of Behavioral Neurobiology,* vol. 10. Plenum, New York, pp. 151–182.

Forbes, J. M. (1992). Metabolic aspects of satiety. *Proc. Nutr. Soc.* 51, 13–19.

Friedman, S., Waldbauer, G. P., Eertmoed, J. E., Naeem, M., and Ghent, A. W. (1991).

Blood trehalose levels have a role in the control of dietary self-selection by *Heliothis zea* larvae. *J. Insect Physiol.* 37, 919–928.

Griss, C., Simpson, S. J., Rohrbacher, J., and Rowell, C. H. F. (1991). Localization in the central nervous system of larval *Manduca sexta* (Lepidoptera: Sphingidae) of areas responsible for aspects of feeding behaviour. *J. Insect Physiol.* 37, 477–482.

Houston, A. I., and Sumida, B. (1985). A positive feedback model for switching between two activities. *Anim. Behav.* 33, 315–325.

Jenkins, A. C., Brown, M. R., and Crim, J. W. (1989). FMRF-amide immunoreactivity and the midgut of the corn earworm (*Heliothis zea*). *J. Exp. Zool.* 252, 71–78.

Kennedy, J. S., and Moorhouse, J. E. (1969). Laboratory observations on locust responses to wind-borne grass odour. *Entomol. Exp. Appl.* 12, 487–503.

Le Magnen, J. (1985). *Hunger.* Cambridge University Press, Cambridge.

Mei, N. (1985). Intestinal chemosensitivity. *Physiol. Rev.* 65, 211–237.

Miles, C. I., and Booker, R. (1994). The role of the frontal ganglion in foregut movements of the moth *Manduca sexta*. *J. Comp. Physiol.* A174, 755–767.

Mitchell, B. K., and Low, R. (1994). The structure of feeding behavior in the Colorado potato beetle, *Leptinotarsa decemlineata* (Coleoptera: Chrysomelidae). *J. Insect Behav.* 7, 707–729.

Montllor, C. B. (1991). The influence of plant chemistry on aphid feeding behavior. In: Bernays, E. A. (ed.), *Insect–Plant Interactions,* vol. III. CRC Press, Boca Raton, FL, pp. 125–173.

Montuenga, L. M., Barrenechea, M. A., Sesma, P., Lopez, J., and Vazquez, J. J. (1989). Ultrastructure and immunocytochemistry of endocrine cells in the midgut of the desert locust, *Schistocerca gregaria* (Forskal). *Cell Tissue Res.* 258, 577–583.

Mordue, W. (1969). Hormonal control of Malpighian tubule and rectal function in the desert locust, *Schistocerca gregaria*. *J. Insect Physiol.* 15, 273–285.

Nakamuta, K. (1985). Mechanism of the switchover from extensive to area-concentrated search behaviour of the ladybird beetle, *Coccinella septempunctata Bruckii*. *J. Insect. Physiol.* 31, 849–856.

Peef, N. M., Orchard, I., and Lange, A. B. (1993). The effects of FMRFamide-related peptides on an insect (*Locusta migratoria*) visceral muscle. *J. Insect Physiol.* 39, 207–215.

Raubenheimer, D., and Bernays, E. A. (1993). Patterns of feeding in the polyphagous grasshopper *Taeniopoda eques,* a field study. *Anim. Behav.* 45, 153–167.

Raubenheimer, D., and Simpson, S. J. (1990). The effects of simultaneous variation in protein, digestible carbohydrate and tannic acid on the feeding behaviour of larval *Locusta migratoria* (L.) and *Schistocerca gregaria* (Forskal). I. Short-term studies. *Physiol. Entomol.* 15, 219–233.

Raubenheimer, D., and Simpson, S. J. (1993). The geometry of compensatory feeding in the locust. *Anim. Behav.* 45, 953–964.

Reynolds, S. E., Yeomans, M. R., and Timmins, W. A. (1986). The feeding behaviour of caterpillars (*Manduca sexta*) on tobacco and on artificial diet. *Physiol. Entomol.* 11, 39–51.

Robb, S., and Evans, P. D. (1990). FMRFamide-like peptides in the locust, distribution, partial characterization and bioactivity. *J. Exp. Biol.* 149, 335–360.

Roessingh, P., and Simpson, S. J. (1984). Volumetric feedback and the control of meal size in *Schistocerca gregaria*. *Entomol. Exp. Appl.* 36, 279–286.

Rohrbacher, J. (1994a). Mandibular premotor interneurones of larval *Manduca sexta*. *J. Comp. Physiol.* A 175, 619–628.

Rohrbacher, J. (1994b). Fictive chewing activity in motor neurones and interneurones of the suboesophageal ganglion of *Manduca sexta* larvae. *J. Comp. Physiol.* A 175, 629–637.

Rowell, C. H. F., and Simpson, S. J. (1992). A peripheral input of thoracic origin inhibits chewing movements in the larvae of *Manduca sexta*. *J. Insect Physiol.* 38, 475–483.

Schachtner, J., and Braunig, P. (1993). The activity pattern of identified neurosecretory cells during feeding behaviour in the locust. *J. Exp. Biol.* 185, 287–303.

Schoofs, L., Vanden Broek, J., and de Loof, A. (1993). The myotropic peptides of *Locusta migratoria*. Structures, distribution, functions and receptors. *Insect Biochem. Mol. Biol.* 23, 859–881.

Schoonhoven, L. M., and Blom, F. (1988). Chemoreception and feeding behaviour in a caterpillar, towards a model of brain function in insects. *Entomol. Exp. Appl.* 49, 123–129.

Sibly, R. M., Nott, H. M. R., and Fletcher, D. J. (1990). Splitting behaviour into bouts. *Anim. Behav.* 39, 63–69.

Simmonds, M. S. J., Simpson, S. J., and Blaney, W. M. (1992). Dietary selection behaviour in *Spodoptera littoralis*, the effects of conditioning diet and conditioning period on neural responsiveness and selection behaviour. *J. Exp. Biol.* 162, 73–90.

Simpson, S. J. (1981). An oscillation underlying feeding and a number of other behaviours in fifth instar *Locusta migratoria* nymphs. *Physiol. Entomol.* 6, 315–324.

Simpson, S. J. (1982). Patterns of feeding, a behavioural analysis using *Locusta migratoria* nymphs. *Physiol. Entomol.* 7, 325–336.

Simpson, S. J. (1983). The role of volumetric feedback from the hindgut in the regulation of meal size in fifth-instar *Locusta migratoria* nymphs. *Physiol. Entomol.* 8, 451–467.

Simpson, S. J. (1990). The pattern of feeding. In: Chapman, R. F., and Joern, A. (eds.), *Biology of Grasshoppers*. John Wiley & Sons, New York, pp. 73–103.

Simpson, S. J. (1992). Mechanoresponsive neurones in the suboesophageal ganglion of the locust. *Physiol. Entomol.* 17, 351–369.

Simpson, S. J. (1994). Experimental support for a model in which innate taste responses contribute to regulation of salt intake by nymphs of *Locusta migratoria*. *J. Insect Physiol.* 40, 555–559.

Simpson, S. J. and Abisgold, J. D. (1985). Compensation by locusts for changes in dietary nutrients, behavioural mechanisms. *Physiol. Entomol.* 10, 443–452.

Simpson, S. J., and Ludlow, A. R. (1986). Why locusts start to feed, a comparison of causal factors. *Anim. Behav.* 34, 480–496.

Simpson, S. J., and Raubenheimer, D. (1993a). The central role of the haemolymph in the regulation of nutrient intake in insects. *Physiol. Entomol.* 18, 395–403.

Simpson, S. J., and Raubenheimer, D. (1993b). A multi-level analysis of feeding behaviour, the geometry of nutritional decisions. *Philos. Trans. R. Soc. London Ser. B* 342, 381–402.

Simpson, S. J., and Simpson, C. L. (1990). The mechanisms of compensation by phytophagous insects. In: Bernays, E. A. (ed.), *Insect–Plant Interactions*, vol. 2. CRC Press, Boca Raton, FL, pp. 111–160.

Simpson, S. J., and Simpson, C. L. (1992). Mechanisms controlling modulation by amino acids of gustatory responsiveness in the locust. *J. Exp. Biol.* 168, 269–287.

Simpson, S. J., Simmonds, M. S. J., Wheatley, A. R., and Bernays, E. A. (1988). The control of meal termination in the locust. *Anim. Behav.* 36, 1216–1227.

Simpson, S. J., Barton Browne, L., and van Gerwen, A. C. M. (1989). The patterning of compensatory sugar feeding in the Australian sheep blowfly. *Physiol. Entomol.* 14, 91–105.

Simpson, S. J., James, S., Simmonds, M. S. J., and Blaney, W. M. (1991). Variation in chemosensitivity and the control of dietary selection behaviour in the locust. *Appetite* 17, 141–154.

Sinoir, Y. (1968). Etude de quelques facteurs conditionnant la prise de nourriture chez les larves du criquet migrateur, *Locusta migratoria migratorioides* (Orthoptera, Acrididae). II. Facteurs internes. *Entomol. Exp. Appl.* 11, 443–449.

Timmins, W. A., and Reynolds, S. E. (1992). Physiological mechanisms underlying the control of meal size in *Manduca sexta* larvae. *Physiol. Entomol.* 17, 81–89.

Timmins, W. A., Bellward, K., Stamp, A. J., and Reynolds, S. E. (1988). Food intake, conversion efficiency, and feeding behaviour of tobacco hornworm caterpillars given artificial diet of varying nutrient and water content. *Physiol. Entomol.* 13, 303–314.

Zitnan, D., Sauman, I., and Sehnal, F. (1993). Peptidergic innervation and endocrine cells of insect midgut. *Arch. Insect Biochem. Physiol.* 22, 113–132.

6

Regulation of a Meal: Blood Feeders

E. E. Davis and W. G. Friend

6.1. Introduction

Hematophagous insects are found in six of the 28 orders of Insecta. They are primarily holometabolous and, except where noted, feed on blood only as adults. Within the order Diptera, there are 12 families that include blood-sucking species, many of which are important transmitters of diseases that affect humans and livestock. The family Calliphoridae includes the only larval blood feeders—for example, *Auchmeromyia senegalensis,* the Congo floor maggot (the adult is coprophagous). Siphonaptera (fleas), all of which are blood-sucking, also includes disease transmitting members. Coleoptera has four families with one species each that feed on ectoderm and only secondarily feed on blood from the resulting skin abrasions; Lepidoptera has a single species of noctuid moth, *Calpe eustrigata,* from South Asia that feeds on mammals; and in Hymenoptera are found the vampire wasps of Vancouver, Canada (Phipps, 1974). The families Reduviidae and Cimicidae, in the order Hemiptera, and the sucking lice of the order Phthiraptera are hemimetabolous, and both sexes feed on blood at all life stages.

In these six orders are a diversity of life styles. There are (a) obligate parasites [e.g., tsetse flies (*Glossina*)] and reduviid bugs (e.g., *Rhodnius*) that must feed on blood for both reproductive and energy requirements, (b) insects that must feed on blood for reproductive purposes, like some mosquitoes, but that can and do feed on plant resources to meet their energy needs (see Chapter 8), and (c) facultative parasites that can feed on either blood or plants for both reproductive and energy requirements; some insects may not express hematophagy until their second gonotrophic cycle. In some groups (e.g., horseflies and mosquitoes), only the female blood-feeds, whereas in others (e.g., fleas and tsetse flies), both sexes feed on blood. Some hematophages feed during the day, some only at night, and others at dawn and/or dusk. Their mouthparts have become specialized for penetration of the host animal's skin and obtaining blood by piercing and sucking, by biting and chewing, or by rasping and abrading (see Lehane, 1991; Smith, 1985); some are vessel feeders, whereas others are pool feeders (see Chapter 3). Unlike phytophagous insects, the hematophages have had to adapt to a food source that is both mobile and often defensive and have evolved various methods of attaining such a host. Some spend their entire life cycle on a single host, while others feed from multiple hosts; some live in the host's domicile and others are ambushers or hunters, often flying several kilometers in search of a blood meal source.

Blood feeding has evolved independently in many groups of the Insecta, in some cases even within the same family—for example, the Culicidae (mosquitoes). Consequently, one might expect great variation in feeding behaviors and their control among various hematophagous insect groups. However, similarities in behaviors and regulatory mechanisms have arisen, probably because the problems associated with blood feeding facing most hematophagous insects do not differ greatly (see Chapter 2). At appropriate times in their life cycles, most blood feeders must detect and orient to the host from a distance, locate and make tarsal contact with the host, probe and pierce the host, locate and ingest blood, direct it to the appropriate part of the digestive tract, cease feeding upon repletion, and leave the host while avoiding host defensive acts in all stages of the process. Following feeding, there usually is a period during which the insect becomes unresponsive to stimuli that would otherwise initiate feeding behaviors in the unfed insect.

Insect feeding behavior involves temporal, often hierarchical sequences of events that many investigations have treated solely as sequences of isolated stimulus–response events, with the output of a prior event merely becoming the input for the next. Although stimulus–response analysis of the feeding process seems to be an essential operational axiom for the development of an understanding of the process, it has its limitations (Bowen, 1991a). Furthermore, there is mounting evidence that once a blood-feeding insect has initiated a particular mode of feeding, the threshold of responses to other appropriate stimuli coming later in the sequence is lowered (Friend and Smith, 1977; Sutcliffe, 1986; Sutcliffe and McIver, 1975). For example, *Tabanus nigrovittatus* is attracted to an artificial feeding apparatus by heat which induces probing and the deployment of mouthparts in the blood-feeding mode as the membrane is pierced. These responses have been shown to lower the threshold of response to nucleotide phagostimulants (Friend and Stoffolano, 1984).

In this review, we summarize the current information on the process of feeding by hematophagous insects focusing on the physiological states, environmental signals, and physical events that mediate and modulate blood-feeding behavior. However, nectar feeding, in which we include feeding on free liquid nectar or water and on dry honeydew, is an integral part of the feeding behavior repertoire of some hematophagous insects such as mosquitoes and blackflies (see Chapter 8). Research on the regulation of feeding by hematophagous insects has been very selective and has concentrated on species that are both medically important and easy to rear in the laboratory; thus the bulk of the literature in this area relates to mosquitoes, and most of that relates to *Aedes aegypti,* the *Culex pipiens* complex, and *Culiseta inornata.* Because of this focus, more is known about the physiology and behavior of mosquito feeding than about the feeding of any other insect group. Consequently, in this review, mosquitoes are used as the basis for understanding the regulation and control of blood feeding. The regulation of blood feeding by other insects will be included to the extent that it is known.

6.2. Model of Feeding Behavior

In order to provide a framework for organizing that which is known and identifying that which is not known about the complex nature of feeding in hematophagous insects, we developed a heuristic model in the form of a decision diagram representing a series of "stages" (numbered rectangles) through which the insect must pass and "decisions" (lettered diamonds) that must be made in the process of feeding. The model is based upon a composite or generic female mosquito that has been inseminated and is in nondiapausing and anautogenous states. (Anautogeny is the dependence of egg production on a blood meal; autogeny is the production of eggs without the ingestion of protein by the adult.) We will also incorporate some of the influences of important factors on these feeding behaviors: the effects of prior feeding as larvae or as adults, circadian/diel periodicities, and other physiological factors.

Decision diagrams 1–4 (Figs. 6.2–6.5) map the various states and conditions that are involved in mosquito feeding, from the initial stage when the female is not responsive to external feeding stimuli to the cessation of feeding and the effect of the meal on subsequent behaviors. The relationships between the individual diagrams is illustrated in Fig. 6.1.

6.3 Initiation of Feeding Behavior [Diagram 1 (Fig. 6.2)]

Initially, newly emerged female *Aedes aegypti* mosquitoes do not respond to external signals that mediate orientation to a plant or animal (Stage 1) (see Davis, 1984b; Laarman, 1955). This nonresponsive condition is also found in diapausing (Bowen et al., 1988), autogenous (O'Meara, 1985), and gravid females (Klowden and Lea, 1979a), as well as in females that have just taken a meal (Gwadz, 1969; Klowden and Lea, 1979b).

Following emergence of adult *Aedes aegypti* female mosquitoes, the antennal receptors for the host attractant, lactic acid, undergo a presumptive maturation process that requires about 3 days to complete at 28°C (Davis, 1984b). This

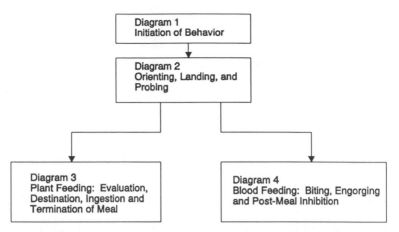

Figure 6.1. Diagram showing the relationships between the diagrams in Figs. 6.2–6.5.

process is concurrent with, but independent of, a juvenile hormone (JH)-dependent follicular development to resting stage (Bowen and Davis, 1989). In contrast, the initiation of biting behavior in *Culex quinquifasciatus* may be JH-dependent (Meola and Petralia, 1980). Furthermore, unlike *Aedes aegypti,* the lactic acid receptors of the autogenous *Aedes atropalpus* reach high sensitivity as early as 10 hr post emergence even though host-locating behavior is not expressed this early (Bowen et al., 1994a,b). Mating occurs in most mosquitoes at this time with resultant humoral and behavioral changes (Hiss and Fuchs, 1972; Jones and Gubbins, 1977; Judson, 1967; Leahy and Craig, 1965) which may affect feeding behavior. For example, uninseminated host-responsive females will continue to be attracted to, and may repeatedly feed from, an animal host (Klowden and Lea, 1979a), whereas inseminated females that mature eggs following a small feed are no longer attracted. This occurs especially in those species that mate in the vicinity of the host.

In newly emerged mosquitoes, the levels of energy reserves are largely dependent upon larval nutrition and will influence the utilization of the insect's first meal (Klowden, 1986, 1990; Klowden et al., 1988). For example, an adult anautogenous *Aedes aegypti* from an underfed larva may require two meals, the first of which may be blood or nectar, before the first clutch of oocytes matures, whereas in an adult from a well-fed larva the oocytes may mature after only a single blood meal (Feinsod and Spielman, 1980).

In older, mature females (more than 5 days post emergence), other physiological states may occur which alter their responsiveness to food-resource signals. For example, during diapause, *Culex pipiens* will not host-seek, but, on termination of diapause, the sensitivity of their lactic acid receptors increases and they switch from feeding solely at plants, on nectar or honeydew, to also seeking a blood meal (Bowen, 1991b). In contrast, female *Anopheles* will seek and feed on both

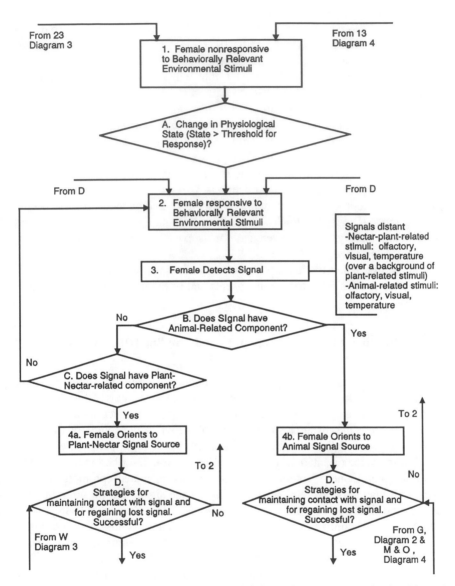

Figure 6.2. Diagram 1: The initiation of feeding behavior. Rectangles in this and subsequent figures represent "stages" through which the insect must pass; diamonds represent "decisions" the insect must make (numbers and letters correspond with those in the text). Diamonds D are repeated at the top of Fig. 6.3.

plants and animals during diapause, but will not use the blood meal for reproductive purposes (Washino, 1977). Gravid females will not host-seek, but may continue to feed at plants (Magnarelli, 1977). However, following oviposition, females that are anautogenous or are autogenous only for the first gonotrophic cycle will seek a blood meal. Females that will not seek a meal because of abdominal distention will do so after the distention is relieved (Klowden, 1990).

The female is now assumed to be fully responsive to relevant environmental signals from potential food resources, plants or animals (Stage 2), so that when she detects an appropriate signal from a plant or animal (Stage 3), she will respond by orienting to the source (Stages 4a or 4b). But, what if she detects both plant and animal signals simultaneously? How will she "select" which feeding mode to engage in? In anautogenous *Aedes aegypti*, there is a dominance in favor of the reproductively more important blood-feeding mode over the nectar-feeding mode (Davis, 1989; Klowden, 1990; Yee and Foster, 1992). Neither the mechanism underlying this preference nor its existence in other mosquitoes has yet been determined.

It should be noted that almost nothing is known about this aspect of the life cycle of most other hematophagous insects and of the initiation of their feeding behaviors.

6.4. Orientation to the Food Resource and Landing [Diagram 2 (Fig. 6.3)]

Female mosquitoes and other flying hematophages orient to either the plant resource (Stage 4a, Fig. 6.2) or the animal resource (Stage 4b) by employing flight pattern strategies for maintaining or regaining contact with the signal emanating from the resource so that they will arrive at the source of the signal from a more-or-less long distance (Decision D). These flight patterns, especially for plant-resource orientation, may be similar to those described in the literature for moth pheromone orientation—that is, an odor-mediated optomotor anemotaxis with interactive patterns for searching for the signal and for orienting to it once found (Baker, 1989; Bell, 1990; Kaissling, 1987; Kennedy, 1986; Visser, 1988). However, unlike insect sex pheromones, an animal providing the signals is not doing so to attract mosquitoes and will often take defensive action against the female as she approaches (Edman and Scott, 1987). The orienting flight patterns for mosquito host location, especially at close range, appear to be aperiodic, nonrandom patterns with low predictability (Davis, *unpublished observations*) rather than the highly predictable self-steered, counterturning patterns of moth casting and orienting flight. Tsetse also appear to employ similar, not readily predictable flight patterns during host orientation (Brady et al., 1989; Gibson et al., 1991). However, a clear picture of the specifics of host-orienting flight strategies of hematophagous insects has yet to be developed.

Many blood-feeding insects attain a host by means other than flying even

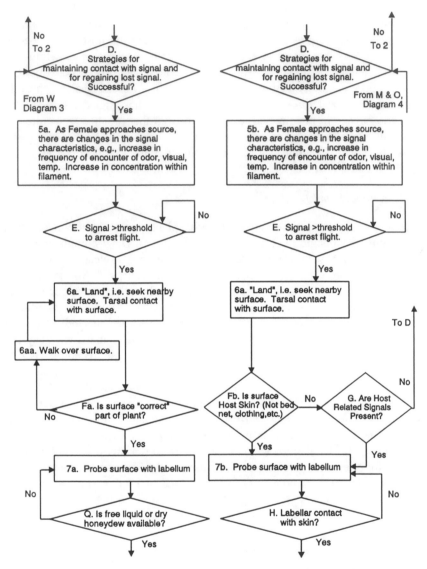

Figure 6.3. Diagram 2: Orienting, probing, and landing. Diamonds D are repeats of those at the bottom of Fig. 6.2; Diamonds Q and H are repeated at the tops of Figs. 6.4 and 6.5, respectively.

though they may be capable of flight. For example, although *Rhodnius prolixus* is a strong flier, no records of triatomine bugs hunting hosts by flight have been found (Gringorten and Friend, 1979). Lice, bedbugs, and the ectoparasitic Coleopterans all walk to a host from locations within the host's nest, bed, or clothing. Insects that orient to and locate a host by walking or crawling seem to do so in a manner similar to flying insects; that is, they rely on odor-mediated anemotactic locomotory activity to bring them to a host. Fleas, on the other hand, do not appear to orient toward the source of host signals. Instead, they appear to engage in nondirected jumping up from their nonhost substrate that, with a little bit of luck and persistence, will land them on the host. These insects that walk, crawl, or jump to a host may also require a reduced series of host signals to find a host (see below).

6.4.1. Approach and Landing

As an insect approaches the odor source, either plant or animal, the characteristics of the odor signal begin to change (Stages 5a and 5b). Other components of the odor blend may now be above detection threshold levels. Near the source, the odor filaments are closer together and less diffuse (Murlis, 1986) so that there may be an increase in both the signal intensity and the frequency of encountering the odor filament. As the insect gets even closer to the odor source, its olfactory receptors may no longer be able to discriminate between the individual odor filaments which will appear to fuse, much like flicker-fusion in the visual system. This may be due to habituation or sensory adaptation of the receptors (Kuenen and Baker, 1981). In moths, when this occurs, the flight velocity decreases and turn angles and turning frequency increase (Baker et al., 1988).

As the insect approaches a surface, it perceives an expanding visual pattern that triggers a forward extension of its legs (Goodman, 1960). Then, when the insect's tarsi make contact with the surface, wing movements cease (Fraenkel, 1932); the insect has landed (Stages 6a and b). Up to this point, the behavior patterns for orientation to a target from a distance and landing on the target have been nearly the same for both plant and animal resources; the principal difference has been in the identity of the signals that mediated the behaviors. From this point on, the feeding behaviors begin to differ markedly.

6.5. Nectar Feeding by Hematophagous Insects [Diagram 3 (Fig. 6.4)]

Stages 6a and 7a and Diagrams 2 and 3 will be briefly discussed and depicted here for several reasons. First, because feeding on plant juices by blood-feeding insects is widespread within hematophagous insects. For example, within the families Ceratopogonidae, Culicidae, Muscidae, Psychodidae, Simuliidae, and Tabanidae, females feed both on plant-associated fluids and on blood (the males feed only on plant-associated fluids). Second, insects that feed on both blood

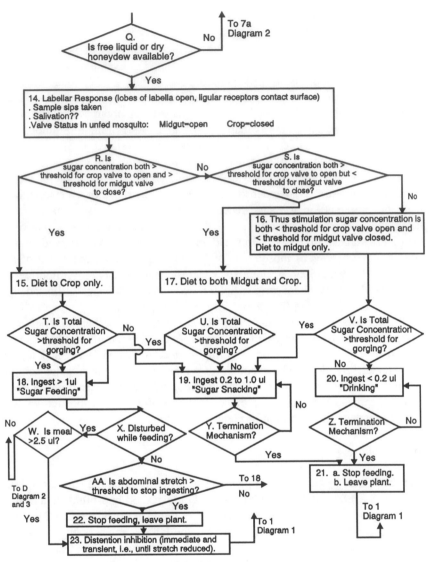

Figure 6.4. Diagram 3: Plant feeding. Diamond Q is a repeat of that at the bottom left of Fig. 6.3.

and plant juices are frequently faced with making a choice between the two feeding modes. Our model for the control of ingestion and diet destination of plant-associated fluids (i.e., nectar, honeydew, or water) is based on a series of studies on *C. inornata* (Friend, 1981; Friend et al., 1988, 1989; Schmidt and Friend, 1991; and see also Chapter 8).

6.5.1. Plant Preference

Mosquitoes exhibit a clear preference for certain plants over others even though the preferred plants may not be the most abundant. For example, in samples of *Aedes sollicitans* collected from vegetation, Magnarelli (1977) found 25% on common dandelion (*Taraxacum officinale*) and king-devil hawkweed (*Hieracium pratense*) even though each had a relative density of about 2% of the total population of flowering plants available in the area. None of the mosquitoes was found on woolly blue violet (*Viola sororia*) despite its relative density of 24%. In another area, he found 20% of *Aedes sollicitans* feeding on yarrow (*Achillea millefolium*), with a relative density of 7%, while none was feeding on red clover (*Trifolium pratense*), relative density 42%. Within the range of acceptable host plants, the preference may change depending on the relative density of acceptable plants, their seasonal availability, and the age of the flowers of a particular host plant (Magnarelli, 1979). Grimstad and DeFoliart (1974) observed similar patterns in 10 species of mosquitoes in Wisconsin. The size and shape of the flower are not recognition or preference factors, and color is only important to the extent that mosquitoes prefer lighter over darker flowers (Grimstad and DeFoliart, 1974; Magnarelli, 1979). On the other hand, there is some evidence that floral odors are attractants. Using floral extracts in the laboratory, Vargo and Foster (1982) demonstrated that *Aedes aegypti* preferred milkweed (*Asclepias syriaca*) over Canada goldenrod (*Solidago canadensis*), and Healy and Jepson (1988) tentatively identified a cyclic or bicyclic monoterpene as a major component of *A. millefolium* that was attractive to *Anopheles arabiensis*. Bowen (1992) described terpene-sensitive receptors in female *Culex pipiens* mosquitoes that probably mediate the attraction to plant floral odors.

6.5.2. Probing

After the female has landed on a plant surface, it will probe the surface with its labellum (Stage 7a) and may walk about and/or take short hop-flights to a nearby surface; that is, it will forage until it detects free liquid or dry honeydew (Decision Q). If free liquid or honeydew is detected, the labellar lobes open, the ligular receptors contact the liquid or surface, and a sample sip is taken (Schmidt and Friend, 1991). Salivation probably occurs with probing because it would be necessary in order to put the dry honeydew in solution to detect it and to take a sample.

While in the nectar-feeding mode, the hematophagous insect must make two decisions. One is where to direct the meal (i.e., to the crop, the midgut, or both), and the other is the amount to ingest. Diet destination and quantity ingested are controlled by two or more separate and distinct mechanisms. The model postulates that the midgut valve is normally open and the crop valve is normally closed in the absence of appropriate stimuli (Stage 14).

6.5.3. Diet Destination [Diagram 3 (Fig. 6.4)]

The destination of the meal in the insect's digestive system is controlled by the molecular species of the ingested sugars and/or their concentration acting on receptors that affect the valves controlling access to the crop and midgut (Clay and Venard, 1972; Day, 1954; Snodgrass, 1960). In the mosquito, *C. inornata* (Schmidt and Friend, 1991), and in the stable fly, *Stomoxys calcitrans* (Ascoli-Christensen et al., 1990a,b), two categories of sugars can be distinguished on the basis of their effect on diet destination; those that induce crop-only filling and those that induce primarily midgut filling. For example, sucrose, maltose, fructose and α-D-glucose, at concentrations above threshold, induce crop-only filling (Stage 15) by causing the crop valve to open and the midgut valve to close. In contrast, lactose, β-D-glucose, trehalose, cellobiose, and raffinose do not induce crop-only filling, some of the diet being found in both the crop and the midgut (Stage 17). These sugars are appropriate stimuli to cause the crop valve to open, but not appropriate stimuli to cause the midgut valve to close even at the highest concentrations tested (Schmidt and Friend, 1991). A third condition occurs when the sugar is not appropriate for changing the status of either the crop or midgut valves, or when the concentration of an appropriate sugar is below the thresholds for changing the status of the crop and/or midgut valves. In these cases, the ingested diet goes to the midgut only (Stage 16). [This is the "default" status of the valves in the unfed mosquito (Stage 14).]

It is noteworthy that diet destination for both blood and sugar solutions in *Tabanus nigrovittatus* is controlled by physical properties of the diet—specifically, its molarity (Friend and Stoffolano, 1991)—and not by the specific molecular constituents dissolved in it as it is in mosquitoes and stable flies.

6.5.4. Meal Size and Termination

The amount of diet ingested is determined by a balance between the stimulus provided by the total concentration of all sugars detected in the diet and the feedback signals resulting from abdominal stretch (Bernays, 1985; Dethier, 1976; Friend, 1989; Friend et al., 1988; Hosoi, 1954) provided that the insect is not disturbed during feeding (Decision X).

If the sugar concentration is adequate for gorging (Decision T) and the insect is undisturbed during feeding, it will terminate feeding and leave the plant (Stage 22) when the abdominal stretch is greater than the threshold to stop gorging (Stage 23). The insect will not seek a feeding resource, either blood or plant, until the distention has been relieved (Stage 23) (Jones and Madhuykar, 1976; Klowden, 1990). If the insect is disturbed during feeding and leaves the plant, it will not return to the plant to continue feeding if the abdominal distention is greater than the threshold for distention inhibition (Decision W). In the case of *Aedes aegypti,* this is about 2.5 μl (Klowden and Lea, 1979b). If the volume of the sugar meal taken before it was disturbed is less than the distention inhibition threshold, the insect will attempt to regain the plant surface and continue feeding (to Stage D), Diagram 2).

When the sugar concentration is not adequate to send the meal to the crop alone (Decision R and Stage 15), the meal may go either to both the midgut and crop (Stage 17) or to the midgut only (Stage 16). In these situations, the insect usually takes a partial meal depending upon whether or not the sugar concentration is above the threshold for gorging (Friend et al., 1989). If it is, the insect will take a small meal, between 0.2 and 1.0 μl (Stage 19), before leaving the plant (Stage 21). If the sugar concentration is below threshold for gorging, the insect will take an even smaller meal (<0.2 μl) (Stage 20). The mechanisms for termination of these smaller meals (Decision Y for Stage 19 and Decision Z for 20) are not known, but most probably do not involve inhibition arising from distention.

Gorging may also be affected by the species of sugar encountered. The most phagostimulatory sugars of the 17 that were tested in the laboratory in descending order of potency were sucrose, melezitose, furanose, raffinose, trehalose, maltose (ED_{50} for gorging <180 mM). A mixture of α-glucose and methyl-α-D-glucopyranoside was less potent, as were the disaccharides isomaltose, melibiose, and cellobiose (ED_{50} values for gorging between 200 and 400 mM). The disaccharides lactose and gentiobiose stimulated little gorging, while β-glucose and methyl-β-D-glucopyranose evoked responses similar to the water controls (Schmidt and Friend, 1991).

Although the foregoing describes the relative efficacy of various sugars in affecting diet destination and gorging, the diagram and the model represent an analysis of the responses of *Culiseta inornata* to *sucrose only* at various concentrations in the diet (Friend et al., 1989). In the field, insects would be responding to plant nectars that contain a complex mixture of sugars, the predominant components of which are sucrose, α-glucose, and fructose (Wykes, 1952). No attempt has been made to insert the responses to sugars other than sucrose into this model.

6.6. Blood Feeding [Diagram 4 (Fig. 6.5)]

The frequency of blood feeding by Glossinidae, Muscidae, fleas, and some other insects differs from mosquitoes. While a mosquito which has taken a sufficiently large blood meal may mature her oocytes after a single feeding, muscids, for example, normally take several blood meals a day (Hopkins, 1964) and require four to six meals to achieve functional reproductive capacity (Spates et al., 1988) as do tsetse (Langley, 1976) and fleas (Prasad, 1986).

6.6.1. Host Preference

This is a poorly understood phenomenon in large part because we do not yet know much about the complex mixture of odors, the bouquet, emanating from

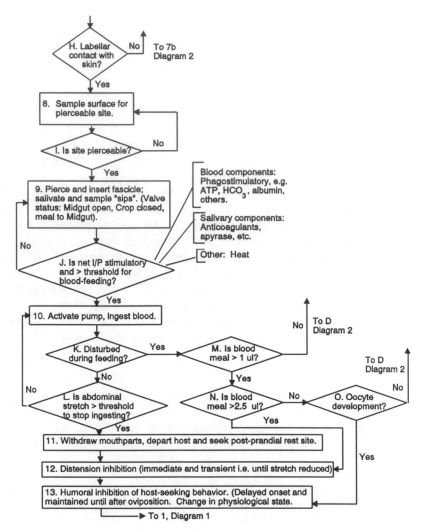

Figure 6.5. Diagram 4: Blood feeding. Diamond H is a repeat of that at the bottom right of Fig. 6.3.

animals that must underlie the ability of a hematophagous insect to recognize and locate its preferred host. That some blood-feeding insects exhibit preferences is clear. For example, certain fleas prefer cats or dogs while some mosquitoes prefer birds or small mammals or humans, and tsetse flies avoid humans. But the bases for these preferences are not known. The best understood system is in tsetse flies. Most species of tsetse prefer cattle and wart hogs over humans; host recognition and discrimination in this insect is part visual and part odor-mediated (Torr, 1989) (see below). In many cases of an apparent host preference, it is a very loose preference because if a nonpreferred host is the only one present, the insect will feed on it. The opportunity to feed is more reproductively critical than is the source of the meal.

6.6.2. Attractants

The olfactory stimuli emanating from a human or from other animals which have been shown to have a role in host seeking by mosquitoes are limited to CO_2, lactic acid, and, for some species such as *Culex quinquefasciatus,* 3-octen-1-ol from cattle breath (Kline et al., 1990). It has been clearly and repeatedly demonstrated that other odors are involved in host seeking and in the initiation of probing (Bar-Zeev et al., 1977; Kline et al., 1990; Schreck et al., 1990; E. E. Davis, *unpublished observations*), but as yet they have not been identified. Lysine, alanine, cadaverine, and estradiol have all been suggested as having attractant effects (see Bos and Laarman, 1975; Brown and Carmichael, 1961), but none has been found very effective upon further testing. In stable flies (Muscidae) (Warnes and Finlayson, 1985) and tsetse flies (Glossinidae) (Torr, 1990), carbon dioxide, cattle and human breath odors, acetone, and acetic acid elicit an increase in spontaneous flight activity and orientation to the host. Antennal receptors of the tsetse fly, *Glossina morsitans morsitans,* that respond to some of these host odors have been characterized by Den Otter and Van der Goes van Naters (1992).

6.6.2.1. LACTIC ACID

In mosquitoes, lactic acid is an important host attractant because there is a strong correlation between the ability of the antennal receptors to detect lactic acid and the presence or absence of host-seeking behavior in *Aedes aegypti* (Davis, 1984a,b; Davis et al., 1987) and *Culex pipiens* (Bowen, 1991a).

6.6.2.2. CARBON DIOXIDE

The role of CO_2 in host orientation is less clear and should be critically viewed. The assignment of the role of "activator" to CO_2 is misleading. According to Gillies (1980), CO_2 acts as an activator only in *still air;* in moving air, CO_2 may be sufficient for orientation to its source. In still air, any "attractive" odor can at best only be an "activator" because there is no directional component to the odor signal toward which the insect can orient. Blackflies (Simuliidae) show orientation to a source of CO_2 (Sutcliffe, 1987). Mosquitoes detect CO_2 via sensilla on the maxillary palps (Kellogg, 1970), and CO_2 has been long thought to be a necessary component of a host-odor blend (Mayer and James, 1969). For tsetse, CO_2 is clearly an important component of a host-attractive odor plume in the field (Torr, 1989), and a receptor for CO_2 has been described on the antennae of tsetse (Bogner, 1992).

The attraction of mosquitoes to a source of CO_2 is a function of the level of

CO_2 emitted (Kline et al., 1990). However, it is not clear whether the behavioral data reflect a dose-dependent effect or merely differences in the active spaces (i.e., the cross-sectional size of the CO_2-containing plumes), because the sizes of the emitting sources (e.g., human versus cow) differ. Perhaps both concentration and active space are important. Furthermore, the efficiency of CO_2 as an attractant has not been demonstrated. That is, what percentage of the available insect population is attracted to a CO_2 source: a large percentage of a moderate to low population or a small percentage of a very large population?

6.6.2.3. VISION

The importance of vision in host detection over a distance is reviewed for biting flies in general by Allen et al. (1987) and Sutcliffe (1987). Diurnal and possibly crepuscular mosquitoes may orient visually to their hosts (Allan et al., 1987), but the role of vision in host seeking by night-biting mosquitoes is unclear and needs further study. It may also be important in avoiding defensive acts of the host at close range. Vision may also be important in plant location.

In addition to this direct involvement in host-finding behavior, the optomotor response is important in all flight behaviors (Kennedy, 1939), and vision is important in landing (Stage 6) (Goodman, 1960). The optomotor response is not discussed here.

6.6.2.4. TEMPERATURE

Heat is also an important host-seeking signal in most hematophagous insects and can be detected over a long distance as well as at close range. Moisture in a host-odor filament can carry thermal information over several meters, and the small temperature fluctuations associated with air compression waves resulting from movement of the animal host can be detected by warm and cold thermal receptor pairs on the mosquito's antennae over a distance of more than 3 meters (Davis and Sokolove, 1975). Thermal convection currents arising from a host also carry host odors while creating a sufficient air flow for anemotaxis. However, electromagnetic properties of infrared black-body radiation are not detected by a mosquito's thermal receptors (Davis and Sokolove, 1975), nor do mosquitoes respond behaviorally to infrared radiation (Peterson and Brown, 1951). Thermal receptors have also been reported in tsetse, specifically cold-sensitive neurons on the tarsi (Reinouts van Haga and Mitchell, 1975) and on the antennae (Den Otter and Van der Goes van Naters, 1992). Preliminary electrophysiological evidence suggests the presence of antennal thermal receptors in *R. prolixus* (Smith and Davis, *unpublished data*) which confirms Wigglesworth's (1941) earlier behavioral observations. Lazzari and Núñez (1989b), using a Y maze, have reported that fifth-instar nymphs of *Triatoma infestans* can orient to an infrared source and differentiate between host (32°C), ambient (18°C), and "hot" (50°C) infrared sources by first turning their antennae toward the source and then walking toward the 32°C source. However, infrared radiation may only play a role in distance orientation because convected heat signals alone, and not radiant heat, will elicit short-range orientation, probing, and salivation (Friend and Smith, 1971; Lazzari and Núñez, 1989a; Wigglesworth and Gillett, 1934). That the odor of mouse skin at room temperature will only cause the antennae to orient toward the source but not cause the insect to approach and probe the mouse skin further illustrates the importance of temperature in host orientation in triatomines (Wigglesworth and Gillett, 1934). Other than pairs of warm- and cold-sensitive neurons on the tarsi of the tick, *Amblyomma variegatum* (Hess and Loftus, 1984), thermal receptors have not yet been reported from other hematophagous arthropods, although they are most certainly present based on the positive behavioral responses of these insects to temperature signals (Friend and Smith, 1977; Hocking, 1971).

6.6.2.5. MOISTURE

Humidity receptors in the antennal grooved-peg sensilla of *Aedes aegypti* were reported by Kellogg (1970), but Davis and Sokolove (1976) found that these receptors had a very low sensitivity to humidity while having a high sensitivity to lactic acid. Furthermore, Price et al. (1979) found that *Anopheles quadrimaculatus* was not behaviorally responsive to changes in humidity when the temperature was held constant. In addition, many species of mosquitoes live and feed in areas of high ambient humidity where it would be unlikely that they could detect small changes in humidity arising from a human or other animal.

6.6.3. Probing

Once the mosquito or other hematophagous insect has landed on a prospective host animal, it will initiate probing (Stage 7b, Diagram 2) with its proboscis or the extension of its haustellum to locate a suitable site to pierce or bite and feed. Probing results in contact by the sensilla at the tip of the proboscis or on the labellar lobes with the substrate surface (Stage 8, Diagram 4) and is influenced by thermal gradients, CO_2, and host airborne and surface chemical signals. Water shortage enhances probing, and moist targets are preferred to dry ones (Khan and Maibach, 1970a,b, 1971). Tarsal chemoreceptor responses and how they affect probing and piercing are very poorly understood (Rutledge et al., 1964; Salama, 1966).

Walker and Edman (1985) analyzed the hierarchy of behaviors involved with labellar probing and feeding site selection by *Aedes triseriatus* on rodents and found that the female did not discriminate between heavily or sparsely furred areas during the initial landing. Immediately after landing, the mosquito would begin "foraging"—that is, walking about while simultaneously tapping its labellum on the host (Stage 8, Decisions H and I). If the female could not find a suitable site—because the hair was too dense, for example—the mosquito would walk or fly to another part of the host and begin foraging again (Stage 5b, Diagram 2). In this way a suitable feeding site was located. These preferred sites were limited to areas where the hair was short and sparse, usually the ears, eyelids, nose, and feet (Walker and Edman, 1985). Coincidentally, these are areas with high subcutaneous vascularization and, hence, a higher skin temperature. Grossman and Pappas (1991) note that it would be an efficient blood-feeding strategy if hematophagous insects could use local skin temperature gradients to locate areas of high blood supply which would enhance blood-feeding success and promote rapid blood ingestion. Unfortunately, their experimental protocol did not allow them to test this hypothesis directly. Although blood-feeding insects can select a region on the host where there is a high probability of finding a blood vessel, the actual location of the vessel appears to be by random probing and piercing (Ribeiro, 1988).

Probing by mosquitoes can be elicited in response to unidentified olfactory signals from a host even though its tarsi are not in contact with skin (Davis, *unpublished observations*). Extension of the proboscis and piercing in *Triatoma infestans* were evoked by the elevated temperature of the feeding surface (Lazzari and Núñez, 1989b). If *Rhodnius prolixus* detects a heat signal, it will extend its proboscis and keep it extended, and will attempt to probe any available surface for a short time after the heat signal is removed. If, during this proboscis extension, an artificial feeding source at room temperature is presented, the insect will feed on it to repletion. Thus, the heat signal appears to act in a switch-like mode at a critical point in the feeding process (Friend and Smith, *unpublished observations*). Temperature is a sufficient stimulus for probing by simuliid blackflies (Smith and Friend, 1982), phlebotomine sandflies (Schlein et al., 1984), and lice (Wigglesworth, 1941).

In stable flies, Gatehouse (1970a,b) found that olfactory stimuli from fresh blood, sweat, carbon dioxide, ammonia, butyric and valeric acid vapors do not induce probing on a live host, but probing can be induced in hungry, water-satiated flies by temperatures associated with mammalian skin. At 35°C there was more probing of surfaces with low reflectance than with high reflectance. Stimulation of tarsal chemoreceptors of *Stomoxys calcitrans* with whole blood or blood components will also induce probing and piercing (Hopkins, 1964).

6.6.4. *Piercing* (Stage 9)

The host-penetrating mouthparts of blood-feeding insects can be divided into two major types. One type is specialized for piercing and sucking. In this type, the mandibles and/or maxillae have been modified to form a long, slender stylet with piercing tips for insertion into the host. The bundle of stylets, including the labrum, is called the *fascicle*. Blood is sucked through a food canal formed by the apposition of two or more of the stylets. Mouthparts of this type are found in the mosquitoes, bugs, lice, and fleas and are generally used for feeding directly from a blood vessel (Griffiths and Gordon, 1952; Lavoirpierre et al., 1959). The other type is designed to rip, tear, or cut the skin and then to suck the blood from the resulting pool. This is found in blackflies, tabanids, and other biting flies. For a recent review of the mouthparts of hematophagous insects, see Lehane (1991).

Mouthpart deployment during blood feeding is very different from that involved in nectar feeding and can affect the subsequent events in the feeding sequence. *Culiseta inornata* will respond to nucleotide phagostimulants only if the mouthparts are deployed in the blood-feeding configuration—that is, unsheathed from the labium—even if this is done artificially (Friend, 1985). In contrast, when feeding on plant secretions the fascicle is not unsheathed, the labellar lobes are spread, and their inner surfaces are applied to the liquid (Owen, 1963). With the mouthparts deployed in this way, the female will not respond to nucleotide stimulants (Friend, 1978). Similarly, *Lutzomyia* spp. require appropriate deployment of the mouthparts in order to respond to blood-feeding signals (Hertig and Hertig, 1927).

6.6.5. Sampling for Blood

Once the host's skin has been pierced (Stage 9), the process of blood finding begins. It "ends when the column of blood in the [insect's] food canal becomes continuous" (Ribeiro, 1988). After a mosquito pierces the skin, the fascicle, which is flexible and anteriorly curved (Gordon and Lumsden, 1939), is alternately pushed into and partially withdrawn from the host in a manner that results in the scanning of a conical segment of host tissue to a depth of about 0.5 mm. As a result of this injury to host tissue, ADP is released which initiates platelet aggregation. Saliva, which is continuously released during this fascicular probing phase (Giffiths and Gordon, 1952), contains apyrase which hydrolyzes the ADP released, thereby preventing clotting and enhancing hemorrhaging and the chances of the insect finding blood (see Chapter 3). Saliva also increases blood volume in the area of the bite (Pappas et al., 1986), probably by causing the release of histamines from the injured tissues (see Chapter 3). When the chemoreceptors at the tip of the fascicle are stimulated by adenine nucleotides (Liscia et al., 1993) within a pool of blood or a blood vessel, its movement stops, the cibarial and pharyngeal pumps usually give a few strokes, and blood is drawn up into the cibarial region (i.e., sample sips of Stage 9, Decision J) (Friend, 1978; Owen, 1963). This brings blood into contact with cibarial chemoreceptors which, a priori, may determine both the acceptability and the destination of the meal (Day, 1954; Friend, 1985; Lee, 1974; Owen, 1963; Salama, 1966). When the presence of the appropriate phagostimulant(s) is detected (see below), the blood-searching phase ends and the engorging phase begins. Successful blood seeking leads to blood ingestion that involves the continuous action of the cibarial and pharyngeal pumps (Stage 10). Unsuccessful blood seeking leads to the termination of probing and fascicular withdrawal (Ribeiro, 1988). The female will then move to a different site on the host either by walking or by short flight, and then she will begin the sequence again at Stage 5b or 7b, Diagram 2. Because the female is in close proximity to the host, she has a high probability of regaining the host and continuing to feed.

6.6.6. Gorging Signals

6.6.6.1. PHAGOSTIMULANTS

Three phagostimulatory regimes have emerged in hematophagous insects that induce gorging once blood has been successfully located. Lehane (1991) described these three regimes based on the source of the phagostimuli; specifically, factors associated with (a) the cellular fraction, (b) the plasma fraction, and (c) intermediate between (a) and (b) (also see Chapter 3). With one possible exception in teneral tsetse, all three regimes require the presence of saline isotonic to blood and all are optimal if the temperature of the diet is near 37°C.

Cellular Factors

The most prevalent and probably the most representative regime requires the presence an adenine-based nucleotide such as ATP, ADP, or, as in blackflies, adenosine (at concentrations similar to those found in blood) in addition to isotonic saline (Friend, 1989; Galun, 1987; Smith, 1979). The other purine and pyrimidine nucleotides are inadequate as phagostimulants (Friend, 1989). That the source of the phagostimulant in this regime is in the cellular fraction is clearly indicated by the induction of gorging by washed intact red blood cells (RBCs), hemolyzed RBCs in saline, or ATP in saline, as well as by the absence of gorging when exposed to RBC-free plasma or RBC ghosts (Galun et al., 1993; Smith, 1979; Smith and Friend, 1982). This system, with some variation, is found in seven of the nine families in which phagostimulation has been studied; specifically, in culicine mosquitoes, Simuliidae, Tabanidae, Glossinidae, Muscidae, Phthiraptera, and Reduviidae (Friend, 1989; Friend and Smith, 1977).

ATP alone is sufficient to evoke gorging in teneral tsetse flies. Neither a particular osmotic pressure nor specific ions such as Na^+ or Cl^- are required (Mitchell and Reinhouts van Haga-Kelker, 1976). However, in post-teneral tsetse two important changes occur: the behavioral threshold for ATP is elevated and other blood-borne factors not related to ATP are now required (Mitchell and Reinhouts van Haga-Kelker, 1976).

The question of how the phagostimulant gets out of the RBCs is unclear (see Chapter 3). Lysis or rupture of the RBCs to release phagostimulant is thought to be too infrequent to account for the consistent and immediate induction of gorging (Haslam and Cusack, 1981). However, distortion of the RBCs, as when they pass through the food canal, may cause them to leak sufficient nucleotide into the fluid medium to stimulate gorging (Turritto and Weiss, 1980; J. J. B. Smith, *personal communication*). If this is the case, then the cibarial receptors are likely to have a role in detecting the stimulant. The detection of ATP by the labral receptors of *Culex pipiens* indicates that they have a role in this process as well (Liscia et al., 1993).

Clearly, there are many gaps in our understanding of cell-derived phagostimuli. Many pieces of the story come from different species in different families and from different laboratories having different approaches. Some of the unresolved issues involve (a) the source and identity of the stimulant and how it gets to its receptor, (b) the specific receptor(s) detecting the stimulant, and (c) precisely what function each receptor type mediates. These questions must be addressed in a comparative manner because of obvious species differences. It is clear that no one system can be generalized to all hematophagous insects, or even to the species in a single family such as the mosquitoes.

Plasma Factors

Saline only, isotonic to blood, is sufficient to induce gorging in phlebotomine flies; adenine nucleotides and other phagostimulants have no effect on gorging (Schlein et al., 1984). Anopheline mosquitoes may be induced to gorge on a diet containing $NaHCO_3$ and saline (Galun et al., 1985). This is in contrast to the culicine and aedine mosquitoes that seem to require the combination of saline and an adenine nucleotide.

Combinations of Cell and Plasma Factors

Fleas will feed on diets that are isotonic with plasma, but will only fully engorge when ATP is also present (Galun, 1966).

6.6.6.2. RECEPTOR MECHANISMS

The nature of the nucleotide phagostimulant receptor system has been investigated primarily in Muscidae, Tabanidae, Triatomidae, and in some mosquitoes. *Stomoxys calcitrans* will gorge on blood plasma and can be induced to gorge by ATP and other nucleotides (Ascoli-Christensen et al., 1990a; Galun, 1967). The plasma-borne phagostimulant appears to be a molecule or molecules larger than 100,000 daltons (Ascoli-Christensen et al., 1990a). ATP and other close analogues appear to directly stimulate feeding in *S. calcitrans* with the potencies ranking as they do for inhibition of the salt-sensitive cell (Ascoli-Christensen et al., 1990b). Later studies involving both feeding and electrophysiological assays showed that the putative "ATP receptor" responds as a P_{2x} type of purinoreceptor (Burnstock and Brown, 1981). Methyl xanthines did not affect feeding, and CH_3-S-ATP was not more potent than ATP. ATP responses were antagonized by

ANAPP$_3$ (Ascoli-Christensen et al., 1991). In addition, tests of 15 purinergic compounds for phagostimulatory activity in *Tabanus nigrovittatus* revealed that ATP had the highest potency and that cyclic AMP induced no gorging. These results indicated that this blood-feeding insect also has a P$_{2x}$ type of purinoreceptor system (Friend and Stoffolano, 1990). Similar approaches have indicated that a P$_{2y}$-type purinoreceptor is present in *Rhodnius prolixus, T. nigrovittatus,* and *Glossina palpalis.* The P$_{2x}$ and P$_{2y}$ receptors differ only in their respective sensitivities to ATP analogues. A third group of insects, including *Culex pipiens, Culista inornata,* and *Simulium venustum,* have been described that have either both a P$_1$ and a P$_2$ type of purinoreceptor or a third receptor type that has yet to be characterized [for a review of this topic, see Friend (1989)].

Changing the solvent system from 150 mM NaCl had significant effects in *T. nigrovittatus.* Addition of 5% bovine albumin and 10 mM NaHCO$_3$ increased the potency of ATP by nearly threefold and the potency of ADP by more than sevenfold. ADP dissolved in choline chloride was as potent as ADP in NaCl. Addition of calcium ions to the solvent reduced the potency of ADP probably because of the activation of apyrase in the saliva (Friend and Stoffolano, 1990).

6.6.6.3. Chemosensory Information

The chemosensory neurons in the largest labellar hairs of *S. calcitrans* that respond positively to NaCl in a concentration-dependent manner are inhibited by ATP and its analogues also in a dose-dependent manner. ATP also inhibits a salt-sensitive cell in *Phormia regina* (Liscia, 1985). Dethier et al. (1965) claim that excitatory inputs from the salt-sensitive cells in *P. regina* inhibit feeding. Thus blocking the input of the salt-sensitive cells might lower the threshold of behavioral responses to other feeding stimulants which act on the chemosensilla of the labrum or cibarium (Ascoli-Christensen et al., 1990b).

6.6.7. Meal Destination

Blood is usually directed to the midgut, the destination being controlled by the chemoreceptors of the labrum and/or cibarium. The destination is largely independent of the mode of feeding (Day, 1954; Galun, 1977; Hosoi, 1959). Trembley (1952) found differences in the control of meal destination among the nine species of mosquitoes she studied; 77% of *Anopheles quadrimaculatus* and 54% of *Culex pipiens* had some blood in the crop after feeding on humans, whereas only 1 of 174 *Aedes aegypti* had any blood in the crop.

In muscid and tabanid flies, blood is normally directed to the midgut, whereas sugar solutions and water go to the crop in *Chrysops vittatus* (Lall, 1969) and *Haematopota lasiophathalma* (Lall, 1970). When blood or dyed sucrose solutions or water in pipettes was used to immerse the haustellum of restrained *Chrysops mlokosieviczi, Hybumnitra peculiaris, Tabanus autumnalis,* or *Haematopota pal-*

lens, all of these fluids went first into the gut and only into the crop when larger amounts were ingested. The authors suggest that crop filling results from rapid absorption of nonblood meals from the midgut, leaving fluid to be found only in the crop upon later observation (Chirov and Alekseyev, 1970). This hypothesis deserves further testing.

Diet destination in *Tabanus nigrovittatus* is affected by the concentration of dissolved solids in the diet. If the molarity is approximately equal to that of blood, all of the diet will be directed to the midgut, as blood would be when the insect was feeding under natural conditions. If the concentration of dissolved solids is hyperosmotic to blood, about half the test insects that ingest large amounts of diet direct it to both the midgut and crop. When the insects are responding to ATP (i.e., they are in the blood-feeding mode), some diet always goes to the midgut regardless of osmolarity. The molar threshold for diet to be partially directed to the crop is above 0.3 M, or about twice the osmotic pressure of vertebrate blood (Friend and Stoffolano, 1991).

6.6.8. Termination of Blood-Feeding Behavior

There are two aspects to be considered here: the termination of a meal and the prevention of host-locating behavior.

6.6.8.1. MEAL TERMINATION

If undisturbed (Decision K, Diagram 4), a feeding insect will ingest blood until it has taken a meal of the appropriate size. As the insect pumps blood into its gut, its abdomen increases in size due to the stretching of the pleural membranes between the cuticular parts of the abdomen. This stretching of the abdomen causes mechanoreceptors in the abdominal or gut wall to signal that a replete meal has been obtained and that the insect should stop pumping blood, withdraw its mouthparts, and depart the host (Stage 11). Abdominal stretch receptor activation has been shown to terminate feeding in *Rhodnius prolixus* (Davey, 1993) (Stage 12). The threshold for the termination of feeding by abdominal stretch appears depend on the level of food deprivation in the insect. In tsetse, starved flies will take larger meals than those that have fed more recently (Moloo and Kutuza, 1970; Tobe and Davey, 1972). Insects that have only taken small amounts of blood, below threshold volume for distention inhibition (Decision N) (Klowden and Lea, 1979b), because of an inadequate feeding site or because they were disturbed during feeding (Decision K) will usually attempt to immediately resume feeding on the host (reentering the model at D or E, Diagram 2).

The pathway(s) for the stretch receptor signals to terminate feeding in Diptera appears to be via the ventral nerve cord and/or the recurrent nerve. Dethier (1976) and Gelperin (1971) obtained hyperphagia in *Phormia regina* with transection of either the ventral nerve cord, which receives input from stretch receptors in the abdominal wall, or the recurrent nerve, which receives similar information from stretch receptors in the foregut of the fly. Opposite results have been obtained in mosquitoes. Bowen et al. (*unpublished observations*) observed hyperphagia in female *Aedes atropalpus* following ventral nerve cord transection at the junction of the abdomen and the thorax, whereas Klowden did not obtain hyperphagia in *Aedes aegypti* when he cut the ventral nerve cord at a similar same place (*unpublished observations;* see Klowden, 1990). Whether the *Phormia*-based model of feeding regulation can be adapted to mosquitoes and other blood-feeding insects remains undetermined.

To accommodate large blood meals, some insects have evolved at least two additional mechanisms, both of which appear to be mediated, in part, by seroto-nin. Released from neurohemal areas of the abdominal nerve by the initiation of a blood meal (Lange et al., 1989), serotonin induces the plasticization of the abdominal pleural membranes of *R. prolixus* nymphs allowing the ingestion of blood up to nine times the insect's unfed weight (Barrett and Orchard, 1990; Orchard et al., 1988). Diuresis may occur just before the insect blood-feeds, as in tsetse (Langley and Pimley, 1973), thus making more room available for the incoming blood, or during blood feeding, as in *Anopheline* mosquitoes, *Culi-coides* spp., triatomines, and probably others, thereby concentrating the blood to allow a greater gross volume to be ingested. Blood-meal-triggered serotonin release may also be involved in the process of diuresis together with diuretic hormone (Barrett et al., 1993; Maddrell et al., 1971).

6.6.8.2. PREVENTION OF HOST-LOCATING BEHAVIOR (Stages 12 and 13)

After an insect obtains a blood meal, it usually will not attempt to locate another source of a blood meal until it has at least partially processed the meal or has laid its complement of eggs. There are two mechanisms by which this may occur. In the mosquito, *Aedes aegypti*, the first is via the same mechanism that terminated the meal—that is, abdominal stretch or distention inhibition (Klowden and Lea, 1979b). However, the threshold volume for inhibition of subsequent host-locating behavior is lower than that for meal termination. Klow-den and Lea (1979b) found that 2.5 µl of blood in the gut would prevent host-locating behavior even though these females, if allowed, would take blood meals up to 5 µl. Distention inhibition of feeding persists until the abdominal stretch is relieved (Klowden, 1990).

A second means of preventing host-locating behavior following a blood meal has been demonstrated in mosquitoes (Stage 13). It is the result of humoral changes that occur during oocyte development induced by the blood meal. This inhibition is due to a hemolymph-borne factor released several hours after the blood meal was taken (Klowden and Lea, 1979a) which causes a reduction in the sensitivity of the lactic acid receptor that mediates host-locating behavior (Davis, 1984b). Oocyte-induced humoral inhibition is maintained until oviposition occurs (Diagram 1, Stage 1). In *Aedes aegypti,* a threshold meal volume between 1 and 2 µl appears to be sufficient (Decision M) to initiate oocyte development (Decision O); humoral inhibition (13) will prevent their seeking to feed again until after oviposition (Diagram 1, Stage 1) (Klowden, 1990). Tsetse flies, which continue to feed throughout the development of their progeny (they are ovoviviparous) (Davey, 1993), and fleas (Prasad, 1986) clearly lack this postprandial humoral inhibition.

6.6.9. Ingestus Interruptus

The ability of a hematophagous insect to feed to repletion on an animal is often related to the host animal's defensive behavior (Diagram 4, Point K) (Klowden and Lea, 1979c). Those hematophagous insects that pierce the skin without causing pain will be less likely to induce a defensive response from the host (e.g., triatomids and mosquitoes) than will tabanids, whose bites are quite painful. If the feeding insect is disturbed during ingestion, it will withdraw its mouthparts and leave the host (Walker and Edman, 1985). Sleeping or otherwise inactive hosts provide the best feeding conditions (Day and Edman, 1984).

6.7. Repeat of Cycle

The patterns of feeding behaviors described here, for both blood feeding and nectar feeding, represent only a single pass through each. In nature these patterns are not one-time events, but are continuously repeated throughout the life cycle of most hematophagous insects. The repetition of blood-feeding behavior is the means by which diseases are transmitted to humans and animals and by which sufficient reproductive capacity is reached to sustain the species, and this is why this group of insects is so important.

Acknowledgments

We wish to thank Drs. M. F. Bowen and J. J. B. Smith for their reading and rereading, discussing, and critiquing of the various forms of this chapter. Some of the research described herein was supported by grants from the Canadian National Science and Engineering Research Council (to WGF) and from the United States National Institutes of Health (grant no. AI21267 to EED).

References

Allan, S., Day, J. F., and Edman, J. D. (1987). Visual ecology of biting flies. *Annu. Rev. Entomol.* 32, 297–316.
Ascoli-Christensen, A., Sutcliffe, J. F., and Straton, C. J. (1990a). Feeding responses

of the stable fly, *Stomoxys calcitrans* (L.), to blood fractions and adenine nucleotides. *Physiol. Entomol.* 15, 249–259.

Ascoli-Christensen, A., Sutcliffe, J. F., and Albert, P. J. (1990b). Effect of adenine nucleotides on labellar chemoreceptive cells of the stable fly, *Stomoxys calcitrans* (L.). *J. Insect Physiol.* 36, 339–344.

Ascoli-Christensen, A., Sutcliffe, J. F., and Albert, P. J. (1991). Purinoreceptors in blood feeding behaviour in the stable fly, *Stomoxys calcitrans. Physiol. Entomol.* 16, 145–152.

Baker, T. C. (1989). Pheromones and flight behavior. In: Goldsworthy, G. J., and Wheeler, C. H. (eds.), *Insect Flight.* CRC Press, Boca Raton, FL, pp. 231–255.

Baker, T. C., Hansson, B. S., Lofstedt, C., and Lofqvist, J. (1988). Adaptation of antennal neurons in moths associated with cessation of pheromone-mediated upwind flight. *Proc. Natl. Acad. Sci. USA* 85, 9826–9830.

Barrett, F. M., and Orchard, I. (1990). Serotonin-induced elevation of cAMP levels in the epidermis of the blood-sucking bug *Rhodnius prolixus. J. Insect Physiol.* 9, 625–633.

Barrett, F. M., Orchard, I., and TeBrugge, V. (1993). Characteristics of serotonin-induced cyclic AMP elevation in the integument and anterior midgut of the blood-feeding bug, *Rhodnius prolixus. J. Insect Physiol.* 39, 581–588.

Bar-Zeev, M., Maibach, H. I., and Khan, A. A. (1977). Studies on the attraction of *Aedes aegypti* to man. *J. Med. Entomol.* 14, 113–120.

Bell, W. J. (1990). Searching behavior patterns in insects. *Annu. Rev. Entomol.* 35, 445–467.

Bernays, E. A. (1985). Regulation of feeding behaviour. In: Kerkut, G. A., and Gilbert, L. I. (eds.), *Comprehensive Insect Physiology, Biochemistry and Pharmacology,* vol. 4. Pergamon Press, Oxford, pp. 1–32.

Bogner, F. (1992). Response properties of CO_2-sensitive receptors in tsetse flies (Diptera, *Glossina palpalis*). *Physiol. Entomol.* 17, 19–24.

Bos, H. J., and Laarman, J. J. (1975). Guinea pig, lysine, cadaverine and estradiol as attractants for the malaria mosquito *Anopheles stephensi. Entomol. Exp. Appl.* 18, 161–172.

Bowen, M. F. (1991a). The sensory physiology of host-seeking in mosquitoes. *Annu. Rev. Entomol.* 36, 139–158.

Bowen, M. F. (1991b). Post-diapausing sensory responsiveness in *Culex pipiens, J. Insect Physiol.* 36, 923–929.

Bowen, M. F. (1992). Terpene-sensitive receptors in female *Culex pipiens* mosquitoes, electrophysiology and behavior. *J. Insect Physiol.* 38, 759–764.

Bowen, M. F., and Davis, E. E. (1989). The effects of allatectomy and juvenile hormone replacement on the development of host-seeking behavior and lactic acid receptor sensitivity in the mosquito, *Aedes aegypti. Med. Vet. Entomol.* 3, 53–60.

Bowen, M. F., Davis, E. E., and Haggart, D. (1988). A behavioral and sensory analysis of host-seeking behavior in the diapausing mosquito *Culex pipiens*. *J. Insect Physiol.* 34, 805–813.

Bowen, M. F., Davis, E. E., Haggart, D., and Romo, J. (1994a). Host-seeking behavior in the autogenous mosquito *Aedes atropalpus*. *J. Insect Physiol.* 40, 511–517.

Bowen, M. F., Davis, E. E., Romo, J., and Haggart, D. (1994b). Lactic acid-sensitive receptors in the autogenous mosquito *Aedes atropalpus*. *J. Insect Physiol.* 40, 611–615.

Brady, J., Gibson, G. A., and Packer, M. J. (1989). Odour movement, wind direction, and the problem of host-finding by tsetse flies. *Physiol. Entomol.* 14, 369–380.

Brown, A. W. A., and Carmichael, A. G. (1961). Lysine and alanine as mosquito attractants. *J. Econ. Entomol.* 54, 317–324.

Burnstock, G., and Brown, C. M. (1981). An introduction to purinergic receptors. In: Burnstock, G. (ed.), *Purinergic Receptors*. Chapman & Hall, London, pp. 1–45.

Chirov, P. A., and Alekseyev, A. N. (1970). On the physiology of feeding in the Tabanidae. *Byull. Mosk. Ova. Ispyt. Prir. Otd. Biol.* 75, 60–67.

Clay, M. E., and Venard, C. E. (1972). The fine structure of the esophageal diverticula in the mosquito *Aedes triseriatus*. *Ann. Entomol. Soc. Am.* 65, 964–975.

Davey, K. G. (1993). How is information about meal size transmitted to the endocrine system in *Rhodnius?* In: Borovsky, D., and Spielman, A. (eds.), *Host Regulated Developmental Mechanisms in Vector Arthropods*. University of Florida, IFAS, Vero Beach, FL, pp. 1–10.

Davis, E. E. (1984a). Regulation of sensitivity in the peripheral chemoreceptor systems for host-seeking behavior by a haemolymph-borne factor in *Aedes aegypti*. *J. Insect Physiol.* 30, 179–183.

Davis, E. E. (1984b). Development of lactic acid-receptor sensitivity and host-seeking behaviour in newly emerged female *Aedes aegypti*. *J. Insect Physiol.* 30, 211–215.

Davis, E. E. (1989). The role of the peripheral receptors in the mediation of behavior in the female mosquito, *Aedes aegypti*. In: Borovsky, D., and Spielman, A. (eds.), *Host Regulated Developmental Mechanisms in Vector Arthropods*. University of Florida, IFAS, Vero Beach, FL, pp. 286–291.

Davis, E. E., and Sokolove, P. G. (1975). Temperature responses of antennal receptors of the mosquito, *Aedes aegypti*. *J. Comp. Physiol.* A96, 223–236.

Davis, E. E., and Sokolove, P. G. (1976). Lactic acid-sensitive receptors on the antennae of the mosquito, *Aedes aegypti*. *J. Comp. Physiol.* A105, 43–54.

Davis, E. E., Haggart, D. A., and Bowen, M. F. (1987). Receptors mediating host-seeking behavior in mosquitoes and their regulation by endogenous hormones. *Insect Sci. Appl.* 8, 637–641.

Day, J. F. (1954). The mechanism of food distribution to the midgut or diverticula in the mosquito. *Aust. J. Biol. Sci.* 7, 515–524.

Day, J. F., and Edman, J. D. (1984). Mosquito engorging on normally host defensive hosts depends on host activity patterns. *J. Med. Entomol.* 21, 732–741.

Den Otter, C. J., and Van der Goes van Naters, W. M. (1992). Single cell recordings from tsetse (*Glossina m. morsitans*) antennae reveal olfactory, mechano- and cold receptors. *Physiol. Entomol.* 17, 33–42.

Dethier, V. G. (1976). *The Hungry Fly.* Harvard University Press, Cambridge, MA.

Dethier, V. G., Solomon, R. L., and Turner, L. H. (1965). Sensory input and central excitation and inhibition in the blowfly. *J. Comp. Physiol. Psychol.* 60, 303–313.

Edman, J. D., and Scott, T. W. (1987). Host-defensive behavior and the feeding success of mosquitoes. *Insect Sci. Appl.* 8, 617–622.

Feinsod, F. M., and Spielman, A. (1980). Nutrient-mediated juvenile hormone secretion in mosquitoes. *J. Insect Physiol.* 26, 113–117.

Fraenkel, G. (1932). Untersuchungen über die Koordination von Reflexen und automatisch-nervösen Rhythmen bei Insekten I. Die Flugreflexen der Insekten und ihre Koordination. *Z. Vergl. Physiol.* 16, 371–393.

Friend, W. G. (1978). Physical factors affecting the feeding responses of *Culiseta inornata* to ATP, sucrose, and blood. *Ann. Entomol. Soc. Am.* 71, 935–940.

Friend, W. G. (1981). Diet destination in *Culiseta inornata* (Williston), Effects of feeding conditions on the response to ATP and Sucrose. *Ann. Entomol. Soc. Am.* 74, 151–154.

Friend, W. G. (1985). Diet ingestion and destination in *Culiseta inornata* (Diptera, Culicidae): effects of mouthpart deployment and contact of the fascicle and labellum with sucrose, water, saline, or ATP. *Ann. Entomol. Soc. Am.* 78, 495–500.

Friend, W. G. (1989). Phagostimulation in vector insects; comparative purinergic responses. In: Borovsky, D., and Spielman, A. (eds.), *Host Regulated Developmental Mechanisms in Vector Arthropods.* University of Florida, IFAS, Vero Beach, FL, pp. 149–158.

Friend, W. G., and Smith, J. J. B. (1971). Feeding in *Rhodnius prolixus*, mouthpart activity and salivation, and their correlation with changes of electrical resistance. *J. Insect Physiol.* 17, 233–243.

Friend, W. G., and Smith, J. J. B. (1977). Factors affecting feeding by bloodsucking insects. *Annu. Rev. Entomol.* 22, 309–331.

Friend, W. G., and Stoffolano, J. G., Jr. (1984). Feeding responses of the horsefly *Tabanus nigrovittatus* to physical factors, ATP analogues and blood fractions. *Physiol. Entomol.* 9, 395–402.

Friend, W. G., and Stoffolano, J. G., Jr. (1990). Feeding responses of the horsefly, *Tabanus nigrovittatus* (Diptera: Tabanidae) to purinergic phagostimulants. *J. Insect Physiol.* 36, 805–812.

Friend, W. G., and Stoffolano, J. G., Jr. (1991). Feeding behaviour of the horsefly *Tabanus nigrovitatus* (Diptera: Tabanidae), effects of dissolved solids on ingestion and destination of sucrose or ATP diets. *Physiol. Entomol.* 16, 35–45.

Friend, W. G., Schmidt, J. M., Smith, J. J. B., and Tanner, R. J. (1988). The effect of sugars on ingestion and diet destination in *Culiseta inornata*. *J. Insect Physiol.* 34, 955–961.

Friend, W. G., Smith, J. J. B., Schmidt, J. M., and Tanner, R. J. (1989). Ingestion and diet destination in *Culiseta inornata:* responses to water, sucrose and cellobiose. *Physiol. Entomol.* 14, 137–146.

Galun, R. (1966). Feeding stimulants of the rat flea *Xenopsylla cheopis. Life Sci.* 5, 1335–1342.

Galun, R. (1967). Feeding stimuli and artificial feeding. *WHO Bull.* 36, 590–593.

Galun, R. (1977). Responses of blood sucking arthropods to vertebrate hosts. In: Shorey, H. H., and McKelvey, J. J., Jr. (eds.), *Chemical Control of Insect Behaviour: Theory and Application.* Wiley–Interscience, New York, pp. 103–115.

Galun, R. (1987). The evolution of purinergic receptors involved in recognition of a blood meal by hematophagous insects. *Mem. Inst. Oswaldo Cruz,* 82, suppl. III, pp. 5–9.

Galun, R., Koontz, L. C., and Gwadz, R. W. (1985). Engorgement response of anopheline mosquitoes to blood fractions and artificial solutions. *Physiol. Entomol.* 10, 145–149.

Galun, R., Vardimon-Friedman, H., and Frankenburg, S. (1993). Gorging of Culicine mosquitoes (Diptera, Culicidae) to blood fractions. *J. Med. Entomol.* 30, 513–517.

Gatehouse, A. G. (1970a). The probing response of *Stomoxys calcitrans* to certain physical and olfactory stimuli. *J. Insect Physiol.* 16, 61–74.

Gatehouse, A. G. (1970b). Interactions between stimuli in the induction of probing by *Stomoxys calcitrans. J. Insect Physiol.* 16, 991–1000.

Gelperin, A. (1971). Regulation of feeding. *Annu. Rev. Entomol.* 16, 364–378.

Gibson, G. A., Packer, M. J., Steullet, P., and Brady, J. (1991). Orientation of tsetse flies to wind, within and outside host odour plumes in the field. *Physiol. Entomol.* 16, 47–56.

Gillies, M. T. (1980). The role of carbon dioxide in host-finding by mosquitoes (Diptera: Culicidae); a review. *Bull. Entomol. Res.* 70, 525–532.

Goodman, L. J. (1960). The landing responses of insects. I. The landing response of the fly, *Lucilia sericata,* and other Calliphorinae. *J. Exp. Biol.* 37, 854–878.

Gordon, R. M., and Lumsden, W. H. R. (1939). A study of the behaviour of the mouthparts of mosquitoes when taking up blood from living tissue. *Ann. Trop. Med. Parasitol.* 33, 259–278.

Griffiths, R. B., and Gordon, R. M. (1952). An apparatus which enables the process of feeding by mosquitoes to be observed in the tissues of a live rodent. *Ann. Trop. Med. Parasitol.* 46, 311–319.

Grimstad, P. R., and DeFoliart, G. R. (1974). Nectar sources of Wisconsin mosquitoes. *J. Med. Entomol.* 11, 331–341.

Gringorten, J. L., and Friend, W. G. (1979). Haemolymph-volume changes in *Rhodnius prolixus* during flight. *J. Exp. Biol.* 83, 325–333.

Grossman, G. L., and Pappas, L. G. (1991). Human skin temperature and mosquito (Diptera: Culicidae) blood feeding rate. *J. Med. Entomol.* 28, 456–460.

Gwadz, R. W. (1969). Regulation of blood-meal size in the mosquito. *J. Insect Physiol.* 15, 2039–2044.

Haslam, R. J., and Cusack, N. J. (1981). Blood platelet receptor for ADP and for adenosine. In: Burnstock, G. (ed.), *Purinergic Receptors*. Chapman & Hall, London, pp. 221–285.

Healy, T. P., and Jepson, P. C. (1988). The location of floral nectar sources by mosquitoes: the long-range responses of *Anopheles arabiensis Patton* (Diptera: Culicidae) to *Achillea millefolium* flowers and isolated floral odors. *Bull. Entomol. Res.* 78, 651–657.

Hertig, A. T., and Hertig, M. (1927). A technique for the artificial feeding of sandflies (Phlebotomus) and mosquitoes. *Science* 65, 328–329.

Hess, E., and Loftus, R. (1984). Warm and cold receptors of two sensilla on the foreleg tarsi of the tropical bont tick *Amblyomma variagatum*. *J. Comp. Physiol.* A155, 187–196.

Hiss, H. A., and Fuchs, M. S. (1972). The effect of matrone on oviposition behavior in the mosquito, *Aedes aegypti*. *J. Insect Physiol.* 18, 2217–2227.

Hocking, B. (1971). Blood-sucking behavior of terrestrial arthropods. *Annu. Rev. Entomol.* 16, 1–26.

Hopkins, B. A. (1964). The probing response of *Stomoxys calcitrans* (L) to vapours. *Anim. Behav.* 12, 513–524.

Hosoi, T. (1954). Mechanism enabling the mosquito to ingest blood into the stomach and sugary fluids into the oesophageal diverticula. *Annot. Zool. Jpn.* 27, 82–90.

Hosoi, T. (1959). Identification of blood components which induce gorging of the mosquito. *J. Insect Physiol.* 3, 191–218.

Jones, J. C., and Madhuykar, B. V. (1976). Effects of sucrose on blood avidity in mosquitoes. *J. Insect Physiol.* 22, 357–360.

Jones, M. D. R., and Gubbins, S. J. (1977). Changes in the circadian flight activity of the mosquito *Anopheles gambiae* in relation to insemination, feeding and oviposition. *Physiol. Entomol.* 268, 731–732.

Judson, C. L. (1967). Feeding and oviposition behavior in the mosquito *Aedes aegypti*. Preliminary studies of physiological control mechanisms. *Biol. Bull.* 133, 369–377.

Kaissling, K.-E. (1987). *R. H. Wright Lectures on Insect Olfaction*. Simon Fraser University, Burnaby, British Columbia, Canada.

Kellogg, F. E. (1970). Water vapour and carbon dioxide receptors in *Aedes aegypti*. *J. Insect Physiol.* 16, 99–108.

Kennedy, J. S. (1939). The visual responses of flying mosquitoes. *Proc. Zool. Soc. London* 109, 221–242.

Kennedy, J. S. (1986). Some current issues in orientation to odour sources. In: Payne, T. L., Birch, M. C., and Kennedy, C. (eds.), *Mechanisms in Insect Olfaction*. Clarendon Press, Oxford, pp. 121–125.

Khan, A. A., and Maibach, H. I. (1970a). A study of the probing response of *Aedes aegypti*. 1. Effect of nutrition on probing. *J. Econ. Entomol.* 63, 974–976.

Khan, A. A., and Maibach, H. I. (1970b). A study of the probing response of *Aedes aegypti*. 2. Effect of desiccation and blood feeding on probing to skin and an artificial target. *J. Econ. Entomol.* 64, 439–442.

Khan, A. A., and Maibach, H. I. (1971). A study of the probing response of *Aedes aegypti*. 4. Effect of dry and moist heat on probing. *J. Econ. Entomol.* 64, 442–443.

Kline, D. L., Takken, W., Wood, J. F., and Carlson, D. A. (1990). Field studies on the potential of butanone, carbon dioxide, honey extract, 1-octen-3-ol, L-lactic acid and phenols as attractants for mosquitoes. *Med. Vet. Entomol.* 4, 383–391.

Klowden, M. J. (1986). Effects of sugar deprivation on the host-seeking behaviour of gravid *Aedes aegypti* mosquitoes. *J. Insect Physiol.* 32, 479–483.

Klowden, M. J. (1990). The endogenous regulation of mosquito reproductive behavior. *Experientia* 46, 660–670.

Klowden, M. J., and Lea, A. O. (1979a). Humoral inhibition of host-seeking behavior in *Aedes aegypti* during oocyte maturation. *J. Insect Physiol.* 25, 231–235.

Klowden, M. J., and Lea, A. O. (1979b). Abdominal distention terminates subsequent host-seeking behavior of *Aedes aegypti* following a blood meal. *J. Insect Physiol.* 25, 583–585.

Klowden, M. J., and Lea, A. O. (1979c). Effect of defensive host behavior on the blood meal size and feeding success of natural populations of mosquitoes. *J. Med. Entomol.* 15, 514–517.

Klowden, M. J., Blackmer, J. L., and Chambers, G. M. (1988). Effects of larval nutrition on the host-seeking behavior of adult *Aedes aegypti*. *J. Am. Mosq. Control Assoc.* 4, 73–75.

Kuenen, L. P. S., and Baker, T. C. (1981). Habituation vs sensory adaptation as the cause of reduced attraction following pulsed and constant sex pheromone pre-exposure in *Tricoplusia ni*. *J. Insect Physiol.* 27, 721–726.

Laarman, J. J. (1955). The host-seeking behavior of the malaria mosquito *Anopheles maculipennis atroparvus*. *Acta Leidensia* 25, 1–144.

Lall, S. B. (1969). Phagostimulants of haematophagous tabanids (Diptera). *Entomol. Exp. Appl.* 12, 325–336.

Lall, S. B. (1970). Loci, structure and function of contact chemical sensilla in haematophagous tabanids (Diptera). *J. Med. Entomol.* 7, 205–222.

Lange, A. B., Orchard, I., and Barrett, F. M. (1989). Changes in haemolymph serotonin levels associated with feeding in the blood-sucking bug, *Rhodnius prolixus*. *J. Insect Physiol.* 35, 393–399.

Langley, P. A. (1976). Initiation and regulation of ingestion by hematophagous arthropods. *J. Med. Entomol.* 13, 121–130.

Langley, P. A., and Pimley, R. W. (1973). Influence of diet composition on feeding and water excretion by the tsetse fly *Glossina morsitans*. *J. Insect Physiol.* 19, 1097–1109.

Lavoipierre, M. M. J., Dickersson, G., and Gordon, R. M. (1959). Studies on the methods of feeding of blood-sucking arthropods. I. The manner in which Triatomine bugs obtain their blood meal as observed in the tissues of the living rodent, with some remarks on the effects of the bite on human volunteers. *Ann. Trop. Med. Parasitol.* 53, 235–250.

Lazzari, C. R., and Núñez, J. A. (1989a). The response to radiant heat and the estimation of the temperature of distant sources in *Triatoma infestans*. *J. Insect Physiol*. 35, 525–529.

Lazzari, C. R., and Núñez, J. A. (1989b). Blood temperature and feeding behavior in *Triatoma infestans* (Heteroptera: Reduviidae). *Entomol. Gen*. 14, 183–188.

Leahy, M. G., and Craig, G. B., Jr. (1965). Accessory gland substance as a stimulant for oviposition in *Aedes aegypti* and *A. albopictus*. *Mosq. News* 25, 448–452.

Lee, R. (1974). Structure and function of the fascicular stylets, and the labral and cibarial sense organs of male and female *Aedes aegypti* (L.) (Diptera, Culicidae). *Quaest. Entomol*. 10, 187–215.

Lehane, M. J. (1991). *Biology of Blood-Sucking Insects*. Harper Collins Academic, London.

Liscia, A. (1985). ATP influences the electrophysiological responses of labellar chemosensilla of *Phormia regina* Meigen (Diptera: Calliphoridae) to salt and sugar stimulation. *Monit. Zool. Ital*. 19, 39–45.

Liscia, A., Crnjar, R., Barbarosa, I. T., Esu, S., Muroni, P., and Galun, R. (1993). Electrophysiological responses of labral apical chemoreceptors to adenine nucleotides in *Culex pipiens*. *J. Insect Physiol*. 39, 261–265.

Maddrell, S. H. P., Pilcher, D. E. M., and Gardiner, B. O. C. (1971). Pharmacology of the Malpighian tubules of *Rhodnius* and *Carausius*, the structure–activity relationship of tryptamine analogues and the role of cyclic AMP. *J. Exp. Biol*. 54, 779–804.

Magnarelli, L. A. (1977). Nectar feeding by *Aedes sollicitans* and its relation to gonotrophic activity. *Environ. Entomol*. 6, 237–242.

Magnarelli, L. A. (1979). Diurnal nectar-feeding of *Aedes cantator* and *A. sollicitans* (Diptera, Culicidae). *Environ. Entomol*. 8, 949–955.

Mayer, M. S., and James, J. D. (1969). Attraction of *Aedes aegypti*, responses to human arms, carbon dioxide, and air currents in a new type olfactometer. *Bull. Entomol. Res*. 56, 629–642.

Meola, R. W., and Petralia, R. S. (1980). Juvenile hormone induction of biting behavior in Culex mosquitoes. *Science* 209, 1548–1550.

Mitchell, B. K., and Reinhouts van Haga-Kelker, H. (1976). Comparison of the feeding behaviour in teneral and post-teneral *Glossina morsitans* (Diptera, Glossinidae) using an artificial membrane. *Entomol. Exp. Appl*. 20, 105–112.

Moloo, S. K., and Kutuza, S. B. (1970). Feeding and crop emptying in *Glossina brevipalpis* Newstead. *Acta Trop*. 27, 356–377.

Murlis, J. (1986). The structure of odour plumes. In: Payne, T. L., Birch, M. C., and Kennedy, C. (eds.), *Mechanisms in Insect Olfaction*. Clarendon Press, Oxford, pp. 27–38.

O'Meara, G. F. (1985). Ecology of autogeny in mosquitoes. *In* Lounibos, L. P., Rey, J. R., and Frank, J. H. (eds.), *Ecology of Mosquitoes: Proceedings of a Workshop*. Florida Medical Entomology Laboratory, Vero Beach, FL, pp. 459–471.

Orchard, I., Lange, A. B., and Barrett, F. M. (1988). Serotonergic supply to the epidermis of *Rhodnius prolixus:* evidence for serotonin as the plasticising factor. *J. Insect Physiol.* 34, 873–879.

Owen, W. B. (1963). The contact chemoreceptor organs of the mosquito and their function in feeding behavior. *J. Insect Physiol.* 9, 73–87.

Pappas, L. G., Pappas, C. D., and Grossman, G. L. (1986). Hemodynamics of human skin during mosquito (Diptera: Culicidae) blood feeding. *J. Med. Entomol.* 23, 581–587.

Peterson, D. G., and Brown, A. W. A. (1951). Studies of the response of the female *Aedes* mosquito, part III: the response of *Aedes aegypti* (L) to a warm body and its radiation. *Bull. Entomol. Res.* 42, 535–541.

Phipps, J. (1974). The vampire wasps of British Columbia. *Bull. Entomol. Soc. Can.* 6, 134.

Prasad, R. S. (1986). A hypothesis of host dependency of rat fleas. In: Payne, T. L., Birch, M. C., and Kennedy, C. (eds.), *Mechanisms in Insect Olfaction.* Clarendon Press, Oxford, pp. 175–177.

Price, G. D., Smith, N., and Carlson, D. A. (1979). The attraction of female mosquitoes (*Anopheles quadrimaculitus*) to stored human emanations in conjunction with adjusted levels of relative humidity, temperature and carbon dioxide. *J. Chem. Ecol.* 5, 383–395.

Reinouts van Haga, H. A., and Mitchell, B. K. (1975). Temperature receptors on the tarsi of the tsetse fly *Glossina morsitans* Westw. *Nature* 255, 225–226.

Ribeiro, J. M. C. (1988). How mosquitoes find blood. The role of vector-host interactions in disease transmission. *Misc. Pub. Entomol. Soc. Am.* pp. 18–23.

Rutledge, L. C., Ward, R. A., Gould, D. J. (1964). Studies on the feeding response of mosquitoes to nutritive solutions in a new membrane feeder. *Mosq. News* 24, 407–418.

Salama, H. S. (1966). The function of mosquito taste receptors. *J. Insect Physiol.* 12, 1051–1060.

Schlein, Y., Yuval, B., Warburg, A., and Mimmick, M. F. (1984). Aggregation pheromone released from the palps of feeding female *Phlebotomus papatasi* (Psychodidae). *J. Insect Physiol.* 30, 153–156.

Schmidt, J. M., and Friend, W. G. (1991). Ingestion and diet destination in the mosquito *Culiseta inornata:* effects of carbohydrate configuration. *J. Insect Physiol.* 37, 817–828.

Schreck, C. E., Kline, D. L., and Carlson, D. A. (1990). Mosquito attraction to substances from the skin of different humans. *J. Am. Mosq. Control Assoc.* 6, 406–410.

Smith, J. J. B. (1979). The feeding response of *Rhodnius prolixus* to blood fractions, and the role of ATP. *J. Exp. Biol.* 78, 225–232.

Smith, J. J. B. (1985). Feeding mechanisms. In: Kerkut, G. A., and Gilbert, L. I. (eds.), *Comprehensive Insect Physiology, Biochemistry and Pharmacology,* vol. 4, Pergamon Press, Oxford, pp. 34–85.

Smith, J. J. B., and Friend, W. G. (1982). Feeding behaviour in response to blood fractions and chemical phagostimulants in the blackfly, *Simulium venustum*. *Physiol. Entomol.* 7, 219–226.

Snodgrass, R. E. (1960). The anatomical life of the mosquito. *Smithson. Misc. Coll.* 139, 1–87.

Spates, G. E., DeLoach, J. R., and Chen, A. C. (1988). Ingestion, utilization and excretion of blood meal sterols by the stable fly, *Stomoxys calcitrans*. *J. Insect Physiol.* 34, 1055–1061.

Sutcliffe, J. F. (1986). Black fly host location. *Can. J. Zool.* 64, 1041–1053.

Sutcliffe, J. F. (1987). Distance orientation of biting flies to their hosts. *Insect Sci. Appl.* 8, 611–616.

Sutcliffe, J. F., and McIver, S. B. (1975). Artificial feeding of simuliids (*Simulium venustum*): factors associated with probing and gorging. *Experientia* 31, 694–695.

Tobe, S. S., and Davey, K. G. (1972). Volume relationships during the pregnancy cycle of the tsetse fly *Glossina austeni*. *Can. J. Zool.* 50, 999–1010.

Torr, S. J. (1989). The host-oriented behavior of tsetse flies (*Glossina*): the interaction of visual and olfactory stimuli. *Physiol. Entomol.* 14, 325–340.

Torr, S. J. (1990). Dose responses of tsetse (*Glossina*) to carbon dioxide, acetone, and octanol in the field. *Physiol. Entomol.* 15, 93–103.

Turritto, V. T., and Weiss, H. J. (1980). Red blood cells: their dual role in thrombus formation. *Science* 207, 541–543.

Trembley, H. L. (1952). The distribution of certain liquids in the esophageal diverticula and stomach of mosquitoes. *Am. J. Trop. Med. Hyg.* 1, 693–710.

Vargo, A. M., and Foster, W. A. (1982). Responsiveness of female *Aedes aegypti* (Diptera: Culicidae) to flower extracts. *J. Med. Entomol.* 19, 710–718.

Visser, J. H. (1988). Host-plant finding by insects, orientation, sensory input and search patterns. *J. Insect Physiol.* 34, 259–268.

Walker, E. D., and Edman, J. D. (1985). Feeding-site selection and blood-feeding behavior of *Aedes triseriatus* (Diptera: Culicidae) on rodent (Sciuridae) hosts. *J. Med. Entomol.* 22, 287–294.

Warnes, M. L., and Finlayson, L. H. (1985). Responses of the stable fly, Stomoxys calcitrans (L.) (Diptera: Muscidae) to carbon dioxide and host odors. I. Activation. *Bull. Entomol. Res.* 75, 519–527.

Washino, R. K. (1977). The physiological ecology of gonotrophic dissociation and related phenomena in mosquitoes. *J. Med. Entomol.* 13, 381–388.

Wigglesworth, V. B. (1941). The sensory physiology of the human louse *Pediculus humanus corporis* De Geer (Anoplura). *Parasitology* 33, 67–109.

Wigglesworth, V. B., and Gillett, J. D. (1934). The function of the antennae in *Rhodnius prolixus* and the mechanism of orienting to the host. *J. Exp. Biol.* 11, 120–139.

Wykes, G. R. (1952). An investigation of the sugars present in the nectar of flowers of various species. *New Phytol.* 51, 210–215.

Yee, W. L., and Foster, W. A. (1992). Diet sugar-feeding and host-seeking rhythms in mosquitoes (Diptera: Culicidae) under laboratory conditions. *J. Med. Entomol.* 29, 784–791.

7

Regulation of Phloem Sap Feeding by Aphids
W. F. Tjallingii

7.1. Introduction

Aphids are phloem feeders. Before feeding can be started, aphids must reach a food plant and select a probing site on the plant. From here their stylets need to penetrate the phloem in a vascular bundle and select a suitable sieve element to puncture. Host plant specificity is extremely well developed in aphids (Hille Ris Lambers, 1979). This has focused the attention of many entomologists on the possible role in host plant selection of allelochemicals, which make plants so specific (Whittaker and Feeny, 1971). Others were intrigued by the phloem specialization of aphids and the nutritional advantage of the phloem sap composition above other plant parts. This stimulated nutritional studies and the development of artificial diets (Mittler, 1988). Both ideas had been combined earlier in the "dual discrimination" hypothesis for food plant selection by aphids (Kennedy and Booth, 1951).

Some studies supported the idea that aphids, like several chewing insects, use allelochemicals as cues for their host plant selection (Nault and Styer, 1972). As phloem feeders, it may be argued that aphids will especially use the peripheral tissues of plants (epidermis and mesophyll) for host selection cues because allelochemicals would not occur in the phloem sap, or are present only at low concentrations (Mullin, 1986). Nevertheless, evidence for the occurrence of some allelochemicals, especially alkaloids and phenols, has been demonstrated in the phloem sap (Hussain et al., 1974; Wink and Witte, 1984). In the peripheral tissues of plants, allelochemicals are compartmentalized, often in the cell vacuoles, and they are mostly stored in a detoxified form, as glucosides or amides. When plant cells become damaged, these stored substrates are mixed with enzymes, occurring in different compartments (extracellular spaces, protoplasts, chloroplasts), which retoxify them (Matile, 1984). Aphids presumably avoid this essential part of the plant's defense strategy by preventing such damage because stylet penetration predominantly follows an extracellular pathway [in contrast, see Pollard (1973) and others] and hence leaves the allelochemicals and retoxifying enzymes in their own compartments. Although the stylets puncture many cells along this pathway, the cells remain intact, at least initially. Cells that were punctured, sometimes repeatedly, looked healthy with their organelles showing normal features in transmission electron micrographs (Tjallingii and Hogen Esch, 1993). These pathway punctures into cells are very characteristically reflected as drops of the electrical potential (pd) in the electrical penetration graph (EPG), a technique which can record the probing activities by aphids (Tjallingii, 1985b, 1988; Fig. 7.1). Intracellular pathway punctures last for about 10 sec each, and protoplasts and vacuoles may both be involved. These punctures may be associated with some watery salivation and ingestion of minute sap samples for chemoreception by the gustatory organs in the cibarial cavity (Wensler and Filshie, 1969). It remains unknown, so far, what the importance of the pathway punctures is for host plant recognition or plant rejection. Aphids on some nonhosts or resistant plants have been shown to continue to the sieve elements first, before aborting probing behavior and leaving the plant (Helden and Tjallingii, 1993). Further research of probing behavior will be needed to establish whether chemical cues from peripheral tissues or phloem sources are really used for food plant selection.

Sieve elements can be involved in the brief pathway punctures, just like the other cells along the stylet path. In addition, sustained punctures, lasting for minutes and sometimes for days (see below, Section 7.5), occur into sieve elements. During a sieve element puncture, only the maxillary stylets are inserted into the living cell (Tjallingii and Hogen Esch, 1993). Two main activities during these sustained punctures can be distinguished: sieve element salivation and sieve element ingestion. Sieve element salivation is represented by the EPG waveform E1 (Fig. 7.1), because this waveform appeared responsible for the inoculation of a phloem-restricted (persistent) virus by viruliferous aphids (Prado and Tjallingii, 1994). Sieve element salivation can be regarded as watery salivation because no sheath material is deposited. There is no ingestion during sieve element salivation (E1). Presumably, the closed cibarial valve in the food canal (Ponsen, 1972) prevents sap from entering the stylets. Sieve element ingestion is reflected as the EPG waveform E2 (Fig. 7.1). The ingestion is principally passive because of the high hydrostatic pressure in the sieve element, as indicated by radiotracers (Tjallingii, 1988). Actually, tracer experiments (radioisotope and virus) have shown that during E2 there is concurrent ingestion of sap and secretion of saliva. This saliva, however, is forced by the sap flow into the food canal and into the foregut immediately. Thus it does not enter the plant (sieve element). In addition to the sap pressure, this process results from the stylet morphology. At the tips of the maxillary stylets, the separate salivary and food canals fuse to a single lumen (Forbes, 1969).

Because sap feeding is mainly passive, it does not seem adequate to refer to

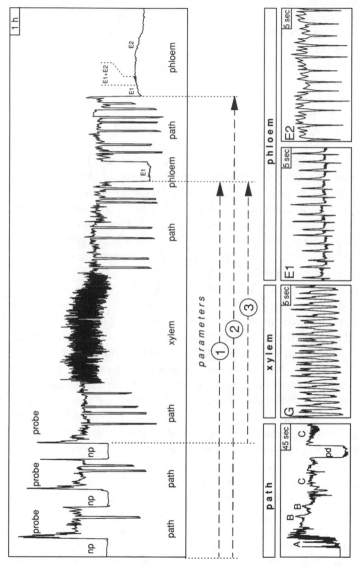

Figure 7.1. Electrical penetration graph (EPG) of an aphid. **Top trace:** Overview of the main features, np, nonprobing; probe, period of stylet penetration; path, pathway phase; xylem, xylem ingestion phase; phloem, phloem sieve element phase. **Parameters** (arrows): 1, time from the beginning of the experiment to the first sieve element puncture; 2, same as 1 but to sustained sieve element puncture with ingestion; 3, same as 1 but from the beginning of the probe instead of the beginning of the experiment. **Bottom traces:** Detailed waveforms of each phase in the top trace. Path: A, first stylet–plant contact; B, sheath material secretion; C, other "pathway activities"; pd, potential drop, intracellular puncture. Xylem: G, active xylem ingestion, "drinking." Phloem: E1, sieve element salivation; E2, sieve element ingestion.

aphids as "sucking insects." Their main mode of feeding is passive ingestion. Sucking is restricted to occasional "drinking" from xylem elements, which is reflected in the EPG as waveform G (Fig. 7.1; Spiller et al., 1990), and, possibly, to sampling during the brief pathway punctures described above.

It may be concluded from the foregoing that food plant selection in aphids is a complicated and largely unknown area (Dreyer and Campbell, 1984; Montllor, 1991; Pollard, 1973), but, in fact, it is a prelude to the question of how feeding is regulated once a plant is accepted as a food plant. The aim of this chapter is to review feeding, or sieve element ingestion, and to contribute to what is known with some new and previously unpublished results.

7.2. Terminology

In the literature on feeding behavior by Homopterans, there is often some terminological confusion. Some authors use feeding for all activities which relate either directly or indirectly to the needs for an insect to feed—in other words, all activities that are related to the "motivational drive" of hunger. In this context, such activities as orientation, dispersion, walking, and flying can be elements of feeding behavior. I would like to adapt this use of "feeding behavior" here, under the condition that "feeding" is only used in combination with the word "behavior."

Feeding, when *not* combined with "behavior," has a more restricted meaning— that is, eating, grazing, or the intake of food in general. Therefore, it largely is a synonym of ingestion, especially when used for sap-feeding Homopterans. Intake of plant fluids also may occur without the direct nutritional purpose to feed. For example, fluid samples can be ingested to make contact to the gustatory organs in the cibarial cavity (Wensler and Filshie, 1969), resulting in behavioral adjustments. If, subsequently, such fluid samples will be ejected, as some authors suggest (Harris, 1977), the term "intake" may be preferred above "ingestion." Ingestion is a general term that will be used here for all substances passing into the gut. Aphids, unlike some other Homopterans, do not ingest cell organelles, and fluids are principally ingested from the sieve elements. Ingestion of xylem sap by aphids seems to occur only under desiccating conditions (Spiller et al., 1990) and, therefore, has been referred to as "drinking." Sap ingestion in aphids is largely synonymous with sap feeding, which more clearly refers to a nutritional purpose. Whenever feeding is used, its context or additional terms will make clear what its meaning is—for example, "feeding behavior," "sap feeding," and so on.

Homopterans insert their mouthparts into the plant to locate the food source (tissue) before sap feeding starts. "Stylet penetration" and "probing" are terms that may include all known and unknown activities in which the stylets are involved while they are in the plant (or artificial diet). Strictly, probing implies exploration or investigations made by a probe—that is, an inserted measuring instrument. Some authors (mainly British) use probing in the restricted sense of "brief test penetrations." However, the word "probing" in itself does not include any time limitation. A long probe may include continuous exploration or monitoring of a number of relevant properties. "Stylet penetration" seems more neutral than "probing," but it may be explained as restricted to the pure mechanical progress of the stylets. However, many activities, such as sap sampling, ingestion, saliva secretion, cell puncturing, and partial withdrawal, are so integrated in the whole process that the use of "stylet penetration" seems to have as many advantages and disadvantages as does the use of "probing." Anyhow, they describe the complete phenomenon much better than, for example, "feeding." Therefore, "stylet penetration" and "probing" will be used here as synonyms, to indicate all activities during stylet contact with the plant (or artificial diet).

Used in combination with "behavior," "probing behavior," or "(stylet) penetration behavior" seems very acceptable to indicate probing in a wider sense—that is, a series of probes including nonprobing intervals on the same plant (see Fig. 7.1, top trace). In this sense, nonprobing normally comprises walking or resting, but not the migratory flights or escape falls from the plant, which may still be included in feeding behavior.

In summary, *feeding behavior* includes all activities related to sap feeding, probing, and exploratory locomotion to find new food plants. *Probing behavior* is a part of feeding behavior, including probing and nonprobing intervals on the same plant. *Probes* and *stylet penetrations* are synonyms used for the periods of stylet contact with the plant. *Ingestion* will refer to all sap intake into the gut. In aphids, it includes *drinking* from xylem elements and *sap feeding* from phloem sieve elements.

7.3. The EPG Technique as a Method to Study Sap Feeding

As discussed above, sap feeding in aphids means passive sap ingestion from phloem sieve elements, which can be recognized during EPG recording as waveform E2 (Fig. 7.1). This makes the EPG technique a good method to study sap feeding. The principle of the EPG technique is that the piercing insect and its feeding substrate—plant or artificial diet—are made part of an electrical circuit. On the insect's dorsum an electrode of thin (10–20 μm) gold wire is attached using conductive silver paint. The feeding substrate is connected by another electrode, which supplies a weak voltage. An AC voltage was used in the original system (McLean and Kinsey, 1964; McLean and Weigt, 1968), still popular in the United States and here referred to as the "AC system," whereas the later modified and improved system (Tjallingii, 1985a, 1988) uses a DC voltage and will be referred to as the "DC system." As soon as the stylets penetrate, the circuit is completed and the EPG can be recorded. The two stylet canals, the food canal, and the saliva canal electrically form a kind of "bottle neck" in the system; that is, phenomena in or near these canals determine the signal, whereas electrical phenomena elsewhere in the insect have no or negligible effects. The leg contacts, also, have too high an electrical resistance to play a role. The probing activities can be recognized as different waveforms or waveform patterns. In the EPG of the DC system, eight waveforms have been distinguished so far and most of them have been identified. First, the EPG can be divided in probing (stylet penetration) and nonprobing (np) periods (Fig. 7.1). Within a probe, three different phases can be distinguished: a pathway phase (path), including waveforms A, B, and C with its drops in potential (pd) due to brief intracellular punctures; a xylem phase, characterized by waveform G; and a phloem sieve element phase (phloem) including waveforms E1 and E2. As stated above, waveform E1 has recently been correlated with saliva secretion into the sieve element (Prado and Tjallingii, 1994) and waveform E2 has been correlated with ingestion and concurrent secretion of saliva that goes with the sap directly into the food canal (Tjallingii, 1990). The only waveform during which honeydew secretion was found is waveform E2. Because the phloem sap is under a high pressure, ingestion during waveform E2 is presumably mainly passive, whereas the ingestion of xylem sap (waveform G) is supposed to be active ingestion by the cibarial pump because xylem sap is generally under negative pressure (see Chapter 2). Several EPG studies support these hypotheses (Spiller et al., 1990; Tjallingii, 1990). Behaviorally, phloem ingestion and xylem ingestion are two clearly separate activities which might be distinguished as "sap feeding" and "drinking," respectively. Xylem drinking only occurs occasionally as a result of water stress. It is not followed by honeydew secretion. Probably, its regulation occurs separately from the nutritional needs, though drinking and sap feeding might be linked at a higher level in the nervous system. Intake of sap samples for sensory purposes, as referred to above, have not been proven so far. Such sap sampling might occur during the intercellular path or during the brief and abundant intracellular punctures (pd, Fig. 7.1). We will concentrate here on sap feeding as the activity to be regulated by the nutritional needs of the organism. Hence, for studying this in aphids, the E2 waveform in EPGs can be used.

There remain, however, some practical problems in using the E2 waveform. Though E1 and E2 are distinctly different, they may show some overlap because, in contrast to most other waveforms, they are not mutually exclusive. During a period of overlap it is difficult to determine where E1 stops and E2 starts, or vice versa. Typically, the overlap period is clear and short, but sometimes, especially during the first few hours after access to a new plant and often on nonhost plants (depending on the aphid species as well), transitional mixtures can be sustained, making quantification of E1 and E2 durations difficult. Signals from AC devices (Backus and Bennett, 1992; Brown and Holbrook, 1976; Kimsey and McLean, 1987; McLean and Weigt, 1968) however, do not allow any distinction between E1 and E2 activities (Tjallingii, 1995). As a consequence, when using results obtained by the AC method, periods defined as "phloem ingestion" or "committed phloem ingestion" (Montllor et al., 1983) will include sieve element salivation (waveform E1) and, as a result, tend to overestimate sap feeding considerably, especially in the first hours.

7.4. Quantifying Ingestion

Early data on ingestion rates using gravimetric or radiotracer methods show a tremendous variation (reviewed in Pollard, 1973). It is very hard to understand why the figures are so different. When one set of data is assumed to reflect realistic figures, it is nearly impossible to explain others, even when the original articles are examined thoroughly. Essential information on experimental setup, procedures followed, or original figures behind the calculated and presented data are often not published. The most uncertain factors, however, are presumably in the behavior of the insects during these experiments. All of these studies used groups of aphids, of which only a fraction had been sap feeding for an unknown part of the experimental time. Data from such experiments seriously mask the real sap feeding by individuals. What seems to be "typical" aphid behavior does, in fact, not exist in any individual aphid, and thus the individual ingestion rate remains unknown.

A few of our later studies, later than Pollard's review (1973), on individual aphids are therefore of interest here. Using plants and sachets of artificial diet, aphids were electrically monitored by EPG recording to quantify different types of ingestion (Tjallingii, 1978, 1987). Other experiments included honeydew recording or stylectomy (stylet cutting) in order to collect sap from sieve elements. Their results may provide at least some tentative figures on sap feeding by aphids which are compatible.

The most reliable ingestion rates were based on ^{32}P-labeled artificial diets. In a first experiment (Tjallingii, 1978), *Brevicoryne brassicae* L. was allowed to penetrate parafilm sachets containing the labeled diet. EPGs were recorded on tape; then the aphids were removed after 2 hr, killed with a droplet of ethanol in a liquid scintillation vial, and their radioactivity was counted directly in water (Cerenkov radiation) for each individual. On these diets, a pathway and ingestion (xylem and phloem) phase were shown in the EPGs. The summed durations of separate waveforms were estimated for each aphid by replaying taped signals to chart strips. Using relevant blanks and standards, the ingested volumes could be established (per aphid, per unit of time for each waveform).

Waveforms A, B, and C were pooled as "path" [Fig. 7.1, pathway phase; this is equivalent to "salivation" in AC devices, McLean and Kinsey (1967)], whereas waveforms E1, E2, and G were pooled as "D" (the common label used at that time, when no distinction was yet made between the xylem and phloem phase). Aphids only showing the pathway phase had an ingestion rate of 0.02 pl $\cdot$ sec^{-1}.

This figure was used as a correction factor for aphids that additionally showed waveform D, in order to calculate the ingestion rate for D. Thus calculated, the ingestion rate for phase D on artificial diet was 1.12 pl · sec^{-1} (Table 7.1, experiment 1).

In a second study (Tjallingii, 1987), we used *Myzus persicae* Sulzer, which is of about the same size but accepts the artificial diet much more readily than *B. brassicae*. The accuracy gained considerably from the new EPG waveforms that were now distinguished and by the use of a small device (Fig. 7.2) that allowed us to control the exposure of the aphid to labeled diet during ingestion. The aphid could penetrate a parafilm membrane stretched over a tiny gully which was cut in the soft plastic surface of the device. At either end of the gully, perpendicular holes were drilled through the surface, connecting the gully to an inlet and an outlet canal formed by syringe needles. A syringe with labeled diet (asterisk in Fig. 7.2) was connected to the inlet. An air bubble separated this diet from an initial amount of unlabeled diet in the gully to prevent the diets from mixing. The aphid started probing on the unlabeled diet until a clear E2 or G waveform was shown. Then the diet was changed by pressure on the syringe. After about 10 min during which the same waveform was produced continuously on the labeled diet, the aphid was removed and killed and the ingested volume was estimated by its counted radioactivity.

Insects exhibiting waveform E2 (Table 7.1, experiment 2) showed an ingestion rate of 1.6 pl · sec^{-1}, about the same order as D (1.1 pl · sec^{-1}) in the previous experiment. In contrast, insects exhibiting waveform G showed a threefold higher ingestion rate (4.5 pl · sec^{-1}). Because E2 is mainly passive sap feeding and G is xylem drinking on plants, the question arises whether this ratio between E2 and G from diet would hold for plants as well. This could not be derived from these experiments.

Some plant data were obtained from honeydew and stylectomy experiments. Tracer studies are unsuitable because of the irregular distribution of an isotope in plants. As for ingestion, however, there are many examples in the literature of honeydew production over periods by groups of aphids (Banks and McCauley, 1964; Cull and Van Emden, 1977; Spiller, 1986) of which only a fraction had been sap feeding. During our EPG recording we collected honeydew from individual aphids. Using *M. persicae* on *Brassica napus* (L.), honeydew was collected in oil to measure the droplet size, and on rotating disks of thin-layer chromatography (TLC) plate (Merck, Silicagel 60F254) as a "honeydew clock" (Banks and MaCaulay, 1964). Droplet volumes from diameters measured in oil were empirically related to the surface diameters (ultraviolet observations by stereomicroscope with ocular micrometer) on the TLC plate, using averages and standard deviations of subsequent measurements by these methods from the same individual aphids. For individual aphids, droplets were fairly constant in volume as were their excretion intervals (Fig. 7.3). Honeydew secretion was observed to coincide only with waveform E2, with a mean secretion rate of 12.25 pl · sec^{-1}

Table 7.1. *Flow rates from different experiments in which it was determined as flow rate based on ingestion of ^{32}P (experiments 1 and 2), honeydew excretion (experiment 3), or sap exudation from severed stylets after stylectomy (experiment 4), during EPG waveforms, or after showing them (experiment 4)*

Experiment number	Experiment type	Insect	Waveform[a]	N[b]	Flow rates[c]		
					Maximum	Average	Standard Deviation
1	^{32}P diet (1978)	*B. brassicae*	Path	24		0.02	
2	^{32}P diet (1987)	*M. persicae*	D	134		1.12	
			E2	19		1.64	0.72
			G	4		4.33	1.31
3	Honeydew	*M. persicae*	E2	7		12.25	1.54
4	Free exudation (stylectomy)	*M. persicae*	E2**	1	90		

[a]Waveforms (see also Fig. 7.1): path—stylet pathway phase (including A, B and C in Fig. 7.1); D—phloem and xylem phase (including E1, E2, and G in Fig. 7.1); E2—sap feeding (on plants from sieve elements); G—sap ingestion (on plants from xylem elements, "drinking"); E2**—waveform before stylectomy.

[b]N, number of replicates.

[c]Flow rates: maximum, average, and standard deviation in picoliters per second (pl·sec^{-1}; 1 pl = 10^{-12} liters).

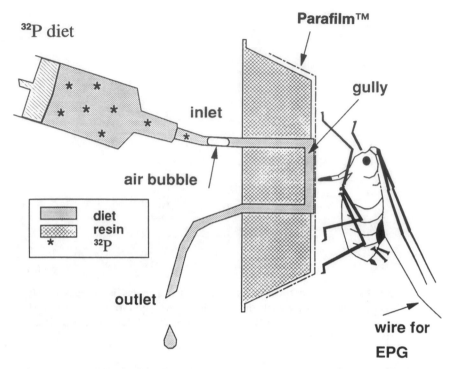

Figure 7.2. Radioactive feeding experiment. Flow chamber made by a small gully in a resin-filled plastic cap. The gully between the inlet and outlet holes is covered by Parafilm™. During ingestion (waveform E2 or G) the unlabeled diet in the chamber was replaced by the ³²P diet from the syringe. After a defined time the aphid was removed, its radioactivity was measured, and the ingestion rate (pl · sec⁻¹) for the waveform could be calculated. The air bubble prevents the two diets to be mixed before the replacement.

(Table 7.1, experiment 3). How can we use this figure as an estimate for the ingestion rate? We should correct for evaporation and other losses due to absorption by the insect. Thus, the related ingestion rate will be higher than the excretion rate, but how much higher? Some work has been done on utilization and conversion suggesting that about 50% of carbon in the intake is used and not excreted (Llewellyn and Leckstein, 1978). There are no figures on water, but if we assume, just as an example, that 50% of the water is also absorbed, then the $12.25 \text{ pl} \cdot \text{sec}^{-1}$ honeydew secretion would be related to an ingestion rate of about $25 \text{ pl} \cdot \text{sec}^{-1}$ during waveform E2.

In another study that may provide some relevant information on sap feeding, we used *M. persicae* on lettuce (cultivar Taiwan). Following stylectomy (i.e., stylet amputation by radio-frequency microcautery), phloem sap was collected for chemical analysis (Mentink et al., 1984). An aphid was wired and its EPG was recorded until waveform E2 occurred, indicating that it had reached the

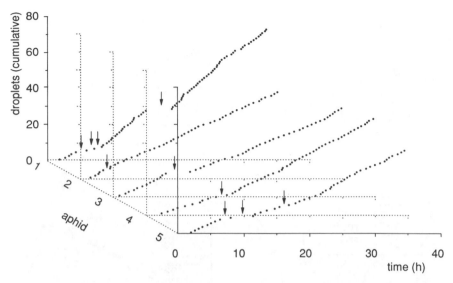

Figure 7.3. Patterns of honeydew excretion by five apterous individuals of *Myzus persicae*. Each dot represents a single droplet. Horizontal gaps (arrows) indicate periods without sap feeding. Though the secretion rate (slope) may differ between aphids, it is fairly constant within an individual.

sieve element and started ingestion. The stylets were cut, and sap exudation occurred immediately in about 25% of the trials due to the sap pressure in the sieve element. Often, exudation stopped after a few minutes and the original rate could not be measured, but, occasionally, sap continued flowing for tens of minutes, an hour, or even longer. These cases allowed us to collect sap and calculate the flow rate. The rate of exudation varied considerably, but maximum values up to about 90 pl $\cdot$ sec^{-1} were measured. More recently (Helden et al., 1994), such experiments were repeated on the same lettuce cultivar with *Naso-novia ribisnigri* and values up to 800 pl $\cdot$ sec^{-1} have been found. Though the figures are very rough, from a different plant than the honeydew data, and from a rather low number of replicates, these are the only figures available. The maximum values are interesting because they may give us at least some idea how fast the phloem sap can be forced into the food canal of the stylets—that is, the plant component of the sap feeding. In addition, the ^{32}P figures from diet may reflect the active contribution of the aphid to the feeding rate. The plant component is passive and it can maximally be as large as the maximal free outflow after stylectomy, but, according to the honeydew figures above, aphids apparently limit the uptake to the estimated rate of 25 pl $\cdot$ sec^{-1}. This considerable reduction of the passive inflow is presumably accomplished by the cibarial valve and by its muscles, allowing the valve to be opened appropriately. Probably, the valve movements controlling the inflow are reflected by the small waves in waveform E2 (Fig. 7.1) at a frequency of about 6 Hz. The sharp downward peaks in the E2 waveform have been related to concurrent salivation by the independently operating salivary pump (Tjallingii, 1978).

On diets where no pressure forces the fluid into the stylets, the ingestion rate of 1.64 pl · sec^{-1} for waveform E2 (Table 7.1) is presumably too low for survival. It has been observed that aphids on artificial diets show a great deal of waveform G, active ingestion, especially when they were born on diet. It may be speculated that this is an adaptive way of feeding to survive the unnatural situation of the artificial diet.

The rate of xylem ingestion on plants is unknown. The data presented here cannot be used. The viscosity of the xylem sap will be much lower than artificial diet which contains 15% sucrose. Also, there are hydrostatic pressure differences between the diet and the plant so that any estimation becomes merely speculative. Even less is known about the rate of intake during sap sampling for gustatory purposes.

7.5. Onset of Sap Feeding in Probing Behavior

Having distinguished sap feeding from other ingestion and estimated its probable rate, we can address the question of its occurrence in relation to other aspects of probing behavior. When does sap feeding start, and for how long is it continued? Is there something like a sap-feeding period which may be considered as a meal?

Recording honeydew secretion and EPGs from individual aphids seem suitable techniques to answer these questions. However, honeydew recording provides rather rough data because the excretion frequency is only one droplet in about 40 min. EPG recording seems more accurate because activity changes within a second can be monitored. But does EPG recording itself not interfere with the natural appearance and duration of sap feeding? Wired aphids have locomotory constraints, and the electrical current through their body may change their behavior. In a separate study (Tjallingii, 1986) it was pointed out that wired aphids did behave differently from free aphids, but the differences were small, especially on host plants. Some aphid species seem to become stressed by tethering and may show an initial delay in probing attempts (Helden and Tjallingii, 1993). On nonhost plants, however, the EPG results tend to overestimate the plant suitability because wired aphids cannot escape from the unsuitable plant and will probe again instead of leaving.

A different criticism comes from Schnorbach and Kunkel (1990), who argue that wired aphids are handled as isolated individuals, which excludes "subsocial behavior" affecting the feeding behavior. These effects, if present, may be due to "crowding" (i.e., contacts between individuals), pheromone (alarm and aggregation) communication, or plant-mediated stimuli elicited by other aphids on the same plant. It may be argued, however, that some of these factors act positively and others negatively on the initiation of, and time spent in, sap feeding.

Though some behavioral effects of locomotory or social constraints on sap feeding cannot be excluded, no indications have been found so far for any effects due to the electrical current. Its value is low, lower than 5 nA. Though this is still significant with respect to the small diameter of the stylet canals, it causes a translocation of charged sap particles (electrophoresis) that does not exceed about 0.1% of the transport by the ingestion rate found for E2 (Table 7.1) (Tjallingii, 1985a); this seems negligible.

Control experiments to investigate the effects of the methods used, especially the consequences for the experimental objectives, are strongly recommended for each new aphid–plant combination. We used honeydew recordings from nonwired individual aphids as a control, for comparison with wired aphids (EPG recording) in respect to the initiation of sap feeding and its duration.

The occurrence of sap feeding over time has been studied preliminarily in *B. brassicae* and *M. persicae,* but more reliable data come from *Aphis fabae* (Scop.) and *Sitobion avenae* (F.) on broad beans (stems and leaves) and cereals (wheat and barley), respectively (Tjallingii and Mayoral, 1992). EPGs recorded for 8 hr from apterous adults, about 15 aphids on stems and on leaves, were stored on a computer hard disk and successively analyzed. The beginning and duration of (new) separate probes and waveforms were scored for each treatment, for each aphid–plant combination, and for each hour. Feeding behavior appeared to be composed of a series of probes (Fig. 7.1; Table 7.2), decreasing in number with time and increasing in duration, separated by periods of nonpenetration. With time, the percentage of probing time spent in sieve element punctures increased. The number of new sieve element punctures increased up to the 4th hour and then decreased gradually due to the increased duration of each puncture (a sieve element puncture contributes to this parameter only in the hour of its appearance). Nonwired individuals were used for honeydew recording in order to check for wire effects.

Table 7.2. Probing and sap feeding of A. fabae *on bean leaves during 8 successive hours[a]*

	Hour since start of experiment							
	1	2	3	4	5	6	7	8
Number of new probes	4.6	2.3	2.4	1.2	1.5	0.6	0.3	0.5
Total time probing (%)	0.75	0.87	0.87	0.91	0.93	0.95	0.97	0.96
Time in phloem/probing time (%)	0.5	1.9	4.5	18.2	43.5	65.1	64.6	78.5
Number of new phloem punctures	0.1	0.3	0.2	0.8	0.5	0.5	0.5	0.5

[a]Mean numbers (per aphid per hour) of initiated probes and sieve element punctures. Mean time spent probing and mean time in phloem sieve elements as part of the probing time (per hour). $N = 15$.

The aphids needed several hours, including numerous probes, to achieve the first puncture in a sieve element (Fig. 7.1 and Table 7.3, parameter 1), and at least 30 min more was needed to achieve sap feeding—that is, a sieve element puncture of more than 10 min with waveform E2 (Fig. 7.1 and Table 7.3, parameter 2). In most cases, sustained (>10 min) sap feeding was maintained for more than 1 hr; some lasted for hours, up to the end of the experiment. Sieve element punctures of less than 10 min mostly occurred early after plant access (start of experiment) and always started with sieve element salivation (E1), occasionally followed by sap feeding (E2). The sustained periods sieve element punctures, however, typically contained 40–60 sec of salivation (E1) followed by sap feeding (E2). This demonstrates the general rule, also found in other aphid species (Montllor, 1991), that reaching the first sieve element does not imply that sap feeding starts immediately. Also, the actual time needed for the aphid stylets to penetrate the tissues from the epidermis to a sieve element (Fig. 7.1 and Table 7.3, parameter 3) is only a small part of the time to the first sieve element puncture and to sap feeding (parameters 1 and 2, respectively). We have to realize, however, that the time to first E waveform in the EPG is not exactly reflecting the time needed to reach the first sieve element (parameter 1), because sieve elements may be involved in intracellular pathway punctures (Fig. 7.1, pd) before the first E waveform is shown (Hogen Esch and Tjallingii, 1992). Thus the first E waveform actually indicates the first observable sieve element puncture, but it may overestimate the time to reach the first sieve element. Though these considerations are very important to understand food plant selection and acceptance by these insects, they do not affect our conclusions here: (a) Aphids may take many hours to start sap feeding on their host plant; (b) many probes are needed, and within these probes a number of short (<10 min) sieve element punctures usually precede sustained sap feeding; and (c) once sustained for more than 10 min, sap feeding tends to last for more than 1 hr and often

Table 7.3. Mean time needed to first observable puncture of a phloem sieve element measured from plant access (parameter 1) or within the probe (parameter 3); time to first sustained (>10 min) sap feeding (parameter 2) is measured from plant access.[a]

| Plant | Part | Time (hours) to first phloem puncture | | Time (hours) to first sustained phloem ingestion in experiment (parameter 2) |
		In experiment (parameter 1)	In probe (parameter 3)	
Broad bean	Leaf	3.4	0.7	5.2
Broad bean	Stem	3.6	0.8	4.2
Wheat	Leaf	2.9	0.4	3.5
Barley	Leaf	5.8	0.6	6.3

[a]For parameters 1, 2, and 3, see also Fig. 7.1.

longer. For an accurate estimation of the sap-feeding periods, however, the 8-hr experiments appeared too short.

The delays between plant access and the onset of honeydew secretion, shown by the nonwired aphids in the control experiment, were similar to the times needed to achieve sustained ingestion found in EPGs by wired aphids (Table 7.3). Therefore, it is unlikely that effects due to wiring or other artifacts biased the results above.

7.6. Maintenance and Termination of Sap Feeding

The relatively short duration (8 hr) of the foregoing experiments allowed no definite conclusions on whether aphid feeding is subdivided into feeding bouts or meals, comparable to chewing insects. In the much-longer-lasting honeydew recordings (*M. persicae* on *B. napus;* see above), we found that many aphids initially produced series of droplets at regular intervals, ranging from one to several hours. The gaps between these series of honeydew droplets (Fig. 7.3) coincided with switches from sap feeding (E2) to pathway and nonprobing wave-forms when an EPG of these aphids was recorded concurrently. Gradually the series of honeydew droplets became longer, from hours to days, up to 10 successive days on one occasion. The honeydew secretion rate, droplet size, and frequency within periods of uninterrupted secretion differed slightly between aphids but was fairly constant within individuals (Fig. 7.3). Thus, the honeydew results support the idea that meals in the sense of chewing insects do not exist in aphids. Sap feeding seems to present a steady inflow, presumably mainly determined by the pressure of the phloem sap. The long honeydew recordings suggest that once a food plant is well accepted, sap ingestion can be sustained for very long periods, if not for the complete duration of a developmental stage (nymphs or adult).

The steady food supply from sieve elements allows the aphids to feed almost continuously. Also, in most adults, feeding is not interrupted by reproductive behavior because most generations are parthenogenetic. Therefore, ingestion of large quantities in a short time may be unnecessary. The continuity of the process also suggests that aphids do not need a separate time for the digestion of their food. Free amino acids occur in the phloem sap, while proteins cannot be used due to the absence from the aphid gut of proteolytic enzymes (Srivastava and Auclair, 1963). On the other hand, the carbohydrate supply is so abundant that limited digestion may satisfy the aphids' needs. The inflow into the foregut from the plant and the outflow to the midgut need no regulation of the blowfly type (Dethier, 1976; also see Chapters 6 and 8). Blowflies take fast meals into a crop, and information from stretch receptors ends the meal; digestion then determines how fast the crop is emptied between the meals.

Using a 16-hr light period per day, we did not observe any significant difference

(α=0.05) between regression coefficients and droplet sizes in series of droplets from light and dark periods within individuals. This is in contrast to what was found by others (Cull and Van Emden, 1977; Spiller 1986). They concluded that there was a clear decrease in honeydew production in the dark and an increase during light. However, they did not observe aphids individually. Thus, the decrease during the dark may have been caused by a smaller number of sap-feeding aphids rather than a decreased ingestion rate by each individual. A reduced carbohydrate concentration in the phloem in darkness, believed to explain the reduced ingestion (Cull and Van Emden, 1977), is not likely to play any role. Winter et al. (1992) showed that the main carbohydrates (sugars) in the phloem hardly fluctuate with photoperiodicity. However, other chemicals (i.e., amino acids and plant hormones) do fluctuate and may have some impact on feeding behavior.

Many allelochemicals have been claimed to reduce ingestion. Azadirachtin, a potent systemic antifeedant, has recently also been found to reduce sap feeding by inducing shorter periods of sieve element ingestion in *M. persicae* (Nisbet et al., 1993). There is no reason why allelochemicals naturally occurring in the sieve tube sap would not act similarly.

Thus, we may conclude that plant factors, especially those in the phloem sieve elements, seem to provide the main cues for aphids in their regulation of sap feeding. That is, they determine the onset, the maintenance, and the termination of sap-feeding periods.

7.7. Aphids and Other Homopterans

We have to be careful before extrapolating our hypotheses on the regulation and control of sap-feeding to all other aphid species and, especially, to other Homopterans. Preliminary results (Tjallingii, *unpublished data*) showed that *Rhopalosiphum padi* and other cereal aphids seem to give less regular honeydew patterns than those presented here. It should be studied further to see whether this deviation is consistent. Also, other taxonomic groups of aphids should be studied first before any definite conclusion can be drawn. Other Homopterans have completely different food sources than the phloem sap. *Nilaparvata lugens* (Stal) secretes honeydew when feeding from phloem as well as from xylem elements (Kimmins, 1989). Many Homopterans are much more mobile than aphids, possibly requiring shorter probes and sap-feeding periods more matching the concept of "meals."

The only comparable data so far come from the phloem-feeding whiteflies (Aleurodidae). We recorded EPGs and honeydew excretion (Lei et al., 1995) from nymphs of the whitefly *Trialeurodes vaporariorum* (Westw.) The young sessile nymphs alternate between an ingestion and a noningestion waveform during a continuous intracellular position of the stylet tips in a phloem sieve element. After molting, the stylets penetrate a new sieve element which is not abandoned before the next moult. In honeydew recordings, the interval between droplets during ingestion showed fluctuations that corresponded with frequency fluctuations within the ingestion waveform in the concurrently recorded EPG. However, we did not measure the droplet volumes in this experiment, and thus we do not exactly know the impact of these figures on the ingestion rate. In addition, first-instar whitefly larvae showed alternating periods of ingestion and noningestion. Whether the periods of noningestion are induced by the food plant or form a way of controlling ingestion remains unknown. If noningestion means salivation, like the E1 waveform in aphids, switching to this waveform may reflect regular attempts to avoid or suppress wound reactions or other sieve element defenses. For the time being, however, this remains mere speculation.

7.8. Conclusions

Sap feeding by aphids seems to be regulated by plant factors which are mainly unknown so far. The ingestion rate is fairly constant per individual insect, and intake control appears to be achieved by changing the duration of an ingestion period.

Initiation of sap feeding takes many hours even on a suitable host plant. Many probes are needed, and within these probes a number of sieve element punctures usually precede sustained sap feeding.

Once sustained for more than 10 min, sap feeding tends to last for more than 1 hr and often longer. Presumably sap feeding will continue during a complete larval or adult stage unless it is terminated by some plant (sieve element) factor. In the feeding behavior of aphids, meals, as shown by chewing insects, do not exist.

Preliminary results from the (sessile) larval life span of the greenhouse whitefly show the same principle even more extremely. It is doubtful if non-phloem feeders control sap feeding in a similar way.

References

Backus, E. A., and Bennett, W. H. (1992). New AC electronic insect feeding monitor for fine-structure analysis of waveforms. *Ann. Entomol. Soc. Am.* 85, 437–444.

Banks, C. J., and Macauley, E. D. M. (1964). Feeding, growth and reproduction of *Aphis fabae* Scop. on *Vicia faba* under experimental conditions. *Ann. Appl. Biol.* 53, 229–242.

Brown, C. M. and Holbrook, F. R. (1976). An improved electronic system for monitoring feeding of aphids. *Am. Potato J.* 53, 457–462.

Cull, D. C., and Van Emden, H. F. (1977). The effect on *Aphis fabae* of diel changes in their food quality. *Physiol. Entomol.* 2, 109–115.

Dethier, V. G. (1976). *The Hungry Fly*. Harvard University Press, Cambridge, MA.

Dreyer, D. L., and Campbell, B. C. (1984). Chemical basis of host-plant resistance. *Plant Cell Environ*. 10, 353–361.

Forbes, A. R. (1969). The stylets of the green peach aphid, *Myzus persicae* (Homoptera, Aphididae). *Can. Entomol*. 101, 31–41.

Harris, K. F. (1977). An ingestion–egestion hypothesis of noncirculative virus transmission. In: Harris, K. F., and Maramorisch, K. (eds.), *Aphids as Virus Vectors*. Academic Press, New York, pp. 165–219.

Helden, M. van, and Tjallingii, W. F. (1993). Tissue localization of lettuce resistance to the aphid *Nasonovia ribisnigri* using electrical penetration graphs. *Entomol. Exp. Appl*. 68, 269–278.

Helden, H. van, Beek, T. A. van, and Tjallingii, W. F. (1994). Phloem sap collection from lettuce (*Lactuca sativa*): methodology and yield. *J. Chem. Ecol*. 20, 3173–3190.

Hille Ris Lambers, D. (1979). Aphids as botanists? *Symp. Bot. Ups*. XXII, 4, 114–119.

Hogen Esch, Th., and Tjallingii, W. F. (1992). Ultrastructure and electrical recording of sieve element punctures by aphid stylets. In: Menken, S. B. J., Visser, J. H., and Harrewijn, P. (eds.), *Proceedings of the 8th International Symposium on Insect–Plant Relationships*. Kluwer Academic Publishers, Dordrecht, pp. 283–285.

Hussain, A., Forrest, J. M. S., and Dixon, A. F. G. (1974). Sugar, organic acid, phenolic acid and plant growth regulator content of the honeydew of the aphid *Myzus persicae* and its host plant *Raphanus sativus*. *Ann. Appl. Biol*. 78, 65–73.

Kennedy, J. S., and Booth, C. O. (1951). Host alteration in *Aphis fabae* Scop. I. Feeding preferences and fecundity in relation to the age and kind of leaves. *Ann. Appl. Biol*. 38, 25–64.

Kimmins, F. M. (1989). Electrical penetration graphs from *Nilaparvata lugens* on resistant and susceptible rice varieties. *Entomol. Exp. Appl*. 50, 69–79.

Kimsey, R. B., and McLean, D. L. (1987). Versatile electronic measurement system for studying probing and feeding behavior of piercing and sucking insects. *Ann. Entomol. Soc. Am*. 80, 118–129.

Lei, H., Tjallingii, W. F., Lenteren, J. C. van, and Xu, R. M. (1995). Stylet penetration by larvae of the greenhouse whitefly on cucumber. *Entomol. Exp. Appl*.

Llewellyn, M., and Leckstein, P. M. (1978). A comparison of energy budgets and growth efficiency for *Aphis fabae* Scop. reared on synthetic diets with aphids reared on broad beans. *Entomol. Exp. Appl*. 23, 66–71.

Matile, P. (1984). Das toxische Kompartiment der Pfanzenzelle. *Naturwissenschaften* 71, 18–24.

McLean, D. L., and Kinsey, M. G. (1964). A technique for electronically recording aphid feeding and salivation. *Nature* 27, 1358–1359.

McLean, D. L., and Kinsey, M. G. (1967). Probing behavior of the pea aphid, *Acyrthosiphon pisum*. I. Definitive correlation of electronically recorded waveforms with aphid probing activities. *Ann. Entomol. Soc. Am*. 60, 400–406.

McLean, D. L., and Weigt, W. A., Jr. (1968). An electronic measurement system to record aphid salivation and ingestion. *Ann. Entomol. Soc. Am.* 61, 181–185.

Mentink, P. J. M., Kimmins, F. M., Harrewijn, P., Dieleman, F. L., Tjallingii, W. F., Rheenen B. van, and Eenink, A. H. (1984). Electrical penetration graphs combined with stylet cutting in a study of host plant resistance to aphids. *Entomol. Exp. Appl.* 35, 210–213.

Mittler, T. E. (1988). Application of artificial feeding techniques for aphids. In: Minks, A. K., and Harrewijn, P. (eds.), *Aphids: Their Biology, Natural Enemies and Control,* vol. 2B. Elsevier, Amsterdam, pp. 145–170.

Montllor, C. B. (1991). The influence of plant chemistry on aphid feeding behavior. In: Bernays, E. A. (ed.), *Insect–Plant Interactions,* vol. 3. CRC Press, Boca Raton, FL: pp. 125–173.

Montllor, C. B., Campbell, B. C., and Mittler, T. E. (1983). Natural and induced differences in probing behavior of two biotypes of the greenbug, *Schizaphis graminum,* in relation to resistance in sorghum. *Entomol. Exp. Appl.* 34, 99–106.

Mullin, C. A. (1986). Adaptive divergence of chewing and sucking arthropods to allelochemicals. In: Brattsten, L. B., and Ahmad, S. (eds.), *Molecular Aspects of Insect–Plant Associations.* Plenum, New York, pp. 175–209.

Nault, L. R., and Styer, W. E. (1972). Effects of sinigrin on host selection by aphids. *Entomol. Exp. Appl.* 15, 423–437.

Nisbet, A. J., Woodford, J. A. T., Strang, R. H. C., and Connolly, J. D. (1993). Systemic antifeedant effects of azadirachtin on the peach–potato aphid *Myzus persicae.* *Entomol. Exp. Appl.* 68, 87–98.

Pollard, D. G. (1973). Plant penetration by feeding aphids (Hemiptera, Aphidoidea), a review. *Bull. Entomol. Res.* 62, 631–714.

Ponsen, M. B. (1972). The site of potato leafroll virus multiplication in its vector, *Myzus persicae,* an anatomical study. *Meded. Landbouw. hogesch.* 16, 1–147.

Prado, E., and Tjallingii, W. F. (1994). Aphid activities during sieve element punctures. *Entomol. Exp. Appl.* 72, 157–165.

Schnorbach, H. J., and Kunkel, H. (1990). Beginn und Verlauf der Nahrungsaufnahme bei Blattläusen (Aphididae, Hemiptera). *Angewandte Zool.* 77, 165–290.

Spiller, N. J. (1986). The probing and feeding behavior of aphids on resistant and susceptible wheat (*Triticum spp.*). Ph.D. thesis, University of London.

Spiller, N. J., Koenders, L., and Tjallingii, W. F. (1990). Xylem ingestion by aphids— a strategy for maintaining water balance. *Entomol. Exp. Appl.* 55, 101–104.

Srivastava, P. N., and Auclair, J. L. (1963). Characteristics and nature of proteases from the alimentary canal of the pea aphid, *Acyrthosiphon pisum* (Harr.) (Homoptera, Aphididae). *J. Insect Physiol.* 9, 469–474.

Tjallingii, W. F. (1978). Electronic recording of penetration behavior by aphids. *Entomol. Exp. Appl.* 24, 721–730.

Tjallingii, W. F. (1985a). Electrical nature of recorded signals during stylet penetration by aphids. *Entomol. Exp. Appl.* 38, 177–186.

Tjallingii, W. F. (1985b). Membrane potentials as an indication for plant cell penetration by aphids. *Entomol. Exp. Appl.* 38, 187–193.

Tjallingii, W. F. (1986). Wire effects on aphids during electrical recording of stylet penetration. *Entomol. Exp. Appl.* 40, 89–98.

Tjallingii, W. F. (1987). Stylet penetration activities by aphids. New correlations with electrical penetration graphs. In: Labeyrie, V., Fabres, G., and Lachaise, D. (eds.), *Insect–Plants*. Junk, Dordrecht, pp. 301–306.

Tjallingii, W. F. (1988). Electrical recording of stylet penetration activities. In: Minks, A. K., and Harrewijn, P. (eds.), *Aphids: Their Biology, Natural Enemies and Control*, vol. 2B. Elsevier, Amsterdam, pp. 95–108.

Tjallingii, W. F. (1990). Continuous recording of stylet penetration activities by aphids. In: Campbell, R. K., and Eikenbary, R. D. (eds.), *Aphid–Plant Genotype Interactions*. Elsevier, Amsterdam, pp. 89–99.

Tjallingii, W. F. (1995). Comparison of AC and DC systems for electrical recording of stylet penetration activitiets by Homopterans. In: Backus, E., and Walker, G. (eds.), *Homopteran Feeding Behavior. Recent Research Advances and Experimental Techniques. Misc. Pap. Entomol. Soc. Am.* (in press).

Tjallingii, W. F., and Hogen Esch, Th. (1993). Fine structure of aphid stylet route in plant tissues in correlation with EPG signals. *Physiol. Entomol.* 18, 317–328.

Tjallingii, W. F., and Mayoral, A. M. (1992). Criteria for host-plant acceptance by aphids. In: Menken, S. B. J., and Visser, J. H., and Harrewijn, P. (eds.), *Proceedings of the 8th International Symposium on Insect–Plant Relationships*. Kluwer Academic Publications, Dordrecht, pp. 280–282.

Wensler, R. J., and Filshie, B. K. (1969). Gustatory sense organs in the food canal of aphids. *J. Morphol.* 129, 473–492.

Whittaker, R. H., and Feeny, P. P. (1971). Allelochemicals, chemical interactions between species. *Science* 171, 757–770.

Wink, M., and Witte, L. (1984). Turnover and transport of quinolizidine alkaloids. Diurnal fluctuations of lupanine in phloem sap, leaves and fruits of *Lupinus albus* L. *Planta* 161, 519–524.

Winter, H., Lohaus, G., and Heldt, H. W. (1992). Phloem transport of amino acids in relation to their cytosolic levels in barley leaves. *Plant Physiol.* 99, 996–1004.

8

Regulation of a Carbohydrate Meal in the Adult Diptera, Lepidoptera, and Hymenoptera

J. G. Stoffolano, Jr.

8.1. Introduction

Few areas of study are more important than the physiological mechanisms regulating feeding behavior. It is their feeding behavior that makes insects pests and vectors of important disease-causing agents of plants, humans, and animals. Unfortunately, their importance as pests and vectors has dramatically influenced the insects studied and the problems addressed. A new look at old problems, or even existing ones, using different insects as potentially new models is always refreshing. When one examines the reviews on the regulation of meal size for carbohydrate feeding by adult insects, it is obvious that two of the major insect orders (i.e., Lepidoptera and Hymenoptera) have not been well studied. The early studies on tarsal chemoreceptors in butterflies (Minnich, 1922) and the honeybee, *Apis mellifera* (Minnich, 1932), were done about the same time as the initial studies on flies (Minnich, 1929), yet the emphasis on future research concerning the physiological mechanisms governing carbohydrate feeding has clearly been on the Diptera and specifically on *Phormia regina* and a few mosquito species (see Chapter 6).

The only adult insect system that is well understood with respect to carbohydrate feeding is the blowfly, *Phormia regina* (Dethier, 1976). Most studies on other species have borrowed the findings directly from research on the blowfly and have used these to explain how meal size is regulated in their insect of choice. Also, with the major focus on grasshoppers and flies (Bernays, 1985), research on other important groups has been neglected. Because of this, distortions of our ideas, hypotheses, and so on, often result because researchers extrapolate from what we know about well-known systems to other groups of insects where information is lacking. An example is that many investigators assume that all Diptera and Lepidoptera must receive tarsal input prior to proboscis extension.

This idea is based on research with *Phormia regina* that shows that proboscis extension is initiated when tarsal contact is made with a stimulating substance. To see how one can be easily misled, one only has to consider the hawk moths (Sphingidae), clearwing moths (Sessiidae), and bee flies (Bombyliidae). Not all members of these families hover in front of the flower, drinking nectar without landing, but those that do this will extend the proboscis without making tarsal contact. Little, if any, information is available on the stimuli causing proboscis extension in these insects. Whether these insects have tarsal chemosensilla and exhibit a tarsal–proboscis response needs to be determined. Potential stimuli eliciting the proboscis response in insects that hover while feeding could be flower color, shape, odor, or any combination of these.

Because several major reviews concerning regulation of meal size already exist (Barton Browne, 1975; Bernays, 1985; Bernays and Simpson, 1982; Gelperin, 1971a; Hsiao, 1985; Langley, 1976), this review will take a somewhat different approach. Because the Diptera, Lepidoptera, and Hymenoptera represent the major feeders on fairly "pure" liquid carbohydrates, we will review the literature on these groups to see what is known; and, based on this, our focus will be on what gaps exist in our knowledge and how one can use *P. regina* as an exemplar research model to design experiments that will help fill in some of the existing gaps. This review will also attempt to place what we already know about the physiological mechanisms regulating carbohydrate feeding, when appropriate, in the context of current ideas about foraging behavior and nutritional ecology.

8.1.1. Insect Diversity and Numbers Linked to Carbohydrates in the Environment

Insects, comprising three-quarters of all known animal species, owe their success to various behavioral, morphological, and physiological adaptations. The most commonly known adaptations are the ability to fly, to fold their wings over the back, holometaboly, diapause, and presence of an ovipositor (Zeh et al., 1989). Evolution of wings was paramount for insect diversity; however, one limitation to aerial success was, and still is, the availability of a somewhat predictable fuel supply. The major source of fuel for most insect flight is exogenous carbohydrates, and floral nectar is not the only carbohydrate source available. It should be noted that some insects use lipids and not carbohydrates as their major fuel for flight (Goldsworthy and Wheeler, 1989).

Few will argue that some of the success of the Diptera, Lepidoptera, and Hymenoptera is linked to the adults having access to exogenous carbohydrates, plus the feeding behaviors and/or mouthpart structures they evolved to exploit this important resource. Finally, the size of insect populations is greatly influenced by carbohydrate availability.

8.1.2. Carbohydrate Resources in Nature

PLANT SAP

Flies, as well as bees, have been seen feeding on plant saps (Downes and Dahlem, 1987; Williams, 1990). Williams (1990) suggests that plant sap may provide an alternative source of carbohydrates during the winter, especially in the temperate regions, while Wellenstein (1952) reported that the ant, *Formica rufa*, feeds on plant sap when other foods are scarce. The fact that some moths and butterflies are attracted to and feed on plant saps is well known to collectors who employ the technique of "sugaring" to collect certain lepidopterans.

NECTAR-LIKE FLUIDS FROM FUNGI

Prior to the presence of angiosperms, insects obtained carbohydrates secreted from the ostioles of the pycnidium of various plant rusts (i.e., Uredinales). These rusts not only produce a nectar-like reward but also lure insects by various odors (Buller, 1950). Numerous citations concerning the "nectar" from rusts, its importance to the rust, and the involvement of insects in the transfer of the two types of rust pycnidia are provided by Buller (1950). Rathay (1883) reported that both ants and flies eagerly seek this rust "nectar." Ingold (1971) stated that the ergot, *Claviceps purpurea*, produces " . . . a copious, stinking 'honey-dew' in which the concentration of sugar may be over 2 molar." Few researchers have considered these fungi as sources of carbohydrates for insects in nature.

HONEYDEW

Probably the most common, nonfloral source of carbohydrates in nature and possibly the first major source of exogenous fuel for insects prior to the presence of flowering plants is that produced by members of the Homoptera. Honeydew is present everywhere the appropriate homopteran producers are found. The hypothesis has been made that early Diptera relied on honeydew as their major source of carbohydrates and that this source of energy greatly influenced dipteran evolution (Downes and Dahlem, 1987). The trophobiotic relationship between ants and homopterans is well known and is reviewed by Hölldobler and Wilson (1990). The contribution of honeydew to the diet of some ant species has been projected at over 50% (Skinner, 1980). The dorsal nectary organs of some lycaenid caterpillars produce a secretion that contains sugars and is often called "honeydew." This secretion is not the same as homopteran honeydew because it is synthesized and produced in special glands of the larvae. It is, however, a choice carbohydrate source for some ants (Huxley and Cutler, 1991).

FLORAL NECTAR

The best-known natural source of carbohydrates is floral nectar. This reward, in exchange for a pollination favor, has been well studied and serves an important energy source for Diptera, Lepidoptera, and Hymenoptera (Bentley and Elias, 1983; Faegri and van der Pijl, 1971; Kevan and Baker, 1983; Real, 1983).

EXTRAFLORAL NECTAR

Extrafloral nectar often rewards the ant bodyguards that protect the plant from herbivores and also serves as a source of carbohydrates for many insects (Bentley, 1977; Koptur, 1992). In addition, extrafloral nectar provides insects in dry environments with water (Koptur, 1992). Extrafloral nectaries are not limited to angiosperms. Lawton and Heads (1984) reported that the bracken fern (*Pteridium aquilinum*) produces a nectar containing glucose, fructose, and sucrose. Both the concentration and types of carbohydrates vary with the type of plant (Koptur, 1992), and during the fruiting season they are actively fed upon by adult dipterans, lepidopterans, and hymenopterans. Koptur (1992) provides the most comprehensive review to date on extrafloral nectar.

PUNCTURED FRUITS, PLANT LEACHATES, AND STIGMATIC EXUDATES

The importance of plant leachates, as a source of carbohydrates, for adults of the three orders discussed here has not been adequately evaluated (Tukey, 1971). Many orchid species deceive the pollinator by failing to provide a reward. Through careful analysis, however, some type of reward or "pseudoreward" is usually found. The orchid, *Dactylorhiza fuschsii* (Druce) Soo, provides the pollinators (i.e., *Bombus* and *Apis*) with a stigmatic exudate that contains glucose and amino acids (Dafni and Woodell, 1986).

The most common carbohydrate in nature is sucrose. Fructose and glucose are next in importance, with maltose, melibiose, melezitose, and raffinose being less frequently encountered (Baker and Baker, 1983a,b).

Later on in the chapter, the importance of these diverse carbohydrate resources to the adults of Diptera, Lepidoptera, and Hymenoptera will be discussed with respect to how individuals in these orders store and handle these important sources of energy.

8.1.3. Importance of Carbohydrates in the Life of Adult Insects

There is no doubt that the abundance of most adult insects in the three major orders discussed here is greatly dependent on an adequate supply of exogenous carbohydrates. Carbohydrates are important for egg development and fecundity, for longevity and flight, as food supplements for both larvae and adults of Hymenoptera, for the production of reserves prior to diapause, and as a source of energy for everyday metabolism (Chippendale, 1978).

8.1.4. Evolution of the Crop and Its Significance

Through specialization of the different regions of the digestive tract, insects increased both the quantity and type of food taken. One of these regions, called the *crop,* developed from the insect foregut. The evolution of the expandable crop, a specific region of the linearly arranged digestive tube, permitted herbivorous insects to consume larger quantities of food. It wasn't until the evolution of the adult diverticulated crop and a mouthpart capable of imbibing liquids that the Diptera and Lepidoptera were able to maximally use exogenous carbohydrates. King (1991) briefly discusses the evolution of the crop in the Diptera and notes that it may be a synapomorphic trait in this group. When studied, the expandable crop being located in a distensible abdomen has been shown to be involved in the regulation of meal size and termination of feeding. In the Hymenoptera it was the evolution of an expandable crop (linearly arranged with other structures) and the relocation of its position from the normal foregut location in the head and thorax of most insects into the abdomen that permitted this group to maximize its use of liquid carbohydrates. Once discovered, the ability to consume as much as possible of an unpredictable carbohydrate resource was dependent on a storage organ. Another advantage of the crop is that it is impermeable to water loss, thus preventing water from entering the hemolymph and upsetting its osmotic pressure. Without a doubt, the evolution of the crop as a storage organ, its location in the insect, and its various modifications were all instrumental in accounting for part of the success and diversification of these three major insect orders.

8.1.5. Nutritional Ecology and Foraging Strategies (Food Handling)

The adaptive significance of feeding on honeydew versus either extrafloral or floral nectar is that the forager obtains a more concentrated carbohydrate resource (maximizing its intake) and, at the same time, is probably less exposed to predation. By feeding on nectar, which is less concentrated than honeydew, the forager reduces the chances of maximizing its diet and also increases the potential of predation. What the insect does gain, however, is time. Feeding on a liquid source of carbohydrates (e.g., extrafloral or floral nectar) takes less time than feeding on a dry honeydew resource or crystal of extrafloral nectar found on various cacti. If the forager leaves the resource (e.g., floral nectar) once feeding has terminated and goes to a less conspicuous habitat to process the meal (i.e., either digesting or concentrating it by bubbling; see Section 8.4.1), it may be further ahead because it has avoided exposure to predators and, depending on the concentration, possibly could have obtained more carbohydrate. Kingsolver and Daniel (Chapter 2) emphasize that in order to understand the energetics of foraging, it is important to know the mechanical determinants involved in nectar feeding. The structure of the mouthparts and the mechanical aspects of feeding for various groups of insects are discussed by Smith (1985). Schmid-Hempel (1986) provides an interesting insight into various foraging models when he says "These general models usually do not make reference to the actual decision-making mechanisms used by the animal." It is hoped that this review will provide the readers with ways that some of these mechanisms can be investigated. Unfortunately, the insect (i.e., *Phormia regina*) for which we know the most about regulation of carbohydrate feeding is also the insect that has been least studied in nature.

8.2. *Phormia regina:* The Model System for Carbohydrate Feeders

Major reviews concerning the physiological mechanisms regulating carbohydrate meals in adult *P. regina* are available (Barton Browne, 1975; Bernays, 1985; Bernays and Simpson, 1982; Dethier, 1976). In this section, the plan is to provide a general overview of what is known about regulation of a carbohydrate meal (i.e., mainly liquid) and to present any pertinent information after the 1985 review of Bernays. This information will be presented so that it will set the stage for discussing information about how carbohydrate feeding is regulated in insects other than *P. regina*.

As long as the concentration of the stimulating sugar is above that determined for the behavioral tarsal acceptance threshold, proboscis extension in *P. regina* is triggered. A hierarchy of decision-making occurs once the tarsal chemosensilla contact the solution. Prior to ingesting the carbohydrate source, decisions are made to drink or not to drink by evaluating receptor input from different locations on the mouthparts or inside the cibarium. In descending order of importance, these include the chemosensilla located on or in the following areas: tarsi, labellar lobes, interpseudotracheal regions on the labellum, and, finally, the cibarium. Based on various internal factors, a decision is finally reached to drink. As the meal enters the fly, a decision is made to direct it either to the crop or to the midgut. Most carbohydrates in nature should be concentrated enough to be directed to the crop.

Abdominal distention, caused by crop enlargement, acts on stretch receptors of the abdominal nerve-net which sends negative feedback to the central nervous system (CNS), where the behavioral tarsal acceptance threshold is somehow elevated. This effect is evaluated, and feeding may continue or terminate. Once a meal has terminated, stretch receptors located in the foregut respond to slugs of fluids moving from the crop into the foregut. The rate of crop emptying— and thus slug release—is influenced by internal factors that regulate the rate of crop emptying (see Section 8.3.5). Information from the stretch receptors located in the foregut region also provides negative feedback to the CNS via the recurrent nerve complex where the input also affects the acceptance threshold. Bowdan and Dethier (1986) suggest that both the recurrent nerve and ventral nerve feedback systems act as temporal fractionators. The system influenced by crop distention is immediate, thus acting to terminate the meal, whereas the one influenced by foregut distention is delayed and cumulative and decays slowly. Research from the laboratory of Murdock raises new questions about the involvement of information from both the recurrent nerve and the median abdominal nerve. Edgecomb et al. (1987) stated that the median abdominal nerve had " . . . no direct role in threshold regulation"; and this was restated in another paper (Sudlow et al., 1987). In the later paper, however, they noted that "Additional, still unidentified, factor(s) may regulate both labellar and tarsal thresholds." More will be said about how negative inputs affect the CNS and what some of these unidentified factors may be in Section 8.7. Internal factors influencing the rate of crop emptying include activity level of the insect, rate of carbohydrate utilization by the cells and tissues, and the carbohydrate concentration of the hemolympyh (see Fig. 8.1). It is extremely important that the early diagrams showing the various factors regulating feeding in *P. regina* (Gelperin, 1971a; Stoffolano,

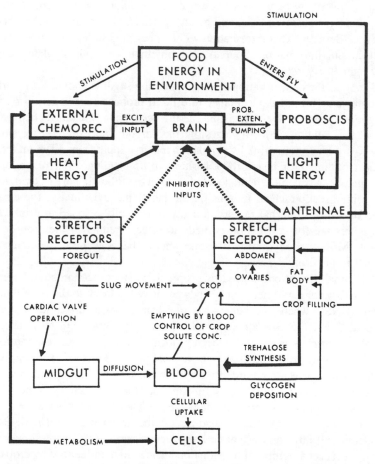

Figure 8.1. A model system based on the blowfly, *Phormia regina*, showing how the fly maintains metabolic homeostasis and how various factors influence feeding behavior. (After Stoffolano, 1974.)

1974) be updated similar to the one developed for mosquitoes by Davis and Friend (see Chapter 6).

8.3. General Regulatory Scheme

Once feeding has commenced, there can be an increase in the central excitatory state; and if phagostimulation (here from carbohydrates) is greater than phagodeterrency, feeding will continue (see Fig. 8.2 and Chapter 5). It is generally accepted that feeding continues only with continued phagostimulatory input. Feeding fails to commence if deterrency is greater than stimulation. If started, however, feeding may cease due to any single factor or a combination of factors listed by Bernays and Simpson (1982): adaptation of chemosensilla; decay of central excitatory state; volumetric factors, mainly by mechanosensillar detection; and nutrient, humoral, and other factors in the hemolymph sending negative feedback to the CNS. The presence of phagodeterrents in nectar (Koptur, 1992) and other natural carbohydrate sources is poorly understood. If these sources are to benefit the plant, it is difficult to imagine that the plant would invest in producing substances that would discourage the insects attracted to them. The general model depicting the various factors involved in the regulation of a meal (see Fig. 8.2) should be useful to investigators interested in modeling how a meal is regulated in other insects. Also, see the model presented by Davis and Friend in Chapter 6.

8.3.1. Hunger

Hunger is a manifestation of food deprivation and can have a major impact on both the amount and quality of food taken. In nature, the various factors influenc-

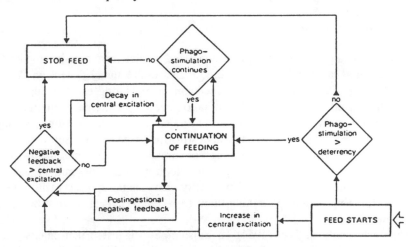

Figure 8.2. A general model of the processes involved and the factors affecting continuation to feed in an insect. (After Bernays and Simpson, 1982.)

ing hunger have not been well studied. What impact do factors such as continued and uninterrupted rain for several days, long periods of unexpected cold weather, an unexpected snow in either the spring or early fall, or extensive flooding have on hunger in insects? Surely, those insects that rely on honeydew will have to increase their foraging following several days of rain which washes off honeydew deposits.

The physiological bases underlying this behavioral state are discussed in previous reviews and focus on the Diptera. The factors responsible for carbohydrate hunger in the Lepidoptera and Hymenoptera have not been well studied. Using the mosquito, *Aedes aegypti*, Wensler (1972) showed that hungry adults became more responsive to an ethyl ether extract of honey as the period of starvation increased. Immediately after feeding, they lost their responsiveness to these odors. One of the main consequences of hunger may be increased searching behavior. This topic is too broad to be covered here, and readers should consult Bell (1991) for an extensive treatment of this topic.

8.3.2. Intake: Quantity and Quality

The size of a carbohydrate meal can be calculated by weighing the insect before and after the meal, measuring the crop volume, or determining the length of the meal. Readers are referred to Baker and Baker (1983a,b) and Kearns and Inouye (1993) for more information on determining the quantity and quality of a carbohydrate meal. Unlike the blood feeders which sample or take a small amount of the meal into the cibarium for monitoring its quality (Langley, 1976), no experimental evidence is available on sampling the meal (i.e., tasting in the cibarium) prior to ingestion in carbohydrate feeders.

8.3.3. Meal Destination

Where the meal goes (i.e., midgut or crop) and its size have a major impact on both the initial meal and future feeding. Because abdominal distension due to crop filling appears to be the major mechanism regulating carbohydrate meal size in flies (Dethier, 1976), meal destination becomes extremely important. In Section 8.3.5, more will be said about the role of the crop in regulating feeding. Little, if any, information exists on meal destination or the mechanisms regulating it in insects that feed exclusively on carbohydrates. Most research has focused on the hematophagous flies, which put the blood meal into the midgut and the carbohydrate meal into the crop. This ability provides these insects with a very efficient foraging option, permitting them to take either meal when it becomes available.

The early study of Day (1954) on mosquitoes suggested that sugars were directed to the crop because they stimulated mouthpart chemoreceptors, which, in turn, caused the midgut sphincters to contract and the crop duct sphincters to open. In the horsefly, *Tabanus nigrovittatus* Macquart, the concentration of dissolved solids in the diet (most likely an osmotic response) also influences diet destination (Friend and Stoffolano, 1991). This latter case may represent a unique situation for blood-feeding insects that usually obtain blood from either a surface wound or pool of blood.

A behavioral model of sugar destination in the mosquito, *Culiseta inornata*, has been developed and shows that two separate receptors are involved in meal destination. Both receptors have separate threshold values, and one regulates crop valve opening whereas the other controls midgut valve closure (Friend et al., 1989). It has also been demonstrated that the carbohydrate, chemosensory-based control of meal destination in the mosquito (*Culiseta inornata*) is separate from that regulating ingestion (Schmidt and Friend, 1991). More information on the regulatory mechanisms controlling crop and midgut valve operation in different insects is needed. Davis and Friend (Chapter 6) provide the most recent model concerning meal destination in insects.

8.3.4. Termination of a Meal

The only known major factor involved in terminating a large carbohydrate meal (i.e., one that causes the abdomen to become distended) appears to be a volumetric one emanating from crop distention and not a mechanism involving feedback from energy reserves (Bernays, 1985). The majority of what we know about the mechanisms regulating a carbohydrate meal in insects was obtained using *Phormia regina*. Very little, if anything, is known about carbohydrate meal regulation in other Diptera, Lepidoptera, or Hymenoptera.

As a fly feeds, the crop becomes distended and increases the size of the abdomen. This increase places pressure on an abdominal nerve-net that is stretched as the crop increases in size (Fig. 8.3). The stretch receptors located at different junctions of the nerve-net provide negative feedback via the ventral nerve cord to the CNS. This negative feedback somehow increases the behavioral acceptance threshold, thus terminating feeding. Information concerning how this feedback influences the acceptance threshold will be discussed in Section 8.7 (also see Chapter 5). An additional source of inhibitory feedback in *P. regina* is from the recurrent nerve complex, which monitors stretch of the foregut. Also, Edgecomb et al. (1989) showed that the water content of the fly can significantly affect the behavioral tarsal acceptance threshold.

8.3.5. Control of Crop and Midgut Emptying

The factors regulating short-term feeding or the taking of a single meal have been extensively studied and reviewed (Barton Browne, 1975; Bernays, 1985; Bernays and Simpson, 1982; Dethier, 1976; Edgecomb et al., 1987; and Sudlow et al., 1987). The area of research that needs to be further investigated is that

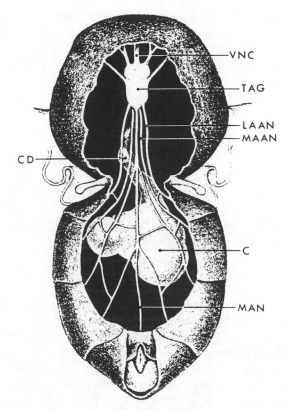

Figure 8.3. Diagram showing the abdominal nervous system of *Phormia* that is involved in providing negative feedback to the CNS about food in the digestive system. Note that the various nerves form a "nerve net" that is ventral to the crop and crop duct. VNC, ventral nerve cord; MAN, median abdominal nerve; C, crop; CD, crop duct; TAG, thoracico-abdominal ganglion; LAAN, lateral accessory abdominal nerve; MAAN, medial accessory abdominal nerve. (After Gelperin, 1971b.)

involved in understanding those factors regulating long-term feeding. Studies to date have used nerve transection experiments and either feeding or injection of various solutes to modify the osmotic pressure of either the diet or hemolymph. What remains to be done is to use information from the vertebrates (Blundell, 1984; Dourish et al., 1989) and other invertebrates (Lent, 1985) to investigate how various neurohormones or neuromodulators impact on crop and midgut emptying, thus influencing the insect's decision to feed or not to feed. This approach is currently being conducted in the laboratory of Ian Orchard on a blood-feeding bug, *Rhodnius prolixus* (Cook and Orchard, 1990; Orchard, 1989). Not until studies similar to the ones mentioned here are done will we understand the underlying basis of "hunger" in insects.

8.4. Carbohydrate Feeding in the Diptera

The Diptera are well adapted for feeding on carbohydrates. Most have chemoreceptors located on their tarsi and are able to perceive carbohydrates whether they are in liquid form (e.g., nectar, plant sap, etc.) or in a dried form (e.g., honeydew). Once located, and if the source is dried honeydew, the flies enter the sugar-feeding mode and use the labella at the tip of the proboscis to obtain the dried sugar. This is done by appressing the labellar lobes onto the carbohydrate source and secreting saliva. When dissolved, the sugar solution is imbibed. Downes (1974) reports that adult midges wander over the surfaces of leaves, "questing" and "scanning" their surface with chemosensilla located on the foretarsi and palps. Both sexes of the tabanid, *Tabanus nigrovittatus*, use their foretarsi to scan leaf surfaces in search of homopteran honeydew (Schutz and Gaugler, 1989; Stoffolano et al., 1990). This scanning behavior, used to locate an exogenous food source, probably evolved early in the evolution of Diptera to locate honeydew. In fact, honeydew was probably the major exogenous source of carbohydrates prior to the appearance of angiosperms that provided either floral and/or extrafloral nectar (Downes and Dahlem, 1987). The ability of flies or other insects to detect dried honeydew results from either the viscous exudate from the pore of the chemosensilla dissolving the sugar or from the hygroscopic nature of the dried carbohydrate source. Thus, dried honeydew will absorb atmospheric water that aids in dissolving the sugar, thus making detection by chemosensilla possible.

As previously stated (see Section 8.1), some flies such as the bee flies do not land to feed. The stimuli causing proboscis extension and the stimuli and receptors involved in turning on ingestion have not been studied in these flies.

8.4.1. Bubbling Behavior

The insect crop is not a major site for water storage. Water, taken in by drinking, is directed to the midgut, and this is probably true for all insects having a diverticulated crop. The main function of the diverticulated crop is to store carbohydrates or to store proteinaceous fluids once the midgut is full. Water, however, enters the crop with either a carbohydrate or protein meal. Energetically, it is not flight-efficient to carry water loads in the crop, but it makes most sense to get rid of it. Some insects such as the hematophagous feeders are able to rapidly excrete water from the blood meal through the anus. Recent studies suggest that the behavior of various dipterans that regurgitate (i.e., form a bubble and then imbibe it) are concentrating the fluid contents of the crop by getting rid of water via evaporation. This behavior is referred to as *bubbling*.

Hewitt (1912) noted that houseflies do not always get rid of the regurgitated fluid droplet at the tip of the labellum but retain it and repeatedly regurgitate and absorb the same droplet (Fig. 8.4a). One fly he observed devoted 1.5 min

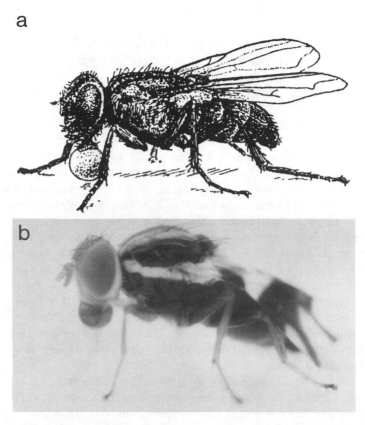

Figure 8.4. Bubbling behavior. (a) Drawing of the housefly, *Musca domestica* L., bubbling. (After Hewitt, 1912). (b) Photograph of apple maggot adult, *Rhagoletis pomonella*, bubbling. (Courtesy of Jorge Hendrichs.)

to this bubbling process. He did not, however, ascribe a function to this behavior. Drew and Lloyd (1987) reported that when fruit flies of the genus *Dacus* filled their crops, they regurgitated several droplets but, rather than retaining droplets on the labellum, they deposited them in a long droplet line on the fruit surface. After a brief interval, they followed the droplet line imbibing all the droplets. Because bacteria are found in this regurgitated material and bacteria have been found to be an important food source for these flies, Drew and Lloyd (1987) stated that this behavior was a way of reinoculating the fruit with bacteria. Reinoculation of the fruit may result from this behavior, but its main function is probably to remove water from the ingested fluids. The study by Hendrichs et al. (1992) concludes that bubbling behavior in *Rhagoletis pomonella* (Fig. 8.4b) is a way to concentrate imbibed fluids by eliminating excess water from the meal. Similar studies need to be conducted using carbohydrate feeders.

Drew and Lloyd (1987) reported that the mechanisms behind the "regurgitation/ reingestion" behavior of *Dacus* had not been investigated. Their study suggests that when the crop is not distended, this behavior does not occur. Thus, crop distention acting on stretch receptors in the abdomen may not only be important for cessation of the meal (Gelperin, 1966) but may elicit the bubbling behavior once feeding has stopped. Not only is it important to know the involvement of the crop in this behavior, but the influence of various sugars and their concentration on this process should be examined. One would anticipate that the amount of time spent bubbling (i.e., food handling) would be shorter for flies with a full crop of concentrated sugar than for those with a less concentrated solution. The dynamics between crop size, crop contents (i.e., type of food and concentration), and the internal state of the fly with respect to bubbling behavior needs to be worked out. The findings of Dethier and Gelperin (1967) on control of crop emptying in the blowfly should serve as a model for studies designed to examine the underlying causes of bubbling.

8.4.2. Hematophagous Insects

Downes (1974) stated that hematophagous dipterans have a dual pattern of feeding (i.e., sugar-feeding mode and blood-feeding mode), which is a basic trait of all dipterans and thus represents a plesiotypic feature of the group. Friend (*personal communication*) considers drinking water as part of the sugar-feeding mode because dilute sugars are treated as water by most dipterans. O'Meara (1987) discusses the role of carbohydrates and carbohydrate feeding in hematophagous dipterans.

The most comprehensive study on the effect of chemical phagostimulants— whether carbohydrates for nectar feeding or purinergic chemicals for blood feeding—on ingestion and diet destination is that of Schmidt and Friend (1991). Gwadz (1969) provided the first detailed study examining the internal regulatory mechanisms controlling food intake in hematophagous dipterans. His study, however, dealt only with blood intake and did not examine how carbohydrate intake is regulated.

Hovering behavior by the males in many dipteran species is extremely important for obtaining a mate. Because these flies use carbohydrates as their main source of fuel for flight, hovering may be energetically expensive, and thus carbohydrates are extremely important. For the tabanid, *Tabanus nigrovittatus*, males were observed feeding on aphid honeydew on the leaves of marsh elder immediately following hovering and mating activity (Schutz and Gaugler, 1989). More studies like the study of Schutz and Gaugler are needed that can correlate a specific biological function with feeding on carbohydrates in nature.

8.4.2.1. INITIATION OF FEEDING, INTAKE, AND MEAL DESTINATION

The various factors involved in the initiation and regulation of blood ingestion have been reviewed (Langley, 1976; also see Chapter 6), but no review is available for sugar feeding in hematophagous dipterans. In most hematophagous dipterans, feeding and the regulation of where the meal goes (i.e., crop or midgut) are controlled by a combination of physical, chemical, and nutritional factors (Chapman, 1985). Lee and Davies (1979) discuss diet destination in the stable fly, *Stomoxys calcitrans*, and present a table concerning the same for over 30 species of hematophagous dipterans. For the mosquito, *Culiseta inornata,* the chemical nature of the meal is the most significant factor influencing meal destination (Schmidt and Friend, 1991), whereas for the tabanid, *Tabanus nigrovittatus,* the total osmotic pressure and not the chemical nature of the meal is more important (Friend and Stoffolano, 1991).

8.4.2.2. MEAL SIZE

The majority of studies on meal size in hematophagous dipterans has focused on blood meals and not carbohydrate intake. Even though they did not directly measure the size of the carbohydrate meal, Schmidt and Friend (1991) showed that the more concentrated the diet, the greater the meal size. This is consistent with the work on *P. regina* (Dethier, 1976).

Most blood-feeding insects can divert the blood meal to the midgut and still take a sugar meal which goes to the crop (Day, 1954; Stoffolano, 1983; Trembley, 1952). Thus, hematophagous dipterans differ considerably from *P. regina* and several muscoid flies which also put liquid proteinaceous material into the crop. Stoffolano (1983) showed that when blood and sugar are taken by *Tabanus nigrovittatus,* regardless of the sequence of intake, the total amount ingested remains the same. It appears that two separate mechanisms exist for regulating sugar and blood intake in the tabanid; otherwise, if only one mechanism existed (e.g., a volumetric one), females should feed to repletion on the first diet presented. In most hematophagous dipterans, a full carbohydrate meal should have its effect on elevating the peripheral tarsal receptor threshold while a full blood meal should affect the labellar or cibarial thresholds and not the tarsal threshold. This is obvious when one considers that, while blood feeding, the tarsi of most hematophagous dipterans usually fail to contact the meal. Only for some pool feeders may the tarsi contact the blood.

The only known study on long-term regulation of a carbohydrate meal in a blood feeding insect is on *Aedes taeniorhynchus* (Nayar and Sauerman, 1974). In order to explain the initially large consumption of sucrose, which then was maintained at a lower but fairly constant level, Nayar and Sauerman focused on the availability of energy reserves (i.e., here glycogen and triglycerides). They proposed that there is a feedback to unidentified centers which somehow measure the changing state of available metabolic reserves. How carbohydrate reserves might be accomplishing this needs to be investigated.

8.4.2.3. TERMINATION OF A MEAL

The first report concerning the regulatory mechanisms terminating a meal in a mosquito was provided by Gwadz (1969). His study, however, only included blood feeding. What he showed was that transection of the ventral nerve cord led to hyperphagia on blood, which is very similar to that reported by Dethier and Gelperin (1967) for *P. regina* fed a sugar diet. Later, Owen and McClain (1981), using a different approach to destroy the ventral nerve cord, produced hyperphagic mosquitoes fed carbohydrates. Neither of these studies on mosquitoes identified the nerves or the stretch receptors involved in preventing hyperphagia. In fact, it is not even known whether mosquitoes have a nerve net similar to that for *Phormia* (see Fig. 8.3). How information on abdominal distention influences peripheral receptor thresholds in hematophagous insects is not known.

The early work of Manjra (1971) demonstrated that the behavioral labellar threshold to sugars decreased with increased starvation, was elevated following a sucrose meal but then decreased with time following the meal. He also showed that the behavioral labellar threshold following a meal was about three times higher for mosquitoes fed only a sucrose meal than that for females fed a mixed meal of blood plus sucrose, which went to the midgut. Why a blood meal should have its inhibitory effect at the tarsal level makes no sense because, while feeding on blood, the tarsi do not even contact blood.

8.4.2.4. CONTROL OF CROP AND MIDGUT EMPTYING

Some evidence suggests that the female tsetse fly, *Glossina austeni*, takes a blood meal that maintains the abdomen at nearly a constant volume during the reproductive cycles (Tobe and Davey, 1972). Little, if any, research has focused on understanding the mechanisms of crop and midgut emptying for hematophagous dipterans fed a sugar meal. Numerous unanswered questions still remain concerning the exact physiological mechanisms regulating both sugar and blood meals in hematophagous insects.

8.5. Carbohydrate Feeding in the Lepidoptera

In many adult Lepidoptera, carbohydrates serve as important energy sources for flying from oviposition sites to different locations where they forage on nectar (Wylie, 1982), for dispersal (Boggs, 1987; Ehrlich, 1984), and for migration (Brown and Chippendale, 1974; Johnson, 1969). There is evidence that adult resources, particularly nectar, may limit population size (Gilbert and Singer, 1975). An unpublished energy budget for *Pieris rapae* (see Gilbert and Singer, 1975) showed that adult feeding accounts for 56% of the adult's budget. Gilbert (1984) states that " . . . it seems unwise to ignore the role of adult resources in shaping community patterns." Yet, despite this obvious omission in our knowledge concerning adult lepidopteran biology, no definitive paper exists on the physiological mechanisms regulating carbohydrate feeding in this group. Boggs (1987) provides an excellent review of the ecological aspects of nectar feeding.

8.5.1. Sources of Carbohydrates

Where did the first adult lepidopterans obtain their exogenous carbohydrates? Common (1975) reported that " . . . it is likely, therefore, that a simple type of proboscis had already evolved by the time angiosperms appeared, although the major radiation of ditrysian forms may well have paralleled that of the angiosperm." The proboscis of the thecline butterfly, *Hypaurotis crysalus,* is shorter than that of flower-feeding Theclini. Kristensen (*personal communication*) stated that the lepidopteran proboscis developed early between the evolution of the Heterobathmiidae and Eriocraniidae and that the coilable proboscis did not evolve until later. Nectar feeding was documented in the Upper Cretaceous, prior to the presence of the Heteroneura, while the diverticulated crop evolved only in subordinate lineages within the Heteroneura.

Sometimes the food sources for the larval and the adult lepidopterans occur in the same habitat. Ehrlich (1984) calls this "co-location" of resources. Scott (1974) reports that the larvae of *Hypaurotis crysalus* feed on *Quercus gambellii;* and, because nectar is lacking from the tree at the time of adult presence, the adults use sap oozing from oak twigs as their source of liquid, which presumably contains carbohydrates.

Johnson and Stafford (1985) stated that "Feeding on homopterous honeydew is a logical extension of the typical nectar-feeding habit of adult Lepidoptera." Because the Lepidoptera appeared prior to the flowering plants (Ronald Hodges, *personal communication*), the statement by Johnson and Stafford should be reevaluated. Honeydew, just as in the Diptera (Downes and Dahlem, 1987), may have played an important role in lepidopteran evolution and served as a natural extension to feeding on floral nectar when it became available. Johnson and Stafford (1985) raise the question and provide some explanations why honeydew is not a major carbohydrate source for more Lepidoptera. They noted that because most butterflies and some moths use vision to find flowers, aphids probably do not provide an adequate visual stimulus for feeding. Not addressed in their paper is whether adult lepidopterans, like the dipterans, can detect dried honeydew and, if detected, whether they are able to process it and imbibe it. In contrast to the idea of Johnson and Stafford, both the general belief amongst amateurs and personal correspondence with Kristensen support the idea that lepidopterans relied on honeydew prior to the presence of floral nectar. More experimental evidence is needed to document whether honeydew is a major carbohydrate food source for adult lepidopterans. *Plusia gamma* is able to utilize concentrated nectars by diluting them with saliva (Procter and Yeo, 1972).

Whether other butterflies and moths use copious amounts of saliva when feeding on nectar is not known. Because both dipterans and lepidopterans use saliva to dissolve and/or dilute dried or concentrated carbohydrate sources, more information should be obtained about the regulatory mechanisms controlling salivation. This is a topic that needs more research and should provide important information about food acquisition and food handling.

In addition to feeding on floral nectar, some moths have been shown to respond to the various sugars present in extrafloral nectaries (Downes, 1968). Clark and Lukefahr (1956) determined that sucrose, glucose, fructose, ribose, rhamnose, and raffinose were the sugars present in Empire cotton (*Gossypium hirsutum*). Even though adult tobacco budworms, a pest of cotton, did not respond behaviorally by proboscis extension to ribose, rhamnose, and raffinose (Ramaswamy, 1987), another noctuid moth, *Spodoptera littoralis,* did (Salama et al., 1984). Some nectars have been shown to contain chemicals that might act as phagodeterrents (Koptur, 1992). It would be informative to determine whether the specific guild of pollinators frequenting plants containing nectar with phagodeterrents have a different tarsal and/or behavioral response to the phagodeterrent chemicals than those that don't frequent these flowers. Such a mechanism could be the underlying basis for the origin and/or maintenance of specificity between some insect–plant relationships.

The importance of amino acids for adult reproduction has been shown in a few butterfly species. *Heliconius* butterflies obtain their requirement of amino acids from pollen (Gilbert, 1972). The presence of amino acids in many types of nectar (Baker and Baker, 1975) suggests that many butterflies and moths obtain this important nutrient requirement from nectar. Both behavioral and electrophysiological studies have shown that moths, and presumably butterflies, are able to detect amino acids (Blaney and Simmonds, 1988). Blaney and Simmonds also reported that even though amino acids were less stimulatory than carbohydrates, the two acted synergistically. Whether adults can discriminate and selectively feed on sugars containing amino acids over sugars lacking them needs to be confirmed. Do butterflies and moths preferentially feed on plants having nectar with higher concentrations of amino acids?

8.5.2. Intake, Destination, and Termination of a Meal

In many flower-feeding adults, whether hovering in front of or landing on the petals of flowers, the tarsi seldom contact nectar yet proboscis extension occurs. What causes proboscis extension in these cases is not known. For those that hover, it may be floral color, odor, or a combination of the two, while for those landing, it may be the phytochemicals in petals that provide important tarsal contact or olfactory stimuli eliciting proboscis extension. Roessingh et al. (1991) demonstrated that the tarsi of adult *Papilio polyxenes* responded electrophysiologically to host-plant extracts. Some moths, especially the sphingid species (Weis,

1930) and possibly the clearwing moths (Sesiidae), which hover while taking nectar, are reported to lack sugar-sensitive chemosensilla on the tarsi. Whether this is true for some butterflies is not known. In many other lepidopterans, tarsal contact with sugars initiates proboscis extension (Anderson, 1932; Kusano and Adachi, 1969a,b; Ramaswamy, 1987). Behavioral and electrophysiological evidence also exists for the presence of sugar-sensitive chemosensilla on their tarsi (Blaney and Simmonds, 1990).

The presence of chemosensilla on the proboscis of lepidopterous adults has mainly been studied in the moths (Blaney and Simmonds, 1988; Städler and Seabrook, 1975; Städler et al., 1974). These studies, both behavioral and electrophysiological, demonstrate that the sensilla of the proboscis respond to sugars as well as to some amino acids. It is obvious from the scanning electron micrograph shown in Fig. 1a of Blaney and Simmonds (1988) that the sensilla are mainly located at the tip of the proboscis, and it is here that they make contact with floral and other forms of carbohydrates in nature. Very few studies have been published on pharyngeal or cibarial receptors in moths and butterflies. Burgess (1880), however, noted that on the convex area of the floor of the hypopharynx in *Danaus plexippus* (Fabr.) there are numerous " . . . little papillae, which possibly may be taste organs." Kristensen (1968) reported the presence of receptors (probably chemoreceptors) on the sclerotized floor of the sucking pump or cibarial plate of *Eriocrania*, noting their early presence in the micropterigids and probably throughout the Lepidoptera. The most detailed study to date (Eaton, 1979) shows contact chemoreceptors in the cibariopharyngeal pump of the adult cabbage looper moth, *Trichoplusia ni*. Whether the pharyngeal sucking pump is started by stimuli contacting these distally located sensilla on the proboscis or whether it is activated by pharyngeal sensilla has not been investigated. If carbohydrate solutions initially enter the proboscis via capillarity, cibarial or pharyngeal chemosensilla would be stimulated and could initiate sucking. More information may be obtained on chemoreceptors in adult lepidopterans by consulting Chapman (1982) and Chapter 4.

The biological significance of proboscis extension occurring when the antennae are stimulated by sugars (Frings and Frings, 1956; Ramaswamy, 1987) remains to be elucidated. Ramaswamy (1987) reported that Frings and Frings (1956) showed that "antennal stimulation increased" when the tarsi were removed from the two moth species they were studying. Thus he postulated that the antennal sugar-receptor system functions as a backup system to the tarsi. Previous work with tabanids (Stoffolano et al., 1990) suggests another possible function. As the adult lepidopteran or tabanid inserts its proboscis deeper into the flower, its head is moved closer to the flower, thus permitting antennal contact with the flower petals. Consequently, sensory input (e.g., from floral chemicals) from the antenna could provide an additional source of input to keep feeding going. The importance of antennal input to feeding could be examined using behavioral observations of intact adults, or using individuals with either the antenna removed or with their chemosensilla inactivated with deleterious chemicals such as HCl, formaldehyde, or NaOH. No function for the antennal contact chemosensilla, which are also responsive to sugars in bees (Marshall, 1935) and tabanids (Stoffolano et al., 1990), has been demonstrated. Only careful behavioral observations or the experimental procedures just described will help uncover the significance of these sensilla in feeding.

Because most hovering lepidopterans feed for such a short time on an individual flower, it seems unlikely that sensory adaptation is the cause of termination of the drink. More likely, however, it is the lack of continued sensory input needed to drive feeding behavior. Sensory input ceases when the nectar supply is depleted. Thus lack of input could terminate the meal, but, before we understand what terminates a meal in adult Lepidoptera, it is imperative that we understand the role of the crop.

8.5.3. Involvement of the Crop

According to Kristensen (*personal communication*), the diverticulated crop evolved only in subordinate lineages within the Heteroneura, whereas the appearance of a proboscis, development of a coilable proboscis, and nectar feeding had already evolved prior to the appearance of the diverticulated crop.

Currently, we have no information on crop filling or emptying in either butterflies or moths. Because both Knight (1962) and Gelperin (1966) have shown that the dipteran gut continues to function in complete isolation, similar studies should be conducted on adult Lepidoptera. Mortimer (1965) mentions a crop sphincter in the adult noctuid, *Plusia,* but additional studies are needed to show how crop filling and emptying are regulated in Lepidoptera. This reviewer was unable to find any references describing muscle sets or valves, as in *P. regina* (Thomson and Holling, 1977), that are involved in either crop filling or emptying in adult lepidopterous insects. Crop functioning (i.e., filling, volume, and emptying) can be studied without sacrificing the adult. By weighing the insect at different times or using the x-ray techniques of Gelperin (1966), researchers can obtain information that is important to providing a more complete understanding of adult butterfly and moth foraging.

Other than a single theoretical paper, one paper on instrumentation used to measure food intake, and two papers on nectar uptake rates, little information exists concerning this order of insects that is so well studied with respect to its ecology and evolutionary biology. Kingsolver and Daniel (1979; also see Chapter 2) developed a theoretical model of feeding energetics in butterflies, which included the rate of nectar extraction. A calibrated microcapillary tube feeder used for investigating rates of sugar removal by three families of butterflies was developed by May (1985a). May (1985b) and Pivnick and McNeil (1985) field-tested some of Kingsolver and Daniel's assumptions and found that nectar concentration does indeed affect rate of uptake. Boggs (1988) examined the effects of sex, size, age, and nectar concentration on the rates of nectar feeding in the grassland, montane, and temperate species, *Speyeria mormonia*. These studies should serve as starting points for future research designed to further test the model of Kingsolver and Daniel and to provide the desperately needed information on the mechanisms regulating meal size, the factors influencing the time intervals between meals, and the effect of internal factors, such as reproductive status, on feeding. How the crop fits into all of this remains to be determined.

An interesting study is that of Dafni et al. (1987) on nectar flow and the pollinators of two co-occurring plant species of *Capparis*. These authors showed that hawk moths preferred *C. ovata*, which had both the higher concentration and volume of nectar, whereas bees preferred *C. spinosa*, which had a lower concentration and volume of nectar. They attributed the higher nectar volume in *C. ovata* to the fact that hawk moths probably have a larger crop than do the bees. This moth and bee system offers an excellent opportunity for researchers to make comparative studies of the various physiological mechanisms underlying meal size and the factors that influence it.

8.5.4. Hunger

Minnich (1922) showed that the tarsal acceptance threshold for the butterfly, *Pyrameis atalanta*, to sugars decreased with increasing deprivation. The significance of the crop size, its contents, and its rate of emptying to the state of "hunger" is not known.

8.5.5. Reproductive, Migratory, and Diapause Effect on Intake

Unlike many Diptera, where a protein meal is required for egg development, there is no evidence for the Lepidoptera that food intake within an individual changes with reproductive status (Boggs, 1987). No one has investigated the importance of carbohydrate feeding with regard to spermatophore formation, accessory gland development, or courtship in male lepidopterans (Boggs, 1987).

No information is available as to whether migratory, as compared to nonmigratory, individuals consume more carbohydrates (Boggs, 1987). Information is needed on the feeding behavior of adult butterflies on carbohydrates prior to winter because several species overwinter in a reproductive diapause. Stoffolano (1974) and Greenberg and Stoffolano (1977) showed that, prior to winter, some insects consume greater quantities of carbohydrates. It would also be important to know whether these carbohydrates are important for producing overwintering reserves, and, if so, what effect the lack of carbohydrate resources in the fall has on overwintering survival.

8.5.6. Soil and Puddle Drinking

Even though drinking from various soils and puddles is not the same as carbohydrate feeding, this situation presents an important paradox to what we know about peripheral stimulation and ingestion in insects in general. Also, nothing is known about where the diet goes, amounts consumed, and what the internal regulatory mechanisms are. Basically, phagostimulation occurs when more impulses from the sugar-sensitive chemoreceptors reach the CNS, whereas phagodeterrency occurs when more impulses from the salt-sensitive chemoreceptors reach the CNS (see Chapter 4). Because this drinking behavior is usually associated with male butterflies and is believed to serve as the mechanism enabling the adult to replenish depleted sodium supplies (Boggs and Jackson, 1991), why do these insects feed on sodium when this solution, in sufficient concentration, normally acts as a deterrent? Interested readers are referred to the paper by Dethier (1977) which focuses on salt appetites in animals, including butterflies, where he makes two important statements pertinent to this topic: "What is surprising is that the ubiquity of salt-sensitive receptors appears to bear only a remote relation to the need for salt and that the taste for salt is not normally and generally involved in regulating its ingestion" and " . . . there is no evidence of a deficit-associated salt appetite."

8.6. Carbohydrate Feeding in the Hymenoptera

Usually the Hymenoptera, more so than the Diptera and Lepidoptera, are known for their nectar feeding on flowers (e.g., bees) and on homopteran honeydew (e.g., ants). In the social insects, unlike the dipterans and lepidopterans, carbohydrate feeding benefits the colony as well as the individual. Also, for most hymenopterans the immobile larvae remain in the nest and require being fed by workers that forage at a distance to bring back their food. The hymenopteran crop is well suited for transporting quantities of liquid carbohydrates from distant sources to the colony and was a major evolutionary adaptation for this insect group. Because carbohydrates are so significant to colony survival, it is surprising that so little information exists on the regulatory mechanisms underlying feeding.

General reviews concerning the nutritional ecology of bees, parasitoids, and ants can be found in Slansky and Rodriguez (1987). Additional information on carbohydrate feeding in the bees, bumblebees, and wasps may be obtained in Breed et al. (1982); Brian (1983), Heinrich (1979), Menzel and Mercer (1987), and Ross and Matthews (1991). Unfortunately, however, no major study has been published about the regulatory mechanisms of feeding in this group.

8.6.1. *Recognition of Food*

Even with their ability to forage for very diverse foods, little information is known concerning the specific cues or chemoreceptors used by ants for food recognition (Stradling, 1987). Using a technique somewhat similar to Dethier (1976) for determining behavioral acceptance thresholds for *P. regina,* Sudd and Sudd (1985) demonstrated that *Formica lugubris* is able to discriminate between sucrose solutions of different concentrations. Schmidt (1938) compared the acceptability of *Lasius niger*, *Myrmica rubra*, and *M. rubida* to various carbohydrates. Ants apparently prefer sucrose and glucose to fructose (Koptur, 1992), but the mechanism for discriminating between different sugars in ants, versus flies (Morita and Shiraishi, 1985), has not been investigated. Von Frisch (1930) determined the behavioral taste threshold of the honeybee, *Apis*, to various sugars; and Minnich (1932), using behavioral techniques, studied the contact chemoreceptors. Ultrastructural and electrophysiological studies have been performed on various mouthpart chemosensilla of honeybees (Whitehead, 1978; Whitehead and Larsen, 1976a,b). Contact chemosensilla sensitive to various sugars have been identified on the antenna of the honeybee (Marshall, 1935). Marfaing et al. (1989) reported on conditioning honeybees using various odorants and the proboscis extension response. As for tabanids and lepidopterans, no biological significance has been given to the proboscis response of honeybees when the chemosensilla on their antenna contact sugars (Minnich, 1932).

8.6.2. Intake, Destination, and Termination of a Meal

Other than descriptive studies, no known information exists on the mechanisms regulating crop filling or emptying in ants. Unlike other insects, some ants have a well-developed proventriculus that functions in both crop filling and emptying (Eisner, 1957). Wheeler (1910) reports that the "pre-stomachal ganglion" innervates the crop, but no studies since then have examined if the crop is under nervous control. Questions like the following should be addressed: What effect does sugar concentration have on crop filling? Does viscosity of the carbohydrate have any effect on crop emptying? Is crop volume involved in termination of a meal? What factors regulate the time interval between meals?

Bees will begin to imbibe liquid if an effective concentration of nectar is encountered. As far as this reviewer knows, the sensilla and the mechanism involved in drinking nectar in bees, bumblebees, and wasps have not been studied (see Chapter 2). Unlike what occurs in the dipterans and lepidopterans, which have a diverticulated crop, carbohydrates in the hymenopterans go directly to the honey-stomach (i.e., crop), which is part of the foregut but located in the abdomen. The physiological factors involved in terminating a meal have not been studied in bees, bumblebees, or wasps. Even though considerable effort has gone into examining the various factors affecting optimal foraging in these hymenopterans (Heinrich, 1979; Kacelnik et al., 1986; Varjú and Núñez, 1991; Waddington, 1983, 1985), little attention has been given to the crop. Schmid-Hempel et al. (1985) and Schmid-Hempel (1986) examined how honeybees made the decision when to leave the patch (i.e., flowers containing the nectar). In his 1986 paper, the author noted that most optimality models of foraging do not include information on how animals actually make decisions. In a series of clever experiments, Schmid-Hempel (1986) attached a rod to the thorax of the honeybee and is able to add small brass nuts to the rod, thus simulating an increased load due to crop filling. Using this approach, he was able to study the effect of weight on the bee's decision to depart from the patch or sources of nectar. One experiment that should have been tried, however, is to see what effect cutting the ventral nerve cord has on crop filling, termination of a meal, and the decision of bees to leave the source of nectar and return to the hive. Also, in order to understand the mechanism(s) involved in between meal regulation, studies need to be conducted to determine what factor(s) influences crop and gut emptying. A whole book (Menzel and Mercer, 1987) has been devoted to the neurobiology and behavior of the honeybee, but no definitive paper discusses the neurophysiological basis for carbohydrate feeding in an insect whose carbohydrate feeding is so important, not only for its own survival and flight but also for the survival of the hive.

8.6.3. Involvement of the Crop

In order to distinguish incoming foragers carrying liquids from those carrying solid foods, Tennant and Porter (1991) classified all foragers with " . . . a visibly enlarged gaster, which appeared striped because the abdomen was distended, exposing the intersegmental membranes" as carrying liquids. The amount of liquid in the crop was determined by weighing incoming ants with distended abdomens and outgoing ants with nondistended abdomens. They also analyzed the crop for the different sugars present. An important finding was that ants foraging exclusively on sucrose solutions contained mainly fructose and glucose in the crop. This suggests enzymatic breakdown of sugars in the crop, which may be possible in ants because invertase has been found in the crop of fire ants (Ricks and Vinson, 1972). The overall conclusion of Tennant and Porter's study was that, based on the high amount and type of liquids taken when compared to solid matter, fire ants are probably not scavengers but primary consumers. Dreisig (1988) developed a model of nectar or honeydew foraging in ants based on the assumption that when an ant finishes feeding on a resource, it has "probably a full crop." He defined ingestion rate as " . . . the amount of resource collected per time, including both waiting and handling times," and reported that the mean ingestion rate for the four species studied was 0.00086 mg/min and that the average crop load was 34% of the unloaded weight. As previously stated, parameters such as concentration and viscosity effects of the carbohydrate resource on rate of ingestion plus crop volume before and after the meal will influence handling time. Such variables must be considered in future models of carbohydrate foraging in ants.

Sudd and Sudd (1985) used a refractometer to determine the quality of the crop contents of ants descending mature pine trees and reported that because of the high concentrations found, further studies are needed to determine whether the honeydew was concentrated by evaporation or was due to other causes (e.g., increased solutes). Whether ants, like dipterans, can concentrate crop contents by bubbling is not known. Traniello (1989) placed emphasis only on forager load capacity and caste efficiency as it relates to noncarbohydrate loads; however, many studies have shown that the major food item taken by most ants is liquid carbohydrate (Tennant and Porter, 1991), thus reinforcing the significance of the crop. As previously suggested for the dipterans and lepidopterans, the crop has an extremely important role in ant biology which previously may have gone unrecognized.

"Bizarre adaptations have proliferated among ants, but none is more unusual than the extreme abdominal distention of certain workers, called repletes, in the honey ants" (Conway, 1977). Truly, this hyperphagic condition is an enigma when one considers existing research on the regulation of meal size in *Phormia regina*. Figure 8.1 shows that one of the major factors terminating a meal is abdominal distention resulting from crop enlargement. Evidence exists stating that repletes are selected from the largest workers (Hölldobler and Wilson, 1990), but how the regulating mechanism(s) normally controlling meal size is "shut off" to produce these "social kegs" remains a mystery. Before one understands this phenomenon, however, the mechanism(s) regulating a single meal in normal workers should be understood.

8.6.4. Hunger

The state of hunger of both workers and larvae have been implicated in influencing foraging and raiding in ants (Stradling, 1987). The physiological mechanisms underlying this behavioral state are not known for ants. More information is needed concerning the possible changes taking place in peripheral sensilla (e.g., sensitivity) associated with carbohydrate feeding to see if sensory modulation is detectable at this level. The same can be said of the bees, bumblebees, and wasps.

8.6.5. New Horizons

The ecology, evolution, and behavior of the hymenopterans has been adequately studied. The amount of information already known about their communication systems is tremendous. What remains to be studied, however, is the sensory aspects of their communication systems as it relates to both foraging and to colony maintenance. At the sensory level, how do workers differentiate between the input provided by larval secretions and that provided by various carbohydrates? If no difference is observed, does the behavioral context in which the input is taking place provide the decisive factor which accounts for the resulting behavioral output? At the same time, more information is needed on the sensory physiology of feeding behavior in a group that exhibits such a rich diversity and repertoire of feeding behavior.

Another aspect of feeding in social insects that needs further exploration is the role played by carbohydrate input in altering the existing physiological state of responsiveness of individual foragers in the colony. Studies exist for both ants (Beckers et al., 1990) and honeybees (Seeley and Levien, 1987) showing that foragers will actually switch foraging on a sucrose source to another that is more concentrated. In ants this behavior depends on when they first encounter the more concentrated diet, whereas in bees the more concentrated diet is always selected regardless of when it is encountered. This infers that the workers encountering the more concentrated solution somehow recognize it as a better source of carbohydrate, remember this while in transit back to the hive or nest, and are able to transmit this information with greater "responsiveness" to the other workers. All of this suggests that sugar solutions of different concentration somehow alter the immediate physiological state of the individual worker, that this "physiological" state does not decay rapidly, and that it affects the worker when it is transmitting information to other members of the colony. What events take place in the foragers of social insects following a meal is a paradox of what occurs in nonsocial insects (Dethier, *personal communication*). In nonsocial insects a period of quiescence usually follows a meal, whereas in social insects the reverse is true. In their mathematical models of cooperative transmission of information (i.e., quality of carbohydrate) versus direct transmission, Beckers et al. (1990) should also include parameters such as physiological state of the worker and crop load.

The factor(s) involved in terminating a carbohydrate meal in insects involved in pollination should be examined. Evolutionarily speaking, the advantage to the plant is to have the insect feed on the nectar source but not have the meal interfere with further foraging. Do pollinating insects stop foraging before they have a full crop? Honeybees probably stop foraging when the honey stomach is full but return to the hive where they unload the nectar. This behavior removes the negative factor(s) (i.e., presumably from the full honey stomach) inhibiting the bee from foraging, and thus the bee continues to forage. The possibility exists that the crop of most pollinators can hold considerably more nectar than is available in several flowers. Certainly, more information on this aspect of meal regulation is needed for various pollinating species.

8.7. Neuropharmacological and Endocrine Involvement in Feeding Regulation

The parabiotic studies of Green (1964a,b) revealed that a humoral factor was involved in changing the activity of the starved fly when it was parabiosed with a fed fly. This implies a humoral factor which still remains unidentified. It is interesting to note that Barton Browne and Evans (1960) concluded that the mechanism regulating locomotor activity following a sugar meal is different from that controlling the tarsal acceptance threshold following a sugar meal. In addition to possible humoral feeding effects on locomotor activity, Barton Browne (1975) hinted at humoral involvement on the tarsal acceptance threshold. In his review he analyzed previous research on *Phormia regina* and attempted to explain elevation of tarsal acceptance thresholds either as being due to immediate feeding effects, thus direct neural effects, or being influenced by humoral events. He concluded by stating that "the increase (tarsal threshold) between 25 and 45 min (following feeding) clearly cannot be explained in terms of the direct neural effects of events during feeding." Thus the question as to the physiological mechanisms underlying elevation of the tarsal threshold is still not completely resolved. Also, Bernays and Simpson (1982) stated that "Hormonal effects following feeding have not been studied in the blowfly. . . ."

The papers of Murdock and his colleagues should serve as an example of one of the directions research on regulation of feeding in insects should be going. In a series of papers (Brookhart et al., 1987; Long and Murdock, 1983; Long et al., 1986), this group provides the most comprehensive study to date on the neuropharmacological control of feeding in any insect. The importance of various biogenic amines (e.g., octopamine, dopamine, and serotonin) in feeding regulation is well established in the vertebrate literature (Blundell, 1984; Fernstrom and Wurtman, 1971; Wurtman and Wurtman, 1979); and because biogenic amines are also present in insects (Evans, 1980, 1985; Mercer, 1987; Orchard, 1982), there is no reason to assume they are not important in insect feeding. Briefly, Murdock and his group showed that octopamine injected into adult *P. regina* caused a significant increase in intake (i.e., hyperphagia) and a significant lowering of the median tarsal acceptance threshold (MTAT). This hyperphagia is similar to that reported by Dethier and Gelperin (1967) when both sources of negative feedback to the CNS are removed by nerve transection. They also showed that both dopamine- and serotonin-injected flies had a significant increase in the MTAT and that these flies took a significantly smaller meal than did the saline-injected flies. When injected with d-amphetamine, a depleter of biogenic amines, flies consumed significantly less sucrose and had highly elevated MTATs. Electrophysiological monitoring of the tarsal chemosensilla showed no difference between the response of d-amphetamine-injected flies and saline-injected controls, thus supporting their hypothesis of central regulation of the MTAT. In conclusion, these authors suggested that octopaminergic receptors in the nervous system positively modulate feeding by increasing responsiveness, which in this case was measured as a low MTAT and increased meal size. On the contrary, the aminergic, serotonergic, and dopaminergic systems are involved in negatively modulating feeding behavior (i.e., decreased responsiveness as measured by a high MTAT and a decreased meal size). The missing link in all the studies to date, however, is the endocrine involvement, which has been previously implicated in regulation of feeding behavior.

The least studied, yet probably the largest, endocrine tissue in the insect is the midgut. Recently our research has shown that a midgut peptide is the "missing link" providing the important communication between the quality of the meal and the endocrine centers of the brain that are involved in regulating various aspects of oogenesis in *P. regina* (Yin et al., 1994). A similar pathway, involving the gut endocrines, is suggested here for hormonal involvement in feeding regulation. The precedent for hormonal involvement in feeding regulation acting via the digestive tract in vertebrate systems already exists (Bado et al., 1989; DelValle and Yamada, 1990; McHugh and Moran, 1986; Weller et al., 1990), and some of the peptides implicated in vertebrate systems have been shown to exist in insects (Cook and Holman, 1985). The blowfly, *P. regina,* provides the best model to investigate hormonal involvement in the regulation of carbohydrate feeding; however, other systems already discussed here should be investigated.

8.8. Concluding Remarks

When carefully analyzed, the contribution of carbohydrates to the total energy budget of the Diptera, Lepidoptera, and Hymenoptera is significant. The ecological importance of the crop as a storage organ for liquid carbohydrates should be reexamined for each of these major orders. The impact of the crop on the evolution of each group should be evaluated—especially in ants, where this important structure has earned the name "social stomach." Finally, the wealth of information on the regulation of carbohydrate feeding in the Diptera should serve to facilitate and stimulate research into the mechanisms operating in the Lepidoptera and Hymenoptera. Only when we understand how these different groups solve the problem of carbohydrate feeding will we be able to take a comparative approach to understanding how the regulation of carbohydrate feeding behavior fits into the evolutionary success of these groups. An example of the comparative approach is given in Schmidt and Friend's (1991) paper where they discuss the diversity in the Diptera of both carbohydrate receptors and sites. It is hoped that this review on what physiological mechanisms regulate carbohydrate feeding in the three major insect groups raises more questions than provides answers, stimulates the reader to ask questions not addressed, and encourages research designed to answer some of these unknowns.

Acknowledgments

I would like to thank Dr. B. J. D. Meeuse, Dr. H. G. Baker, and the late Dr. I. Baker for suggestions on obscure references concerning plant production of non-nectar exudates. Appreciation also goes to the following individuals for their suggestions on the manuscript and also for specific information for their specialty area: Dr. C. L. Boggs (Lepidoptera), Dr. N. P. Kristensen (Lepidoptera), and Dr. G. Schumann (rusts). Finally, I would like to express gratitude to my mentor and long-time friend Dr. V. G. Dethier for providing continued moral support, suggestions, and comments on this manuscript.

References

Anderson, A. L. (1932). The sensitivity of the legs of common butterflies to sugars. *J. Exp. Zool.* 63, 253–259.

Bado, A., Roze, C., Lewin, M. J. M., and Dubrasquet, M. (1989). Endogenous opioid peptides in the control of food intake in cats. *Peptides* 10, 967–971.

Baker, H. G., and Baker, I. (1975). Studies of nectar-constitution and pollinator-plant coevolution. In: Gilbert, L. E., and Raven, P. R. (eds.), *Coevolution of Animals and Plants*. University of Texas Press, Texas, pp. 100–140.

Baker, H. G., and Baker, I. (1983a). Floral nectar sugar constituents in relation to pollinator type. In: Jones, C. E., and Little, R. J. (eds.), *Handbook of Experimental Pollination Biology*. Scientific and Academic Editions, New York, pp. 117–141.

Baker, H. G., and Baker, I. (1983b). A brief historical review of the chemistry of floral nectar. In: Bentley, B., and Elias, T. (eds.), *The Biology of Nectaries*. Columbia University Press, New York, pp. 126–152.

Barton Browne, L. (1975). Regulatory mechanisms in insect feeding. *Adv. Insect Physiol.* 11, 1–116.

Barton Browne, L., and Evans, D. R. (1960). Locomotor activity of the blowfly as a function of feeding and starvation. *J. Insect Physiol.* 4, 27–37.

Beckers, R., Deneubourg, J. L., Goss, S., and Pasteels, J. M. (1990). Collective decision making through food recruitment. *Insectes Sociaux Paris* 37, 258–267.

Bell, W. J. (1991). *Searching Behaviour*. Chapman and Hall, New York.

Bentley, B. (1977). Extrafloral nectaries and protection by pugnacious bodyguards. *Annu. Rev. Ecol. Syst.* 8, 407–427.

Bentley, B., and Elias, T. (eds.) (1983). *The Biology of Nectaries*. Columbia University Press, New York.

Bernays, E. A. (1985). Regulation of feeding behaviour. In: Kerkut, G. A., and Gilbert, L. I. (eds.), *Comprehensive Insect Physiology, Biochemistry, and Pharmacology*, vol. 4. Pergamon Press, Oxford, pp. 1–32.

Bernays, E. A., and Simpson, S. J. (1982). Control of food intake. *Adv. Insect Physiol.* 16, 59–118.

Blaney, W. M., and Simmonds, M. S. J. (1988). Food selection in adults and larvae of three species of Lepidoptera: a behavioural and electrophysiological study. *Entomol. Exp. Appl.* 49, 111–121.

Blaney, W. M., and Simmonds, M. S. J. (1990). A behavioural and electrophysiological study of the role of tarsal chemoreceptors in feeding by adults of *Spodoptera, Heliothis virescens* and *Helicoverpa armigera*. *J. Insect Physiol.* 36, 743–756.

Blundell, J. E. (1984). Serotonin and appetite. *Neuropharmacology* 23, 1537–1551.

Boggs, C. L. (1987). Ecology of nectar and pollen feeding in Lepidoptera. In: Slansky, F., Jr., and Rodriguez, J. G. (eds.), *Nutritional Ecology of Insects, Mites, and Spiders.* John Wiley & Sons, New York, pp. 369–391.

Boggs, C. L. (1988). Rates of nectar feeding in butterflies: effects of sex, size, age and nectar concentration. *Funct. Ecol.* 2, 289–295.

Boggs, C. L., and Jackson, L. A. (1991). Mud puddling by butterflies is not a simple matter. *Ecol. Entomol.* 16, 123–127.

Bowdan, E., and Dethier, V. G. (1986). Coordination of a dual inhibitory system regulating feeding behaviour in the blowfly. *J. Comp. Physiol.* A158, 713–722.

Breed, M. D., Michener, C. D., and Evans, H. E. (1982). *The Biology of Social Insects.* Westview Press, Boulder, CO.

Brian, M. V. (1983). *Social Insects, Ecology and Behavioural Biology.* Chapman and Hall, London.

Brookhart, G. L., Edgecomb, R. S., and Murdock, L. L. (1987). Amphetamine and reserpine deplete brain biogenic amines and alter blow fly feeding behavior. *J. Neurochem.* 48, 1307–1315.

Brown, J. J., and Chippendale, G. M. (1974). Migration of the monarch butterfly, *Danaus plexippus:* energy sources. *J. Insect Physiol.* 20, 1117–1130.

Buller, A. H. R. (1950). *Researches on fungi,* vol. 7. *The Sexual Process in the Uredinales.* University of Toronto Press, Toronto, Canada.

Burgess, E. (1880). Contributions to the anatomy of the milk-weed butterfly *Danais archippus* (Fabr.). *Anniversary Memoirs of the Boston Society of Natural History,* pp. 1–18.

Chapman, R. F. (1982). Chemoreception: the significance of receptor numbers. *Adv. Insect Physiol.* 16, 247–356.

Chapman, R. F. (1985). Coordination of digestion. In: Kerkut, G. A., and Gilbert, L. I. (eds.), *Comprehensive Insect Physiology, Biochemistry, and Pharmacology,* vol. 4. Pergamon Press, Oxford, pp. 213–240.

Chippendale, G. M. (1978). The functions of carbohydrates in insect life processes. In: Rockstein, M. (ed.), *Biochemistry of Insects.* Academic Press, New York, pp. 1–55.

Clark, E. W., and Lukefahr, M. J. (1956). A partial analysis of cotton extrafloral nectar and its approximation as a nutritional medium for adult pink bollworms. *J. Econ. Entomol.* 49, 875–876.

Common, I. F. B. (1975). Evolution and classification of the Lepidoptera. *Annu. Rev. Entomol.* 20, 183–203.

Conway, J. R. (1977). Analysis of clear and dark amber repletes of the honey ant, *Myrmecocystus mexicanus hortideorum. Ann. Entomol. Soc. Am.* 70, 367–369.

Cook, B. J., and Holman, G. M. (1985). Peptides and kinins. In: Kerkut, G. A., and Gilbert, L. I. (eds.), *Comprehensive Insect Physiology, Biochemistry and Pharmacology,* vol. 11. Pergamon Press, Oxford, pp. 531–593.

Cook, H., and Orchard, I. (1990). Effects of 5,7-DHT upon feeding and serotonin content of various tissues in *Rhodnius prolixus. J. Insect Physiol.* 36, 361–367.

Dafni, A., and Woodell, S. R. J. (1986). Stigmatic exudate and the pollination of *Dactylorhiza fuchsii* (Druce) Soo. *Flora* 178, 343–350.

Dafni, A., Eisikowitch, D., and Ivri, Y. (1987). Nectar flow and pollinators' efficiency in two co-occurring species of *Capparis* (*Capparaceae*) in Israel. *Plant Syst. Evol.* 157, 181–186.

Day, M. F. (1954). The mechanism of food distribution to midgut or diverticulum in the mosquito. *Aust. J. Biol. Sci.* 7, 515–524.

DelValle, J., and Yamada, T. (1990). The gut as an endocrine organ. *Annu. Rev. Med.* 41, 447–455.

Dethier, V. G. (1976). *The Hungry Fly*. Harvard University Press, Cambridge, MA.

Dethier, V. G. (1977). The taste of salt. *Am. Sci.* 65, 744–751.

Dethier, V. G., and Gelperin, A. (1967). Hyperphagia in the blowfly. *J. Exp. Biol.* 47, 191–200.

Dourish, C. T., Rycroft, W., and Iversen, S. D. (1989). Postponement of satiety by blockade of brain cholecystokinin (CCK-B) receptors. *Science* 245, 1509–1511.

Downes, J. A. (1968). A nepticulid moth feeding at the leaf-nectaries of poplar. *Can. Entomol.* 100, 1078–1079.

Downes, J. A. (1974). The feeding habits of adult Chironomidae. *Entomol. Tidskr.* 95, 84–90.

Downes, W. L., Jr., and Dahlem, G. A. (1987). Keys to the evolution of Diptera: role of Homoptera. *Environ. Entomol.* 16, 847–854.

Dreisig, H. (1988). Foraging rates of ants collecting honeydew or extrafloral nectar, and some possible constraints. *Ecol. Entomol.* 13, 143–154.

Drew, R. A. I., and Lloyd, A. C. (1987). Relationship of fruit flies (Diptera: Tephritidae) and their bacteria to host plants. *Ann. Entomol. Soc. Am.* 80, 629–636.

Eaton, J. L. (1979). Chemoreceptors in the cibario-pharyngeal pump of the cabbage looper moth, *Trichoplusia ni* (Lepidoptera: Noctuidae). *J. Morphol.* 160, 7–16.

Edgecomb, R. S., Murdock, L. L., Smith, A. B., and Stephen, M. D. (1987). Regulation of tarsal taste threshold in the blowfly, *Phormia regina*. *J. Exp. Biol.* 127, 79–94.

Edgecomb, R. S., Pyle, A. R., and Murdock, L. L. (1989). The role of water in tarsal taste thresholds to sugar in the blowfly *Phormia regina*. *J. Exp. Biol.* 142, 245–255.

Ehrlich, P. R. (1984). The structure and dynamics of butterfly populations. In: Vane-Wright, R. I., and Ackery, P. R. (eds.), *The Biology of Butterflies*. Academic Press, London, pp. 25–40.

Eisner, T. (1957). A comparative morphological study of the proventriculus of ants (Hymenoptera: Formicidae). *Bull. Mus. Comp. Zool.*, Harvard 116, 439–490.

Evans, P. D. (1980). Biogenic amines in the insect nervous system. *Adv. Insect Physiol.* 15, 317–473.

Evans, P. D. (1985). Octopamine. In Kerkut, G. A., and Gilbert, L. I. (eds.), *Comprehensive Insect Physiology, Biochemistry and Pharmacology*, vol. 11. Pergamon Press, Oxford, pp. 499–530.

Faegri, K., and van der Pijl, L. (1971). *The Principles of Pollination Ecology*. Pergamon Press, New York.

Fernstrom, J. D., and Wurtman, R. J. (1971). Brain serotonin content: increase following ingestion of carbohydrate diet. *Science* 174: 1023–1025.

Friend, W. G., and Stoffolano, J. G., Jr. (1991). Feeding behaviour of the horsefly *Tabanus nigrovittatus* (Diptera: Tabanidae): effects of dissolved solids on ingestion and destination of sucrose or ATP diets. *Physiol. Entomol.* 16, 35–45.

Friend, W. G., Smith, J. J. B., Schmidt, J. M., and Tanner, R. J. (1989). Ingestion and diet destination in *Culiseta inornata:* responses to water, sucrose and cellobiose. *Physiol. Entomol.* 14, 137–146.

Frings, H., and Frings, M. (1956). The loci of contact chemoreceptors involved in feeding reactions in certain lepidoptera. *Biol. Bull.* 110, 291–299.

Gelperin, A. (1966). Control of crop emptying in the blowfly. *J. Insect Physiol.* 12, 331–345.

Gelperin, A. (1971a). Regulation of feeding. *Annu. Rev. Entomol.* 16, 365–378.

Gelperin, A. (1971b). Abdominal sensory neurons providing negative feedback to the feeding behavior of the blowfly. *Z. Vergl. Physiol.* 72, 17–31.

Gilbert, L. E. (1972). Pollen feeding and reproductive biology of *Heliconius* butterflies. *Proc. Natl. Acad. Sci. USA* 69, 1403–1407.

Gilbert, L. E. (1984). The biology of butterfly communities. In: Vane-Wright, R. I., and Ackery, P. R. (eds.), *The Biology of Butterflies*. Academic Press, London, pp. 41–54.

Gilbert, L. E., and Singer, M. C. (1975). Butterfly ecology. *Annu. Rev. Ecol. Syst.* 6, 365–397.

Goldsworthy, G. J., and Wheeler, C. H. (1989). *Insect Flight*. CRC Press, Boca Raton, FL.

Green, G. W. (1964a). The control of spontaneous locomotor activity in *Phormia regina* Meigen. I. Locomotor activity patterns in intact flies. *J. Insect Physiol.* 10, 711–726.

Green, G. W. (1964b). The control of spontaneous locomotor activity in *Phormia regina* Meigen. II. Experiments to determine the mechanism involved. *J. Insect Physiol.* 10, 727–752.

Greenberg, S. L., and Stoffolano, J. G., Jr. (1977). The effect of age and diapause on the long-term intake of protein and sugar by two species of blowflies *Phormia regina* (Meig.) and *Protophormia terraenovae* (R.D.). *Biol. Bull.* 153, 282–298.

Gwadz, R. W. (1969). Regulation of blood meal size in the mosquito. *J. Insect Physiol.* 15, 2039–2044.

Heinrich, B. (1979). *Bumblebee Economics*. Harvard University Press, Cambridge, MA.

Hendrichs, J., Cooley, S. S., and Prokopy, R. J. (1992). Post-feeding bubbling behavior in fluid-feeding Diptera: concentration of crop contents by oral evaporation of excess water. *Physiol. Entomol.* 17, 153–161.

Hewitt, C. G. (1912). *House-flies and How They Spread Disease*. Cambridge University Press, London, England.

Hölldobler, B., and Wilson, E. O. (1990). *The Ants*. The Belknap Press of Harvard University Press, Cambridge, MA.

Hsiao, T. H. (1985). Feeding behavior. In: Kerkut, G. A., and Gilbert, L. I. (eds.), *Comprehensive Insect Physiology, Biochemistry, and Pharmacology*, vol. 9. Pergamon Press, Oxford, pp. 471–512.

Huxley, C. R., and Cutler, D. F. (1991). *Ant–Plant Interactions*. Oxford University Press, New York.

Ingold, C. T. (1971). *Fungal Spores-Their Liberation and Dispersal*. Clarendon Press, Oxford, England.

Johnson, C. G. (1969). *Migration and Dispersal of Insects by Flight*. Methuen & Co., London.

Johnson, J. B., and Stafford, M. P. (1985). Adult noctuidae feeding on aphid honeydew and a discussion of honeydew feeding by adult Lepidoptera. *J. Lepid. Soc.* 39, 321–327.

Kacelnik, A., Houston, A. I., and Schmid-Hempel, P. (1986). Central-place foraging in honey bees: the effect of travel time and nectar flow on crop filling. *Behav. Ecol. Syst.* 19, 19–24.

Kearns, C. A., and Inouye, D. W. (1993). *Techniques for Pollination Biologists*. Turris Ebora Books, Hyattsville, Maryland.

Kevan, P. G., and Baker, H. G. (1983). Insects as flower visitors and pollinators. *Annu. Rev. Entomol.* 28, 407–453.

King, D. G. (1991). The origin of an organ: phylogenetic analysis of evolutionary innovation in the digestive tract of flies (Insecta:Diptera). *Evolution* 45, 568–588.

Kingsolver, J., and Daniel, T. (1979). On the mechanics and energetics of nectar feeding in butterflies. *J. Theor. Biol.* 76, 167–179.

Knight, M. R. (1962). Rhythmic activities of the alimentary canal of the black blowfly, *Phormia regina*. *Ann. Entomol. Soc. Am.* 55, 380–382.

Koptur, S. (1992). Extrafloral nectary-mediated interactions between insects and plants. In: Bernays, E. A. (ed.), *Insect–Plant Interactions*. vol. 4. CRC Press, Boca Raton, FL, pp. 81–129.

Kristensen, N. P. (1968). The anatomy of the head and the alimentary canal of adult *Eriocraniidae* (Lep., Dacnonypha). *Entomol. Medd.* 36, 239–313.

Kusano, T., and Adachi, H. (1969a). Proboscis extending time on distilled water, sugars and salts and their nutritive value in the cabbage butterfly (*Pieris rapae crucivora*). *Kontyu* 36, 427–436.

Kusano, T., and Adachi, H. (1969b). Relation between proboscis extending time and ingested amount of sugars, and the regulation of sucking of sucrose in the white cabbage butterfly. *Pieris rapae crucivora* Boisduval (Lepidoptera: Pieridae). *Appl. Entomol. Zool.* 4, 161–170.

Langley, P. A. (1976). Initiation and regulation of ingestion by hematophagous arthropods. *J. Med. Entomol.* 13, 121–130.

Lawton, J. H., and Heads, P. A. (1984). Bracken, ants and extrafloral nectaries. I. The components of the system. *J. Anim. Ecol.* 53, 995–1014.

Lee, R. M. K. W., and Davies, D. M. (1979). Feeding in the stable fly, *Stomoxys calcitrans* (Diptera: Muscidae). I. Destination of blood, sucrose solution and water in the alimentary canal, the effects of age on feeding, and blood digestion. *J. Med. Entomol.* 15, 541–554.

Lent, C. M. (1985). Serotonergic modulation of the feeding behavior of the medicinal leech. *Brain Res. Bull.* 14, 643–655.

Long, T. F., and Murdock, L. L. (1983). Stimulation of blowfly feeding behavior by octopaminergic drugs. *Proc. Natl. Acad. Sci. USA* 80, 4159–4163.

Long, T. F., Edgecomb, R. S., and Murdock, L. L. (1986). Effects of substituted phenylethylamines on blowfly feeding behavior. *Comp. Biochem. Physiol.* 83C, 201–209.

Manjra, A. A. (1971). Regulation of threshold to sucrose in a mosquito, *Culiseta inornata* (Williston). *Mosq. News* 31, 387–390.

Marfaing, P., Rouault, J., and Laffort, P. (1989). Effect of the concentration and nature of olfactory stimuli on the proboscis extension of conditioned honey bees *Apis mellifica ligustica. J. Insect Physiol.* 35, 949–955.

Marshall, J. (1935). On the sensitivity of the chemoreceptors on the antenna and fore-tarsus of the honey-bee, *Apis mellifica* L. *J. Exp. Biol.* 12, 17–26.

May, P. G. (1985a). A simple method for measuring nectar extraction rates in butterflies. *J. Lepid. Soc.* 39, 53–54.

May, P. G. (1985b). Nectar uptake rates and optimal nectar concentrations of two butterfly species. *Oecologia* 66, 381–386.

McHugh, P. R., and Moran, T. H. (1986). The stomach, cholecystokinin, and satiety. *Fed. Proc.* 45, 1384–1390.

Menzel, R., and Mercer, A. (eds.). (1987). *Neurobiology and Behavior of Honeybees.* Springer-Verlag, New York.

Mercer, A. R. (1987). Biogenic amines in the insect brain. In: Gupta, A. P. (ed.), *Arthropod Brain: Its Evolution, Development, Structure, and Function.* John Wiley & Sons, New York, pp. 399–414.

Minnich, D. E. (1922). A quantitative study of tarsal sensitivity to solutions of saccharose in the red admiral butterfly, *Pyrameis atalanta* L. *J. Exp. Zool.* 36, 445–457.

Minnich, D. E. (1929). The chemical sensitivity of the legs of the blow, *Calliphora vomitoria* Linn. to various sugars. *Z. Vergl. Physiol.* 11, 1–55.

Minnich, D. E. (1932). The contact chemoreceptors of the honey bee *Apis mellifera* L. *J. Exp. Zool.* 61, 375–393.

Morita, H., and Shiraishi, A. (1985). Chemoreception physiology. In: Kerkut, G. A., and Gilbert, L. I. (eds.), *Comprehensive Insect Physiology, Biochemistry, and Pharmacology,* vol. 6. Pergamon Press, Oxford, pp. 133–170.

Mortimer, T. J. (1965). The alimentary canals of some adult Lepidoptera and Trichoptera. *Trans. R. Entomol. Soc. London* 117, 67–93.

Nayar, J. K., and Sauerman, D. M., Jr. (1974). Long-term regulation of sucrose intake by the female mosquito, *Aedes taeniorhynchus*. *J. Insect Physiol*. 20, 1203–1208.

O'Meara, G. F. (1987). Nutritional ecology of blood-feeding Diptera. In: Slansky, F., Jr., and Rodriguez, J. G. (eds.), *Nutritional Ecology of Insects, Mites, Spiders and Related Invertebrates*. John Wiley & Sons, New York, pp. 741–764.

Orchard, I. (1982). Octopamine in insects: neurotransmitter, neurohormone and neuromodulator. *Can. J. Zool*. 60, 659–669.

Orchard, I. (1989). Serotonergic neurohaemal tissue in *Rhodnius prolixus:* synthesis, release and uptake of serotonin. *J. Insect Physiol*. 35, 943–947.

Owen, W. B., and McClain, E. (1981). Hyperphagia and the control of ingestion in the female mosquito, *Culiseta inornata* (Williston) (Diptera: Culicidae). *J. Exp. Zool*. 217, 179–183.

Pivnick, K. A., and McNeil, J. N. (1985). Effects of nectar concentration on butterfly feeding: measured feeding rates for *Thymelicus lineola* (Lepidoptera: Hesperiidae) and a general feeding model for adult Lepidoptera. *Oecologia* 66, 226–237.

Proctor, M., and Yeo, P. (1972). *The Pollination of Flowers*. Taplinger Publishing Company, New York.

Ramaswamy, S. B. (1987). Behavioural responses of *Heliothis virescens* (Lepidoptera: Noctuidae) to stimulation with sugars. *J. Insect Physiol*. 33, 755–760.

Rathay, E. (1883). Untersuchungen uber die Spermogonien der Rostpilze. *Denkschrift Kais. Akad. Wissensch*. Wien 46, 1–51.

Real, L. (ed.) (1983). *Pollination Biology*. Academic Press, New York.

Ricks, B. L., and Vinson, S. B. (1972). Digestive enzymes of the imported fire ant, *Solenopsis richteri* (Hymenoptera: Formicidae). *Entomol. Exp. Appl*. 15, 329–334.

Roessingh, P., Städler, E., Schöni, R., and Feeny, P. (1991). Tarsal contact chemoreceptors of the black swallowtail butterfly *Papilio polyxenes:* responses to phytochemicals from host- and non-host plants. *Physiol. Entomol*. 16, 485–495.

Ross, K. G., and Matthews, R. W. (eds.) (1991). *The Social Biology of Wasps*. Comstock Publishing Associates, Ithaca, NY.

Salama, H. S., Khalifa, A., Azmy, N., and Sharaby, A. (1984). Gustation in the lepidopterous moth, *Spodoptera littoralis* (Boisd.). *Zool. Jb. Physiol*. 88, 165–178.

Schmid-Hempel, P. (1986). Do honeybees get tired? The effect of load weight on patch departure. *Anim. Behav*. 34, 1243–1250.

Schmid-Hempel, P., Kacelnik, A. and Houston, A. I. (1985). Honeybees maximize efficiency by not filling their crop. *Behav. Ecol. Sociobiol*. 17, 61–66.

Schmidt, A. (1938). Geschmacksphysiologische Untersuchungen an Ameisen. *Z. Vergl. Physiol*. 25, 351–378.

Schmidt, J. M., and Friend, W. G. (1991). Ingestion and diet destination in the mosquito *Culiseta inornata:* effects of carbohydrate configuration. *J. Insect Physiol*. 37, 817–828.

Schutz, S. J., and Gaugler, R. (1989). Honeydew-feeding behavior of salt marsh horse flies (Diptera: Tabanidae). *J. Med. Entomol*. 26, 471–473.

Scott, J. A. (1974). The interaction of behavior, population biology, and environment in *Hypaurotis crysalus*. *Am. Midl. Nat.* 91, 383–394.

Seeley, T. D., and Levien, R. A. (1987). Social foraging in honeybees: how a colony tracks rich sources of nectar. In: Menzel, R., and Mercer, A. (eds.), *Neurobiology and Behavior of the Honeybee*. Springer-Verlag, Berlin, pp. 38–53.

Skinner, G. J. (1980).The feeding habits of the wood-ant, *Formica rufa* (Hymenoptera: Formicidae). *Bull. Los Angeles City Mus. Nat. Hist. Sci.* 24, 1–163.

Slansky, F., Jr., and Rodriguez, J. G. (1987). *Nutritional Ecology of Insects, Mites, Spiders, and Related Invertebrates*. John Wiley & Sons, New York.

Smith, J. J. B. (1985). Feeding mechanisms. In: Kerkut, G. A., and Gilbert, L. I. (eds.), *Comprehensive Insect Physiology, Biochemistry and Pharmacology*, vol. 4. Pergamon Press, Oxford, pp. 33–85.

Städler, E., and Seabrook, W. D. (1975). Chemoreceptors on the proboscis of the female eastern spruce budworm: electrophysiological study. *Entomol. Exp. Appl.* 18, 153–160.

Städler, E., Städler-Steinbrüchel, M., and Seabrook, W. D. (1974). Chemoreceptors on the proboscis of the female eastern spruce budworm. *Morphol. Histol. Mitt. Schweiz. Entomol. Ges.* 47, 63–68.

Stoffolano, J. G., Jr. (1974). Central control of feeding and drinking in diapausing insects. In: Barton Browne, L. (ed.), *Experimental Analysis of Insect Behaviour*. Springer-Verlag, New York, pp. 32–47.

Stoffolano, J. G., Jr. (1983). Destination of the meal and the effect of a previous sugar or blood meal on subsequent feeding behavior in female *Tabanus nigrovittatus* (Diptera: Tabanidae). *Ann. Entomol. Soc. Am.* 76, 452–454.

Stoffolano, J. G., Jr., Angioy, A. M., Crnjar, R., Liscia, A., and Pietra, P. (1990). Electrophysiological recordings from the taste chemosensilla of *Tabanus nigrovittatus* (Diptera: Tabanidae) and determination of behavioral tarsal acceptance thresholds. *J. Med. Entomol.* 27, 14–23.

Stradling, D. J. (1987). Nutritional ecology of ants. In: Slansky, F., Jr., and Rodriguez, J. G. (eds.), *Nutritional Ecology of Insects, Mites, Spiders, and Related Invertebrates*. John Wiley & Sons, New York, pp. 927–969.

Sudd, J. H., and Sudd, M. E. (1985). Seasonal changes in the response of wood-ants (*Formica lugubris*) to sucrose baits. *Ecol. Entomol.* 10, 89–97.

Sudlow, L. C., Edgecomb, R. S., and Murdock, L. L. (1987). Regulation of labellar and tarsal taste thresholds in the black blowfly, *Phormia regina*. *J. Exp. Biol.* 130, 219–234.

Tennant, L. E., and Porter, S. D. (1991). Comparison of diets of two fire ant species (Hymenoptera: Formicidae): solid and liquid components. *J. Entomol. Sci.* 26, 450–465.

Thomson, A. J., and Holling, C. S. (1977). A model of carbohydrate nutrition in the blowfly *Phormia regina* (Diptera: Calliphoridae). *Can. Entomol.* 109, 1181–1198.

Tobe, S. S., and Davey, K. G. (1972). Volume relationships during the pregnancy cycle of the tsetse fly *Glossina austeni*. *Can. J. Zool.* 50, 999–1010.

246 / J. G. Stoffolano, Jr.

Traniello, J. F. A. (1989). Foraging strategies of ants. *Annu. Rev. Entomol.* 34, 191–210.

Trembley, H. L. (1952). The distribution of certain liquids in the oesophageal diverticula and stomach of the mosquito. *Am. J. Trop. Med. Hyg.* 1, 693–710.

Tukey, H. B. (1971). Leaching of substances from plants. In Preece, T. F., and Dickinson, C. H. (eds.), *Ecology of Leaf Surface Micro-organisms*. Academic Press, New York, pp. 67–80.

Varjú, D., and Núñez, J. A. (1991). What do foraging honeybees optimize? *J. Comp. Physiol.* A169, 729–736.

von Frisch, K. (1930). Veruche uber den Geschmackssin der Bienen. *Naturwissenschaften* 18, 169–174.

Waddington, K. D. (1983). Foraging behavior of pollinators. In: Real, L. (ed.), *Pollination Biology*. Academic Press, New York, pp. 213–239.

Waddington, K. D. (1985). Cost-intake information used in foraging. *J. Insect Physiol.* 31, 891–897.

Weis, I. (1930). Versuch uber die Geschmachsrezeption durch die tarsen des admirals, *Pyrameis atalanta* L. *Z. Vergl. Physiol.* 12, 206–248.

Wellenstein, G. (1952). Zur Ernahrungsbiologie der Roten Waldemeise. *Z. Pflanzenkr. Pflanzenschutz.* 59, 430–451.

Weller, A., Smith, G. P., and Gibbs, J. (1990). Endogenous cholecystokinin reduces feeding in young rats. *Science* 247, 1589–1591.

Wensler, R. J. (1972). The effect of odors on the behavior of adult *Aedes aegypti* and some factors limiting responsiveness. *Can. J. Zool.* 50, 415–420.

Wheeler, W. M. (1910). *Ants: Their Structure, Development and Behavior*. Columbia University Press, New York.

Whitehead, A. T. (1978). Electrophysiological response of honey bee labial palp contact chemoreceptors to sugars and electrolytes. *Physiol. Entomol.* 3, 241–248.

Whitehead, A. T., and Larsen, J. R. (1976a). Ultrastructure of the contact chemoreceptors of *Apis mellifera* L. (Hymenoptera: Apidae). *Int. J. Insect Morphol. Embryol.* 5, 301–315.

Whitehead, A. T., and Larsen, J. R. (1976b). Electrophysiological responses of galeal contact chemoreceptors of *Apis mellifera* to selected sugars and electrolytes. *J. Insect Physiol.* 22, 1609–1616.

Williams, C. E. (1990). Late winter foraging by honeybees (Hymenoptera: Apidae) at sapsucker drill holes. *Great Lakes Entomol.* 23, 29–32.

Wurtman, J. J., and Wurtman, R. J. (1979). Drugs that increase central serotonergic transmission diminish elective carbohydrate consumption by rats. *Life Sci.* 24, 895–904.

Wylie, F. R. (1982). Flight patterns and feeding behavior of adult *Milionia isodoxa* Prout at Bulolo, Papua New Guinea (Geometridae). *J. Lepid. Soc.* 36, 269–278.

Yin, C-M., Zou, B-X., Li, M-F., and Stoffolano, J. G., Jr. (1994). Discovery of a midgut peptide hormone which activates the endocrine cascade leading to oogenesis in *Phormia regina* (Meigen). *J. Insect Physiol.* 40, 283–292.

Zeh, D. W., Zeh, J. A., and Smith, R. L. (1989). Ovipositors, amnions and eggshell architecture in the diversification of terrestrial arthropods. *Q. Rev. Biol.* 64, 147–168.

Long-Term Regulation
of Feeding

9

The Mechanisms of Nutritional Homeostasis

S. J. Simpson, D. Raubenheimer, and P. G. Chambers

9.1. Introduction

Insects have evolved means of using an extraordinary variety of food sources, many of which are, to say the least, nutritionally unpromising. In addition to possessing an impressive array of mechanisms to deal with the physical, chemical, and other challenges posed by such foods, it has become evident in recent years that insects can also cope with two additional problems: the heterogeneity of available diets and their own changing nutritional needs. Compensatory responses are exhibited which help to alleviate the deleterious effects of such nutritional discrepancies. Such responses can be both behavioral (through food selection and regulation of amounts eaten) and postingestive. The properties of foods which play a role include not only nutrients, both singly and interactively (Simpson and Raubenheimer, 1993a,b; Simpson and Simpson, 1990; Slansky, 1993; Slansky and Wheeler, 1991; Waldbauer and Friedman, 1991; Wheeler and Slansky, 1991), but also water (Bernays, 1990; Raubenheimer and Gade, 1993, 1994) and the combination of nutrients and allelochemicals (Raubenheimer, 1992; Raubenheimer and Simpson, 1990; Slansky, 1992; Slansky and Wheeler, 1992).

The behavioral and underlying physiological mechanisms with which insects achieve nutritional homeostasis have been studied intensively in recent years. The subject was reviewed for phytophagous insects by Simpson and Simpson (1990), while Waldbauer and Friedman (1988, 1991) provided a detailed overview of dietary selection. Following the publication of these reviews there have been advances in our understanding of the proximate mechanisms controlling nutritional homeostasis and also in the way in which nutrition is viewed in functional terms (Raubenheimer, 1992; Raubenheimer and Simpson, 1993; Simpson and Raubenheimer, 1993a,b).

In this chapter we review nutritional homeostasis in the light of the more recent work. In particular, we address the multidimensionality of nutritional needs and the interactions occurring between the physiological components which simultaneously regulate intake and utilization of a range of different nutrients.

9.2. The Geometry of Compensation

9.2.1. Targets

Much of this and the chapter by Barton Browne (Chapter 11) is based on the idea that meeting the nutritional requirements of an animal can be viewed as a problem of multidimensional geometry (Raubenheimer and Simpson, 1993; Simpson and Raubenheimer, 1993b). At any instant there is a particular quantity and blend of nutrients which, if provided to the tissues, would maximize the animal's fitness. This is defined as the *nutritional target*. It is a "global" optimum (i.e., the best possible outcome, without further evolution) in the interaction between the genotype and the environment in which that genotype evolved (Raubenheimer and Simpson, 1994). The target lies in an *n*-dimensional "nutrient space," where *n* is the number of different nutrients needed.

The position of the nutritional target moves as the requirements of the tissues for nutrients change (see Chapter 11). Just as the needs of the tissues change with time, so too may the sources of nutrients available to the animal. The insect is therefore faced with the problem of matching the uncertainty of nutrient availability with changing nutritional requirements, both of which may move in *n* dimensions.

Approaching the nutritional target involves two stages: feeding and postingestive processing of food. The *intake target* is defined as the amount and blend of nutrients that must be ingested for postingestive processing to act at optimal efficiency, and thus reach the nutritional target.

There is also a third target: the *growth target*. This is the point in nutrient space which represents the optimal quantity and mix of nutrients required to build new tissue (somatic, reproduction, storage). The growth target will be separated from the nutritional target by the amount of nutrients needed to fuel metabolism.

9.2.2. Reaching the Intake Target: Nutritional Rails, Points of Best Compromise, and Behavioral Rules

A single food item, consisting of a fixed proportion of various nutrients, can be thought of as providing a "rail" in multidimensional nutrient space. The analogy of the rail illustrates the fact that an animal eating one food can only move in nutrient space along the trajectory representing the ratio of nutrients in that food. If the food contains two nutrients, A and B, in the ratio 1:2, then every mouthful ingested will provide twice as much B as A. The animal can "decide" how far along the rail to go (i.e., how much to eat), but it can only leave the rail by differentially utilizing ingested nutrients (i.e., through postingestive compensation; see Section 9.4) or by choosing a different food.

Unless the rail passes through the intake target (i.e., the food is perfectly balanced), the intake target is unreachable for an animal restricted to that food. Even if, as is usually the case, there is a choice of different foods available, the intake target will still be unreachable if it lies outside the nutrient space bounded by the rails for the various food items (Fig. 9.1). When the intake target cannot be reached, there will be a *point of best compromise,* which is the amount and ratio of nutrients which, within the nutrient space available, provides maximal fitness when ingested. Each rail will bear its own point of best compromise (intra-rail optimum). The array of such points across a selection of rails within the nutrient space will provide a pattern that can be described by a particular behavioral rule (Fig. 9.2).

The point of best compromise is determined not only by the nutritional composition of the food, but also by such factors as the energetic and ecological costs of acquiring and processing food and the presence of allelochemicals (Hinks et al., 1993; Martin and Van't Hof, 1988; Raubenheimer, 1992; Slansky, 1992, 1993).

9.2.3. Reaching the Nutritional and Growth Targets: Leaving the Rails

Although an insect with a restricted choice of foods is only able to reach certain parts of the multidimensional nutrient space by altering feeding behavior, it is, in theory, able to move over a much wider region by differentially utilizing ingested nutrients. This could occur at the level of the gut, or once nutrients have entered the hemolymph. In this way the animal may be able to reach the growth target, despite not having been able to reach the intake target. Such postingestive mechanisms are discussed further in Section 9.4.

9.2.4. Experimental Data

The concepts described so far were developed and tested in an experiment on fifth-instar nymphs of the locust *Locusta migratoria* (Raubenheimer and Simpson, 1993; Simpson and Raubenheimer, 1993b). This was similar to the hypothetical experiment shown in Fig. 9.2. Locusts were provided with one of 25 artificial foods, containing one of five levels each (7%, 14%, 21%, 28%, and 35%) of protein and digestible carbohydrate in an otherwise nutritionally similar mix. Consumption and growth were measured over the stadium.

The position of the growth target was estimated from plots in which growth over the stadium was separated into two components: that derived from ingested protein and that from ingested carbohydrate (Fig. 9.3). The position of the intake target was independently estimated in three ways, by: (a) using performance

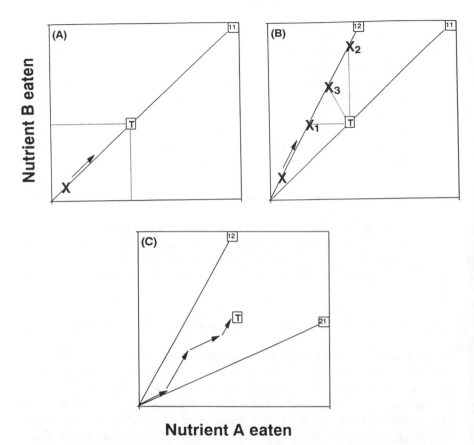

Nutrient A eaten

Figure 9.1. Nutrient planes for intake of two nutrients, A and B. Amounts eaten of A and B are plotted against each other, with the ratio of A to B in each food being shown in the boxes at the end of each rail. The first and second digits of this number give the level in the food of A and B, respectively. The intake target (T) lies on the rail-bearing foods containing an A-to-B ratio of 1:1. (**A**) Graph showing how an animal on a food containing the optimal A-to-B ratio can reach the target from its current position (X). The dry weight of food which must be eaten to reach T will depend on how dilute the food is. (**B**) The animal is given a food containing an A-to-B ratio of 1:2, in which case T is unreachable. Instead the animal might (a) move along the rail until the target coordinate for B is reached (X_1), and thus suffer a shortfall in A, (b) eat until the coordinate for A is reached (X_2), and thus overeat B, or (c) move to an intermediate point (X_3). (**C**) Graph showing how an animal given a choice of foods 1,2 and 2,1 is able to move freely within the space bounded by these rails, and thus reach the target. (After Simpson and Raubenheimer, 1993b.)

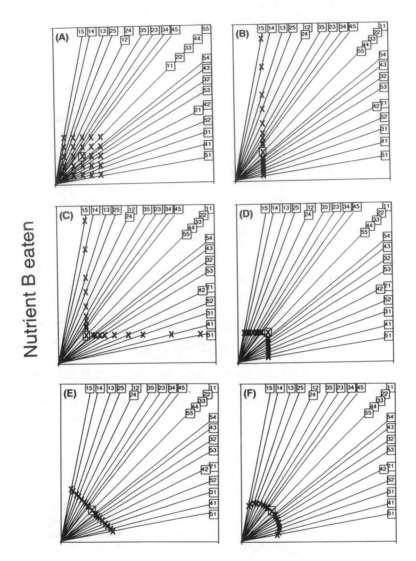

Nutrient A eaten

Figure 9.2. Nutrient planes from a hypothetical study in which animals were fed one of 25 foods, containing one of five levels each of two nutrients, A and B. Each of the panels indicates the array arising from a particular functional feeding rule. The intake target is shown as a square in each panel.

(A) **Rule 1:** Eat the same volume of food, irrespective of how much A and B it contains. (B) **Rule 2:** Eat until the level of the intake target is reached for A, irrespective of how much B is consumed. (C) **Rule 3:** Eat until at least the intake target level of both A and B is reached. (D) **Rule 4:** Eat until the intake target level of either A or B is reached. (E) **Rule 5:** Eat until the sum of A and B ingested equals the sum at the intake target. (F) **Rule 6:** Eat until the geometrically closest distance to the intake target is reached on the rail ("closest distance optimization"). (After Simpson and Raubenheimer, 1993b.)

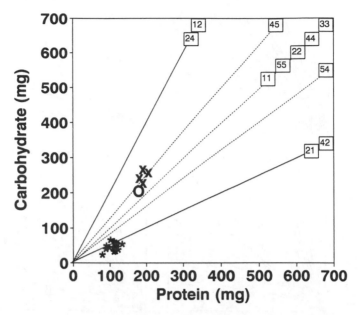

Figure 9.3. Nutrient plane for protein and carbohydrate in fifth-instar nymphs of *Locusta migratoria*. The asterisks indicate growth points (bicoordinate points of growth derived from ingested protein and carbohydrate) reached by locusts fed a wide range of artificial foods varying five-fold (7–35%) in protein and carbohydrate content in no-choice experiments. Note how they are tightly clustered, indicating the position of the growth target. Some of the 25 foods used are shown in the boxes at the end of the rails. The percentage of protein and carbohydrate in the foods can be derived by multiplying the first and second digits by 7. The crosses indicate the amounts of nutrients consumed by locusts provided with a choice of foods (protein, carbohydrate), 1,2 or 2,4, with 2,1 or 4,2 in a separate experiment (see Chambers et al., 1995). The closely grouped points indicate the likely position of the intake target. The circle shows another estimate of the intake target derived from the growth target and published values for metabolic costs and digestive asymmetries. (After Simpson and Raubenheimer, 1993b.)

criteria (minimum mortality and development time), (b) calculating back from the growth target, using published values for respiration and digestive efficiencies, and (c) allowing insects to select and defend a point on the nutritional plane. The latter was achieved by providing nymphs with one of a series of different two-diet choices (Chambers et al., 1995; Simpson et al., 1988). There was very close agreement between the different estimates for the position of the intake target for protein and carbohydrate, with the target lying close to the rail representing a protein-to-digestible-carbohydrate ratio of 45:55 (Fig. 9.3).

Plots of intake of carbohydrate and protein over the main feeding period in the stadium showed markedly arc-like arrays (Fig. 9.4), very similar to that described by the behavioral rule closest distance optimization (Fig. 9.2F). To

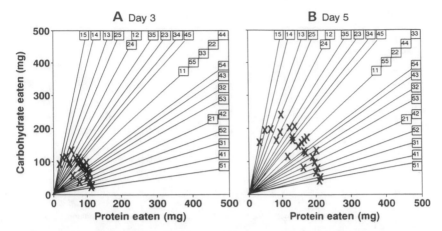

Figure 9.4. Nutrient planes from an experiment in which fifth-instar locust nymphs were given one of 25 artificial foods, containing one of five levels each of protein and digestible carbohydrate (7%, 14%, 21%, 28%, and 35%). The cumulative intake of digestible carbohydrate and protein is shown over the first 3 days (**A**) and the first 5 days (**B**) of the stadium. Each cross is the mean of eight locusts. Given that the intake target lies close to the mid-rail (Fig. 9.3), this array is closely similar to that described by functional rule 6, the closest distance optimization (see Fig. 9.2F). (After Raubenheimer and Simpson, 1993.)

provide such patterns the insects had eaten considerably different dry weights of the various foods, this being especially evident where foods lay on the same rail but were more or less diluted by addition of cellulose. Reaching the same growth point from the disparate points of protein and carbohydrate intake meant that locusts on different foods had differentially utilized protein and carbohydrate postingestively (see Section 9.4).

An analysis of published data from larvae of a number of other insect species (Simpson and Raubenheimer, 1993b) showed that the position of the intake target for protein and carbohydrate differs widely according to life-history characteristics such as possession of mycetocyte symbionts and whether or not the insects fed as an adult. All data were nevertheless consistent with the same behavioral rule: closest distance optimization (Fig. 9.5). A study of first-instar pea aphids, *Acyrthosiphon pisum,* was also compatible with this rule (Fig. 9.6), although, because the intake target lay very close to the carbohydrate axis (a dry-weight amino-acids-to-sugar ratio of 6:94 provided maximum performance) and only a narrow range of possible combinations of sugar and amino acids was physiologically reasonable to test, the rule shown in Fig. 9.2B described the recorded intake pattern equally well (Abisgold et al., 1994). The latter rule also represented the relationship between dietary salt and nonmineral nutrients in locust nymphs, where the intake target salt-to-nonmineral-nutrients ratio was 6:94 (Trumper and Simpson, 1993).

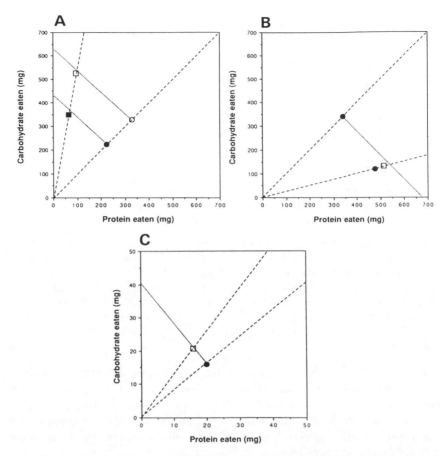

Figure 9.5. A reanalysis of published data for protein and carbohydrate intake. (**A**) Nymphs of the brown-banded cockroach, *Supella longipalpa*. (Data from Cohen et al., 1987.) (**B**) Caterpillars of *Helicoverpa zea*. (Data from Waldbauer et al., 1984). (**C**) Larval *Tribolium confusum*. (Data from Waldbauer and Battacharya, 1973). In each case the squares indicate the average point reached by a cohort of insects given a choice of two foods, while the circles show the points reached when the insects were fed only one food [in the case of *H. zea* there are two such points, one for a food containing (protein, carbohydrate) 1,1 and another 8,2 (the self-selected ratio)]. In the graph for *S. longipalpa*, data are shown for insects tested over the entire larval period (open symbols) and the last half of larval development (closed symbols). Note how the position of the self-selected point varies between the species, being protein-biased for the caterpillar (which feeds for much of its adult as well as its larval protein needs), carbohydrate-biased for the cockroach (which possesses mycetocyte symbionts implicated in upgrading dietary nitrogen), and close to the mid-rail for the beetle (which feeds as an adult and lacks mycetocyte symbionts). Additionally, note how in each case a line drawn at right angles to the rails at the points reached by insects given no choice of food passes close to the selected point. This is consistent with closest distance optimization, if it is assumed that the selected point is the intake target. (After Simpson and Raubenheimer, 1993b.)

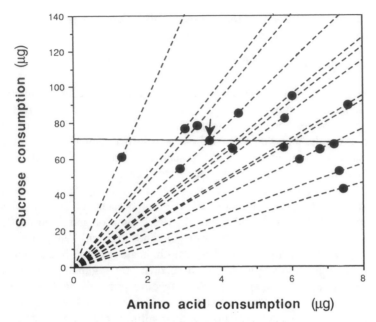

Amino acid consumption (µg)

Figure 9.6. Plot of data from first-instar larvae of the pea aphid *Acyrthosiphon pisum* fed a wide range of artificial foods varying in amino acid and sugar content. Note that, unlike Fig. 9.5, the axes are scaled differently from each other. The position of the intake target (derived from performance criteria) is arrowed and is highly carbohydrate-biased (aphids, like cockroaches, have mycetocyte symbionts). The array of intakes is consistent with closest distance optimization (the expected line being virtually indistinguishable from the line of best fit through the data, which is shown on the figure). Because of the closeness of the intake target to the y-axis (when the two axes are similarly scaled), the rule shown in Fig. 9.2B also describes the data equally well. (After Abisgold et al., 1994.)

9.3. The Physiological Control of Nutrient Intake

Approaching the intake target or point of best compromise in nutrient space involves regulating the type of food selected and also the amounts of each food eaten. These are not alternatives but are different stages in the regulation of nutrient intake, and as such they share certain controlling mechanisms. These mechanisms direct the insect through nutrient space by integrating information about the current nutritional state of the animal and the nutritional quality of the food (Simpson and Raubenheimer, 1993a).

While insects can gain information about the nutritional quality of food by using chemoreceptors innately tuned to various nutrients (Bernays and Simpson, 1982; Simpson and Simpson, 1990; Trumper and Simpson, 1993; also see Chapter 4), and perhaps even chemoreceptors in the gut (Champagne and Bernays, 1991; Timmins and Reynolds, 1992; see also Deutsch, 1990), these inputs alone cannot provide a mechanism for nutritional homeostasis unless they are integrated with information about the insect's current nutritional state.

9.3.1. Assessing Nutritional State

Data from a range of insect species have indicated a central role for the hemolymph in nutritional homeostasis (reviewed by Simpson and Raubenheimer, 1993a). Hemolymph composition varies with time since a meal, with quantity and quality of previous meals, and with the metabolic and growth demands of the tissues. As a result, the hemolymph has the potential to provide constantly updated information about the insect's nutritional state. That such information is used in nutritional homeostasis is evidenced by the fact that feeding behavior is affected by blood titers of specific nutrients, notably amino acids and sugars (Abisgold and Simpson, 1987, 1988; Friedman et al., 1991; Simpson and Simpson, 1992), and also by hemolymph osmolality (Abisgold and Simpson, 1987; Bernays and Chapman, 1974; Gelperin, 1966). The latter indicates the combined solute concentration of the hemolymph. Although adequate for regulating feeding in an insect such as a male blowfly, which is feeding primarily to meet energetic demands (i.e., has a one-dimensional nutrient space), general properties such as osmolality are not adequate indicators of the multiple needs of a growing insect.

9.3.2. Linking Nutritional State to Behavior

Feeding involves a sequence of behavioral components: finding food, accepting or rejecting a potential food item, and then ingestion. Hemolymph composition is known to influence each of these in several groups of insect, including blowflies, grasshoppers, and caterpillars (reviewed by Simpson and Raubenheimer, 1993a). How, then, is hemolymph composition linked to behavior?

9.3.3. Finding Food

Nutrient deficiency results in increased incidence of locomotion, which, in turn, enhances the probability of finding appropriate food. This is the case for insects which are deprived entirely of food (Barton Browne, 1975; Simpson and Simpson, 1990), fed nutritionally diluted diets (Naeem et al., 1992; Simpson et al., 1990), or given foods which lack adequate quantities of single nutrients within an otherwise complete mix (Cohen et al., 1987; Naeem et al., 1992; Trumper and Simpson, 1994). Similarly, insects which are provided with a nutritionally dilute food locomote sooner, on average, after a meal than when fed more concentrated foods (Simpson and Abisgold, 1985; Simpson et al., 1990).

Osmolality and levels of specific nutrients in the hemolymph play a role in such responses, along with declining inhibition from stretch receptors on the gut or body wall and falling titers of neurohormones released as a result of gut or body distention (Abisgold and Simpson, 1987; Bernays and Simpson, 1982;

Friedman et al., 1991). How these control locomotion is as yet unclear. Cohen et al. (1988) suggested that perturbations in levels of neurotransmitters in the central nervous system (CNS) may result from feeding on nutritionally imbalanced diets, perhaps as a result of low levels of precursors coming from the hemolymph, and lead to enhanced locomotion.

Not only is the likelihood of locomoting affected by nutritional state, so too is the direction of locomotion, and instances are known where orientation occurs toward stimuli emanating from foods containing specific nutrients required at that time (e.g., Robacker, 1992; Simpson and White, 1990). In nymphs of *Locusta migratoria*, learning is involved here (Simpson and White, 1990; see Section 9.3.6).

9.3.4. Accepting or Rejecting Food

In most insects the initiation of feeding is dependent upon appropriate input to the CNS from chemoreceptors responding to the food (Bernays and Simpson, 1982). It is now well known for locusts and caterpillars (but disputed for blowflies) that the responsiveness of such receptors is not fixed but varies with the nutritional state of the insect (Abisgold and Simpson, 1988; Blaney et al., 1986; Simmonds et al., 1992; Simpson et al., 1990b; Simpson and Simpson, 1992; Simpson et al., 1991). This means that insects can anticipate the nutritional suitability of a newly contacted food without first having to ingest it and then rely on postingestive feedbacks.

How such responses are mediated is best understood in nymphs of *Locusta migratoria* (Abisgold and Simpson, 1987, 1988; Simpson et al., 1990a; Simpson and Simpson, 1992; Simpson et al., 1990, 1991). In this animal the responsiveness of mouthpart taste receptors to stimulation with amino acids and sugars changes with nutritional state, with chemosensitivity to the two nutrient groups being regulated independently. Similar, independent modulation of gustatory responsiveness to sugars and amino acids has also been described in *Spodoptera littoralis* caterpillars (Simmonds et al., 1992).

In locusts, peripheral responsiveness to amino acids is modulated by levels of eight key amino acids in the hemolymph (Abisgold and Simpson, 1988; Simpson et al., 1990a). Central neural or hormonal feedbacks are not required. Instead, blood amino acids modulate maxillary palp taste receptors at the periphery (Simpson and Simpson, 1992). It would seem that the taste receptors somehow measure the difference between levels of amino acids in the hemolymph and in (or on) the food.

Taste receptors do not exist for all required nutrients, in which cases the type of mechanism which contributes to regulation of amino acid and carbohydrate intake cannot operate. However, levels of amino acids and sugars in the hemolymph could still provide a means of ensuring that adequate quantities of other nutrients are eaten. Shortage or absence in the food of any required nutrient will sooner or later cause a metabolic bottleneck within the insect and lead to elevated concentrations in the hemolymph of other nutrients which are present in the food at adequate concentrations.

For instance, shortage of a vitamin may eventually result in a buildup in amino acids and sugars which will "jam" feeding control systems, partly by causing desensitization of taste receptors (Friedman et al., 1991; Simpson and Rauben-heimer, 1993a; Simpson and Simpson, 190; Fig. 9.7). As the effect of the bottleneck becomes more pronounced, the likelihood that the insect will reject the food will increase. Rejection and subsequent locomotion will enhance the probability that the animal will move away and contact other, perhaps more suitable foods which will be accepted once hemolymph nutrient levels have declined following excretion, respiration, and so on.

In addition to direct modulation of gustatory receptors, other mechanisms are likely to link nutritional state with acceptance behavior by specifically varying central thresholds to chemosensory input. For instance, Okajima et al. (1989) discovered an internal receptor responding to hemolymph concentrations of treha-lose in larval *Mamestra brassicae,* and central neurones responding to nutrients are well known in vertebrates (e.g., Gietzen, 1993; Karadi et al., 1992; Oomura, 1988).

9.3.5. Ingestion: Regulation of Meal Size

The size of meal eaten of a given food is affected by an insect's nutritional state. This has been demonstrated in larval *L. migratoria* and *S. littoralis* following pretreatment on artificial foods lacking either protein or carbohydrate. Here meals, if they began at all (Section 9.3.4), were considerably smaller on foods lacking the deficient nutrient than on those containing it (Simmonds et al., 1992; Simpson et al., 1988, 1990, 1991). Experiments with these two species, and with larvae of two other caterpillars, *Helicoverpa zea* (Friedman et al., 1991) and *Manduca sexta* (Timmins and Reynolds, 1992; Timmins et al., 1988), indicate a central role for hemolymph nutrients in the regulation of meal size, although the results from *M. sexta* could also be explained by chemoreceptors in the gut, if these exist. In male sheep blowflies, *Lucilia cuprina,* fed *ad libitum* on one of two concentrations of sugar solution, larger meals were taken of the dilute than of the concentrated solution (Simpson et al., 1990). This was explica-ble in terms of a nonspecific effect of the concentration of sugar on hemolymph osmolality (Gelperin, 1966; Simpson et al., 1990).

Nutritional state, as represented by osmolality or nutrient levels in the hemo-lymph, could potentially influence the amount of food ingested during a meal in one or both of two ways. First, high levels of nutrients remaining from previous meals might provide inhibition which results in the meal being terminated sooner than if blood nutrient levels were low when the meal began. Second, nutrient levels might rise during the current meal and augment other inhibitory inputs which terminate feeding, such as volumetric feedbacks. Both are likely to occur, and in the case of *L. migratoria* and *S. littoralis* they are known to involve modulation of gustatory responsiveness (Simmonds et al., 1992; Simpson et al., 1991).

9.3.6. Learning

The emphasis in the discussion so far has been on the direct feedback mechanisms which regulate nutrient intake. However, it is becoming increasingly apparent that, as in vertebrates (Booth, 1991; Rozin, 1976), learning also plays a major role and that a combination of learned responses and direct feedbacks allows considerable behavioral flexibility (Figs. 9.7 and 9.8; see Chapter 10). The balance of the two in the control of behavior is likely to vary according to the type of insect and the nature of its nutritional environment (Bernays and Bright, 1993).

Learned responses described to date involving nutrients have been of several types: positive associations (Raubenheimer and Blackshaw, 1994; Simpson and White, 1990), food aversions (Bernays, 1993; Champagne and Bernays, 1991) and nonassociative responses, such as neophilia (Bernays and Raubenheimer, 1991; Geissler and Rollo, 1988; Trumper and Simpson, 1994). Learning has been implicated in orientation toward nutritionally appropriate food cues, food acceptance, and the control of meal size. The unconditioned stimuli in learned associations are as yet unclear, but are likely at some level to involve hemolymph composition (Simpson and Raubenheimer, 1993a).

9.3.7. A Model for Nutritional Compensation

Figure 9.7 shows a graphical summary of a model for the control of compensatory feeding, based on data from locusts and caterpillars. The *y*-axes of all graphs are to an arbitrary scale of 1–9.

The top line of graphs shows the state of the insect immediately before it contacts a potential food item. The left graph shows the concentration of nutrients in the hemolymph. For a–d there is gustatory responsiveness. Nutrients e+ are those for which there is no gustatory responsiveness. The middle graph shows the level of gustatory responsiveness to nutrients in the food. The level for each nutrient is a reciprocal function of the blood titer of that nutrient (except for e+, which are not detected). The right-hand graph indicates the requirements of the tissues for the nutrients at that time. Tissue demands are a function of developmental programs and environmental conditions and vary with time. In a full model incorporating a time axis, tissue demand would be represented as mol/time.

The second line shows the nutrient content of three foods. Food A is balanced with respect to the present demands of the tissues. Food B is imbalanced but

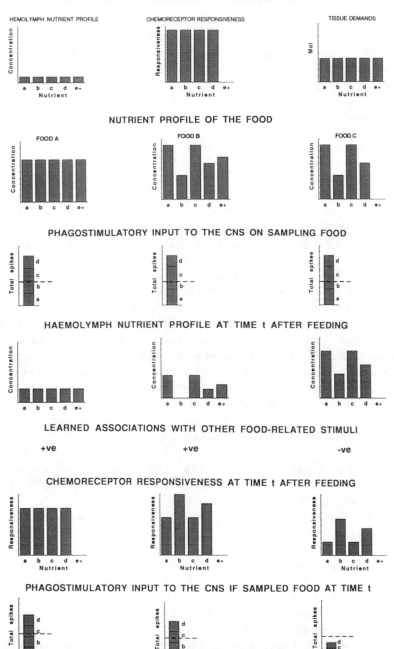

Figure 9.7. A graphical summary of a model for regulation of nutrient intake in insects. See text for explanation. (After Simpson and Raubenheimer, 1993a.)

not seriously deficient in any nutrient group. Food C has the same profile of nutrients a–d as does Food B, but lacks e+ entirely.

The third line shows the phagostimulatory input to the CNS when the insect samples one of the three foods. The number of spikes elicited by each of the nutrients a–d is a function of chemoreceptor responsiveness × concentration in the food. The assumption is that spikes for a–d are of equal weighting and that they are simply summed in the CNS. Although other non-nutrient phagostimulants or deterrents are not shown, they could readily be incorporated. The dashed line indicates the threshold of input needed to trigger the initiation of feeding. Meal size is proportional to the excess above the threshold (see Chapter 5). Because in this instance foods A–C elicit the same spike total, meal size on all three is the same.

The fourth line shows the nutrient profile of the blood at time t after having taken a meal of foods A–C. The concentration of each nutrient is a function of amount ingested in the meal, minus tissue demands, minus amount egested and excreted. Both of the latter have been considered to be zero in the figure, but nonzero values could easily be incorporated. In the case of food C, the tissue demands could not be met for any nutrient, despite the presence of sufficient of a–d in the blood. This is because the absence of e+ in the food created a metabolic bottleneck preventing the nutrients in the blood from being utilized. As a result, nutrients a–d accrued in the hemolymph.

The fifth line indicates whether or not a learned association was made with other food-related stimuli, following feeding on foods A–C. A positive association was made on A and B because the tissue demands were met, whereas a negative association was made on the deficient food, C. The unconditioned stimulus is some function of blood nutrients.

The sixth line indicates chemoreceptor responsiveness at time t after feeding on foods A–C, calculated as described above. It can be seen here how feeding on the imbalanced, but nutritionally complete, food B leads to an asymmetrical pattern of responsiveness, which is the reciprocal of the levels of the nutrients in the food. Provided that the levels of all nutrients are high enough to support the needs of the tissues, this reciprocal pattern of responsiveness will result in the insect continuing to feed over a period of time on the food if given no other choice, rather than rejecting it as occurs for food C, which is seriously deficient in certain nutrients (see next line). It also means that if the insect encounters a food with higher levels of the more limiting nutrients, it will choose that food in preference to food B.

The bottom line of graphs shows the phagostimulatory input to the CNS if the insect was to sample the same foods again at time t. It can be seen that foods A and B would still elicit feeding, whereas food C would be rejected before ingestion.

This is a highly simplified set of relationships. Nevertheless, it shows how complex nutritional decisions can be based on simple mechanisms. The same components can be used to explain behavior under both choice and no-choice situations. It can explain, for instance, why more of a nutritionally dilute food is ingested over time than of a nutrient-rich food, yet why the insect will prefer the nutritionally rich food in a choice test against the dilute one.

Figure 9.8 summarizes the known and hypothetical mechanisms involved in nutritional homeostasis.

9.4. Movement in Nutrient Space Through Postingestive Processes

Differential utilization of ingested nutrients enables an insect to move across nutrient space from the point of intake toward the nutritional and growth targets. That this occurs has been demonstrated experimentally in locust nymphs (Raubenheimer, 1992; Raubenheimer and Simpson, 1993; Simpson and Raubenheimer, 1993b; Fig 9.3) and larval aphids (Abisgold et al., 1994) and can be inferred from a number of other studies [see, e.g., Karowe and Martin (1989) and references in Simpson and Simpson (1990) and Slansky (1993)].

9.4.1. Nutrient Budgets

An understanding of postingestive processes involves the derivation of nutrient budgets. Budgets which investigate individual nutrient groups (e.g., proteins or carbohydrates, or individual compounds) have been termed discriminatory budgets (Raubenheimer and Simpson, 1994, 1995). Unitary budgets, by contrast, treat the sum of all nutrients as a single variable. Commonly, such budgets include all ingesta, even non-nutritional material such as cellulose. Unitary budgets are of limited interest because they fail to discriminate among physiological processes (Simpson and Simpson, 1990).

Raubenheimer and Simpson (1994) present a structure and terminology for nutrient budgets which allows the basic terms to be grouped according to either methodological or functional criteria. Using a functional classification, the discriminatory budget for a given nutrient, n, over time t can be stated in its most condensed form as:

$$I_n = U_n + W_n \qquad [1]$$

where I is the amount of n ingested, U is the amount utilized by the animal for maintenance, behavior, growth, and so on, and W is wastage, due to ingested nutrients either being inaccessible to digestion and absorption or being in excess of metabolic requirements.

The terms W and U can be expanded to varying degrees, depending on the interests of a specific study [see Raubenheimer and Simpson (1994) for a full discussion]. The component terms of W and U are labeled using a system of

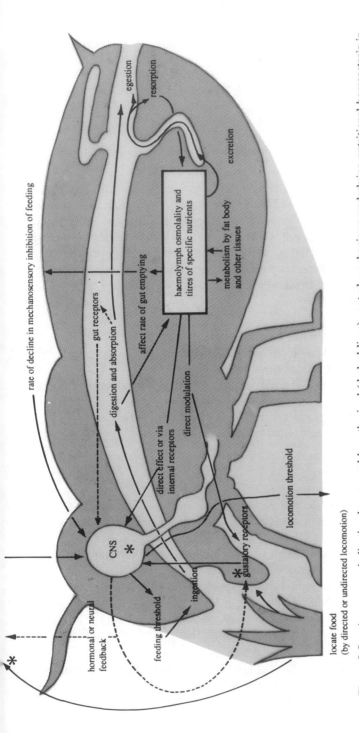

Figure 9.8. A summary indicating known and hypothesized (dashed lines) control mechanisms underlying nutritional homeostasis in a composite insect (based on acridids, caterpillars, and blowflies). At the center lies the composition of the hemolymph, which is influenced by nutrients coming from the gut, excretion, resorption from the hindgut, and metabolism by the fat body and other tissues. Hemolymph osmolality affects the rate of gut emptying, and hence the rate at which volumetric inhibition of feeding and its consequent effects (e.g., stretch-induced release of hormones affecting locomotion and gustatory sensitivity) decline after the last meal. Additional information regarding food in the gut may come from chemoreceptors in the gut lumen. Hemolymph osmolality, and perhaps also concentrations of specific nutrients, influences central neural thresholds for locomotion and feeding, either by acting via internal receptors or by affecting the balance of neurotransmitters in the CNS. Specific nutrient feedbacks for amino acids and sugars directly and independently modulate gustatory responsiveness, thereby altering the probability of accepting a food or terminating a meal. Further nutrient-specific modulation of sensory receptors may come from hormonal or centrifugal neural feedbacks. Asterisks indicate possible loci for learning associations between food-related stimuli and nutritional state; these could be central or peripheral.

267

symbols which indicates their logical status in the budget. For example, the terms from equation [1] might be expanded as:

$$W_n = (Deu + Ded + Dex + {}^wDm)_n + ({}^wRs + {}^wRp)_n \qquad [2]$$

and

$$U_n = ({}^uDm + Dc)_n + (Rc + Rg + Rr + {}^uRs + {}^uRp + Ri)_n \qquad [3]$$

where D denotes a component which is dissociated from the animal's body, and R is a component which is retained within the relevant time period. The second letter in each term distinguishes various subcategories of dissociated and retained nutrient (e.g., e is egested nutrient, m is metabolized, s is stored, and p is utilized by parasites and symbionts, etc.), while the third letter distinguishes subsets nested within these terms. Superscripted w and u denote, respectively, the wastage and utilized components of terms which occur both in W and U.

Thus, from equation [2]:

$Deu = n$ egested without having been digested;
$Ded = n$ egested having been digested but not absorbed across the gut;
$Dex = n$ catabolized and excreted;
${}^wDm = n$ respired as a means of removal (i.e., a wastage component of respiration);
${}^wRs =$ stored n which does not contribute to fitness (e.g., obesity);
${}^wRp = n$ lost to parasites associated with the animal.

And, from equation [3]:

${}^uDm = n$ respired to meet the energetic requirements for maintenance;
$Dc = n$ incorporated into secretions which are dissociated from the animal (e.g., peritrophic membranes, defensive secretions, silk, etc.);
$Rc = n$ incorporated into secretions which are retained by the animal;
$Rg = n$ incorporated into somatic growth;
$Rr = n$ incorporated into reproductive growth;
${}^uRs = n$ stored for later utilization;
${}^uRp = n$ allocated to the maintenance of symbionts;
$Ri = n$ interconverted to meet shortfalls in the budgets of other nutrients.

The sum of the term R_i across discriminatory budgets for all nutrients constitutes a common metabolic pool from which individual budgets draw (Raubenheimer and Simpson, 1994, 1995).

Relating these terms to nutritional targets (Section 9.2.1), we obtain

$$NT_n = GT_n + M_n \qquad [4]$$

where NT_n is the nutritional target coordinate (i.e., optimal) value for n in nutrient space, GT_n is the coordinate for n of the growth target, and M_n is the amount of n needed to fuel metabolism when feeding on a nutritionally ideal food under optimal environmental conditions.

Thus, when feeding on a nutritionally and environmentally optimal food:

$$I_n = IT_n \text{ (the coordinate level of } n \text{ at the intake}$$
$$\text{target)}$$
$$(R_g + Rr + {}^uRs)_n = GT_n$$
$$({}^uDm + Dc + Rc + {}^uRp)_n = M_n$$

W_n will be minimized and this will be true simultaneously for all nutrients.

If the amount of n eaten (I_n) is less than that required to reach the nutritional target level, then the insect will not be able to reach the growth target coordinate for n, unless an excess of other nutrients can be converted to make good the deficiency. For instance, deamination of ingested amino acids may provide metabolic fuel or structural carbohydrate/lipid when amounts of carbohydrate eaten are low (Hinks et al., 1993; van Loon, 1988; Raubenheimer and Simpson, 1993). Nutrients are often not mutually interconvertible, however, as seen, for example, between carbohydrate and protein, or between amino acids within protein (e.g., see Dadd, 1985; van Loon, 1988).

Provided that all the ingested n is available for digestion, it would be expected that potential wastage (W_n) would be minimal when intake is less than or equal to requirements, where requirements are the combined sum of GT, M, and any extra energetic costs associated with obtaining and processing suboptimal diets [Raubenheimer and Simpson (1994); see also Karowe and Martin (1993), Martin and Van't Hof (1988), and Slansky (1993) for detailed discussion of such potential costs]. On the other hand, if intake exceeds requirements, then there will be wastage. Because, by definition, there are fitness costs associated with exceeding the nutritional target (just as there are for not reaching it), it would be expected that amounts ingested above requirements should be removed from the insect or converted for use elsewhere. Fitness costs may stem from an excess of a particular nutrient being toxic, or else jamming the feeding control systems (Section 9.3.4) and thereby preventing intake of sufficient amounts of other nutrients [i.e., cause incidentally restricted intake of others, *sensu* Raubenheimer (1992)]. Other ecological consequences of overeating may also reduce fitness.

9.4.2. Utilization Plots

The efficiency with which nutrients are processed may be clearly visualized in bicoordinate plots of uptake (intake or absorption) and the various components of nutrient budgets [termed "utilization plots" by Raubenheimer and Simpson (1994)]. Such plots are the geometrical representation of analysis of covariance (ANCOVA) designs. ANCOVA provides a more statistically powerful and valid means of analyzing such data than do the widely used nutritional indices (Wald-bauer, 1968), with which there are not only statistical (Raubenheimer and Simpson, 1994), but also potentially severe methodological problems (Farrar et al., 1989; van Loon, 1991, 1993; Schmidt and Reese, 1986; Simpson and Simpson, 1990).

A hypothetical utilization plot is shown in Fig. 9.9. It indicates the pattern of utilization for a given nutrient which would be expected across a range of intakes. For intakes below requirements, wastage is close to zero, while ingested nutrient above this level is removed from the animal (see equation [2]).

Patterns very similar to this were found for utilization of ingested nitrogen and carbohydrate in experiments on nymphs of *L. migratoria* (Zanotto et al., 1993; Fig. 9.10). Locust nymphs removed nitrogen in excess of needs principally postabsorptively, by increasing uric acid production (i.e., Dex_n in equation [2]). Catabolism and oxidation of excess amino acids have also been reported in larval

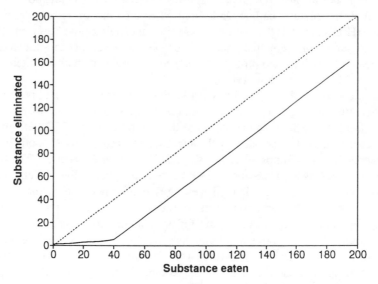

Figure 9.9. A hypothetical utilization plot for a nutrient. The dotted line represents the expected pattern for a non-nutrient (all that is eaten is eliminated as wastage, with none being used for maintenance, behavior, or growth). The solid line shows the case where virtually all of a nutrient ingested is used up to the level required for maintenance, growth, and behavior. Intake beyond this level (40 arbitrary units on the figure) is eliminated from the animal (by egestion, excretion, or wastage respiration). (Adapted from Rauben-heimer and Simpson, 1994.)

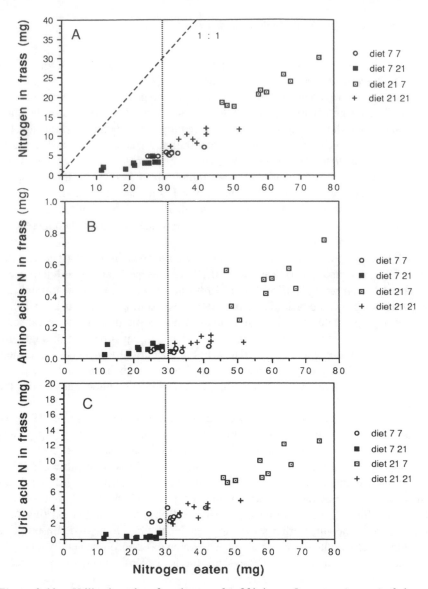

Figure 9.10. Utilization plots for nitrogen for fifth-instar *Locusta migratoria* fed one of four foods varying in protein and digestible carbohydrate content. Note the resemblance to Fig. 9.9. Numbers to the right of each graph show percent protein and digestible carbohydrate in the diets. (**A**) Graph showing the total nitrogen in the frass. When the insects consumed less than 30 mg of nitrogen, relatively little was excreted/egested, whereas when intake exceeded 30 mg, the rate of removal was much higher. (**B**) Graph indicating the amount of free amino acids in the frass. A similar pattern to total nitrogen is evident, although amino acids contributed only a small proportion of the total. (**C**) Graph showing uric acid excreted. When consumption of nitrogen exceeded 30 mg, uric acid excretion was a major component of total nitrogen removal. Notice differences in the scales of *y*-axes between graphs A and C. (After Zanotto et al., 1993.)

Bombyx mori (Horie and Watanabe, 1983), *Pieris brassicae* (van Loon, 1988) and *Spodoptera eridania* (Karowe and Martin, 1989).

Within the nitrogen budget for locust nymphs there was evidence for selective removal of lysine (Zanotto et al., 1994). This is the most effective of the suite of amino acids whose combined presence is required in the hemolymph at elevated concentrations to cause reduction in behavioral and gustatory responsiveness to amino acids in the food (Simpson et al., 1990a; Section 9.3.4). Selective removal of lysine would be expected to alleviate inhibition of feeding for more limiting nutrients. The site of control of lysine removal is unknown but could occur via low rates of absorption from the gut (Deu_n in equation [2]), high rates of excretion, and/or low rates of resorption from the rectum (either of the last two increasing Dex_n in equation [2]).

In locust nymphs, excess carbohydrate was removed mainly by wastage respiration (^wDm_n in equation [2]) (Zanotto et al., 1993), as has been suggested to occur in larval *Samea multiplicalis* (Taylor, 1989) and is known to occur through the activities of brown adipose tissue in vertebrates (Rothwell and Stock, 1979).

Although possible, there is little or no evidence as yet in insects for preabsorptive removal of excess nutrients (Deu_n and Ded_n in equation [2]) through, for instance, increasing gut emptying rate (without concurrently decreasing intermeal intervals), declining enzyme:substrate ratio (due to there being a fixed or decreasing rate of digestive enzyme secretion), or lowered absorption rates across the gut.

Gut emptying rate either does not change (Simpson and Abisgold, 1985) or varies negatively with the concentration of nutrients in ingested food (e.g., Yang and Joern, 1994a), in part as a result of hormonal and osmotic effects on gut motility (e.g., Cazal, 1969; Gelperin, 1966). These effects on intrinsic rate of emptying are enhanced by increased consumption of nutritionally dilute foods (this being driven by the more rapidly emptying gut and low hemolymph nutrient titers; see Section 9.3), so that incoming food physically pushes existing food backwards in the gut, further shortening residence time. Increased consumption of nutritionally dilute foods has also been shown to be accompanied by a rapidly invoked increase in gut size (Yang and Joern, 1994b), which might increase digestive and absorptive efficiencies.

Levels secreted of digestive enzymes either do not vary with dietary nutrients or are positively correlated [Broadway and Duffey, 1986; Lemos et al., 1992 (for trypsin)], although there is a report of lowered production of aminopeptidase in larvae of *Ceratitis capitata* fed high- versus low-protein foods (Lemos et al., 1992). In theory, egestion of excess nutrient will occur if only enough digestive enzymes are produced to provide the tissues with required levels of a nutrient; any ingested nutrients above this required level will pass undigested through the gut (Raubenheimer and Simpson, 1993). At least in locust nymphs this was found not to be the case for both protein and carbohydrate (Zanotto et al., 1993), where virtually all ingested nutrients were digested and absorbed over a wide range of intakes. This is perhaps not surprising, given that digestive enzymes hydrolyze macromolecules into their constituents; and it is these constituents (e.g., amino acids, monosaccharides), not the dietary macromolecules (whose composition is variable), whose levels must be regulated in the body. Similarly, the digestive yield from a given titer of enzyme will vary with, for instance, gut passage time, temperature, and other dietary constituents (Karowe and Martin, 1993; Slansky, 1992, 1993; Yang and Joern, 1994a).

Such regulation could occur via altered rates of absorption of nutrients from the gut. Changes in concentration gradient between the gut lumen and hemolymph might influence uptake of nutrients which are absorbed passively, while active absorption processes found for other nutrients might also vary (Dow, 1986; Turunen, 1985; see also Kasarov and Diamond, 1988). Evidence is lacking for such effects in insects, however.

9.5. Summary

The multidimensional geometric approach to insect feeding and nutrition has provided a framework for exploring and uniting mechanistic, functional, ontogenetic, and evolutionary questions (Simpson and Raubenheimer, 1993a). The present review has highlighted the mechanistic aspect of the framework.

Meeting nutritional needs (approaching the nutritional target) involves regulation of feeding and also postingestive utilization of food. Regulating intake (reaching the intake target or point of best compromise) requires that the animal integrate information about the nutritional quality of its food and its own current nutritional state. The former information is provided by sensory inputs whose behavioral efficacy is to some extent genetically hard-wired, but is also modulated by previous nutritional experience. Information regarding nutritional state comes in large part from the hemolymph and is linked to foraging, dietary selection, and control of meal size through various mechanisms, including direct modulation of taste receptor responsiveness and learning. Such multiple mechanisms provide the basis for considerable behavioral flexibility.

Insects can move from their point of intake in nutrient space to approach their growth target by differentially utilizing nutrients postingestively. Such postingestive compensation can involve a range of responses and is usefully displayed by "utilization plots," which are the graphical representations of analysis of covariance designs.

References

Abisgold, J. D., and Simpson, S. J. (1987). The physiology of compensation by locusts for changes in dietary protein. *J. Exp. Biol.* 129, 329–346.

Abisgold, J. D., and Simpson, S. J. (1988). The effect of dietary protein levels and haemolymph composition on the sensitivity of the maxillary palp chemoreceptors of locusts. *J. Exp. Biol.* 135, 215–229.

Abisgold, J. D., Simpson, S. J., and Douglas, A. E. (1994). Responses of the pea aphid (*Acyrthosiphon pisum*) to simultaneous variation in dietary amino acid and sugar levels. *Physiol. Entomol.* 19, 95–102.

Barton Browne, L. (1975). Regulatory mechanisms in insect feeding. *Adv. Insect Physiol.* 11, 1–116.

Bernays, E. A. (1990). Water regulation. In: Chapman, R. F., and Joern, (eds.), *Biology of Grasshoppers.* John Wiley and Sons, New York, pp. 129–141.

Bernays, E. A. (1993). Aversion learning in feeding. In: Papaj, D. R., and Lewis, A. C. (eds.), *Insect Learning.* Chapman and Hall, New York, pp. 1–17.

Bernays, E. A., and Bright, K. L. (1993). Mechanisms of dietary mixing in grasshoppers, a review. *Comp. Biochem. Physiol.* A104, 125–131.

Bernays, E. A., and Chapman, R. F. (1974). The effects of haemolymph osmotic pressure on the meal size of nymphs of *Locusta migratoria* L. *J. Exp. Biol.* 61, 473–480.

Bernays, E. A., and Raubenheimer, D. (1991). Dietary mixing in grasshoppers: changes in acceptability of different plant secondary compounds associated with low levels of dietary protein (Orthoptera: Acrididae). *J. Insect Behav.* 4, 545–556.

Bernays, E. A., and Simpson, S. J. (1982). Control of food intake. *Adv. Insect Physiol.* 16, 59–118.

Blaney, W. M., Schoonhoven, W. M., and Simmonds, M. S. J. (1986). Sensitivity variations in insect chemoreceptors; a review. *Experientia* 42, 13–19.

Booth, D. A. (1991). *Integration of internal and external signals in intake control. Proc. Nutr. Soc.* 51, 21–28.

Broadway, R. M., and Duffey, S. S. (1986). The effect of dietary protein on the growth and digestive physiology of larval *Heliothis zea* and *Spodoptera exigua. J. Insect Physiol.* 32, 673–680.

Cazal, M. (1969). Actions d'extraits de corpora cardiaca sur le peristaltisme intestinal de *Locusta migratoria. Arch. Zool. Exp. Gen.* 110, 83–89.

Chambers, P. G., Simpson, S. J., and Raubenheimer, D. (1995). Behavioural mechanisms of nutrient balancing in *Locusta migratoria* nymphs. *Anim. Behav.*

Champagne, D. E., and Bernays, E. A. (1991). Phytosterol suitability as a factor mediating food aversion learning in the grasshopper *Schistocerca americana. Physiol. Entomol.* 16, 391–400.

Cohen, R. W., Waldbauer, G. P., Friedman, S., and Schiff, N. M. (1987). Nutrient self-selection by *Heliothis zea* larvae: a time-lapse film study. *Entomol. Exp. Appl.* 44, 65–73.

Cohen, R. W., Friedman, S., and Waldbauer, G. P. (1988). Physiological control of nutrient self-selection in *Heliothis zea* larvae: the role of serotonin. *J. Insect Physiol.* 34, 935–940.

Dadd, R. H. (1985). Nutrition: organisms. In: Kerkut, G. A., and Gilbert, L. I. (eds.),

Comprehensive Insect Physiology, Biochemistry and Pharmacology, vol. 4. Pergamon Press, Oxford, pp. 313–390.

Deutsch, J. A. (1990). Gastric factors. In: Stricker, E. M. (ed.), *Handbook of Behavioral Neurobiology*, vol. 9. Plenum, New York, pp. 151–180.

Dow, J. A. T. (1986). Insect midgut function. *Adv. Insect Physiol.* 19, 187–328.

Farrar, R. R., Barbour, J. D., and Kennedy, G. G. (1989). Quantifying food consumption and growth in insects. *Ann. Entomol. Soc. Am.* 82, 592–598.

Friedman, S., Waldbauer, G. P., Eertmoed, J. E., Naeem, M., and Ghent, A. W. (1991). Blood trehalose levels have a role in the control of dietary self-selection by *Heliothis zea* larvae. *J. Insect Physiol.* 37, 919–928.

Geissler, T. G., and Rollo, C. D. (1988). The influence of nutritional history on the response to novel food by the cockroach, *Periplaneta americana* (L.). *Anim. Behav.* 35, 1905.

Gelperin, A. (1966). Control of crop emptying in the blowfly. *J. Insect Physiol.* 12, 331–345.

Gietzen, D. W. (1993). Neural mechanisms in the response to amino acid deficiency— critical review. *J. Nutr.* 123, 610–625.

Hinks, C. F., Hupka, D., and Olfert, O. (1993). Nutrition and the protein economy in grasshoppers and locusts. *Comp. Biochem. Physiol.* A104, 133–142.

Horie, Y., and Watanabe, K. (1983). Effects of various kinds of dietary protein and supplementation with limiting amino acids on growth, haemolymph components and uric acid excretion in the silkworm *Bombyx mori. J. Insect Physiol.* 29, 187–189.

Karadi, Z., Oomura, Y., Nishino, H., Scott, T. R., Lenard, L., and Aou, S. (1992). Responses of lateral hypothalamic glucose-sensitive and glucose-insensitive neurons to chemical stimuli in behaving rhesus monkeys. *J. Neurophysiol.* 67, 389–400.

Karowe, D. N., and Martin, M. M. (1989). The effects of quality and quantity of diet nitrogen on the growth, efficiency of food utilization, nitrogen budget, and metabolic rate of fifth-instar *Spodoptera eridania* larvae (Lepidoptera: Noctuidae). *J. Insect Physiol.* 35, 699–708.

Karowe, D. N., and Martin, M. M. (1993). Determinants of diet quality: the effects of diet pH, buffer concentration and buffering capacity on growth utilization by larvae of *Manduca sexta* (Lepidoptera: Sphingidae). *J. Insect Physiol.* 39, 47–52.

Karasov, W. H., and Diamond, J. M. (1988). Interplay between physiology and ecology in digestion. *BioScience* 38, 602–611.

Lemos, F. J. A., Zucoloto, F. S., and Terra, W. R. (1992). Enzymological and excretory adaptations of *Ceratitis capitata* (Diptera: Tephritidae) larvae to high protein and high salt diets. *Comp. Biochem. Physiol.* A102, 775–779.

Martin, M. M., and Van't Hof, H. M. (1988). The cause of reduced growth of *Manduca sexta* larvae on a low-water diet: increased metabolic processing costs or nutrient limitation? *J. Insect Physiol.* 34, 515–525.

Naeem, M., Waldbauer, G. P., and Friedman, S. (1992). *Heliothis zea* larvae respond to diluted diets by increased searching behaviour as well as by increased feeding. *Entomol. Exp. Appl.* 65, 95–98.

Okajima, A., Kumagai, K., and Watanabe, N. (1989). The involvement of interoceptive chemosensory activity in the nervous regulation of the prothoracic gland in a moth, *Mamestra brassicae. Zool. Sci.* 6, 859–866.

Oomura, Y. (1988). Chemical and neuronal control of feeding motivation. *Physiol. Behav.* 44, 555–560.

Raubenheimer, D. (1992). Tannic acid, protein, and digestible carbohydrate: dietary imbalance and nutritional compensation in locusts. *Ecology* 73, 1012–1027.

Raubenheimer, D., and Blackshaw, J. (1994). Locusts learn to associate visual stimuli with drinking. *J. Insect Behav.* 7, 569–575.

Raubenheimer, D., and Gade G. (1993). Compensatory water intake in locusts (*Locusta migratoria*), implications for the mechanisms regulating drink size. *J. Insect Physiol.* 39, 275–281.

Raubenheimer, D., and Gade, G. (1994). Hunger–thirst interactions in the locust, *Locusta migratoria. J. Insect Physiol.* 40, 631–639.

Raubenheimer, D., and Simpson, S. J. (1990). The effects of simultaneous variation in protein, digestible carbohydrate and tannic acid on the feeding behaviour of larval *Locusta migratoria* (L.) and *Schistocerca gregaria* (Forskal). I. Short-term studies. *Physiol. Entomol.* 15, 219–223.

Raubenheimer, D., and Simpson, S. J. (1993). The geometry of compensatory feeding in the locust. *Anim. Behav.* 45, 953–964.

Raubenheimer, D., and Simpson, S. J. (1994). The analysis of nutritional budgets. *Funct. Ecol.* 8, 783–791.

Raubenheimer, D., and Simpson, S. J. (1995). The construction of nutrient budgets. *Entomol. Exp. Appl.*

Robacker, D. C. (1992). Specific hunger in *Anastrepha ludens* (Diptera, Tephritidae)— effects on attractiveness of proteinaceous and fruit-derived lures. *Environ. Entomol.* 20, 1680–1686.

Rothwell, N. J., and Stock, M. J. (1979). A role for brown adipose tissue in diet-induced thermogenesis. *Nature* 281, 31–35.

Rozin, P. (1976). The selection of foods by rats, humans and other animals. *Adv. Study Behav.* 6, 21–76.

Schmidt, D. J., and Reese, J. C. (1986). Sources of error in nutritional index studies of insects on artificial diets. *J. Insect Physiol.* 32, 193–198.

Simmonds, M. S. J., Simpson, S. J., and Blaney, W. M. (1992). Dietary selection behaviour in *Spodoptera littoralis:* the effects of conditioning diet and conditioning period on neural responsiveness and selection behaviour. *J. Exp. Biol.* 162, 73–90.

Simpson, C. L., Simpson, S. J., and Abisgold, J. D. (1990a). An amino acid feedback and the control of locust feeding. *Symp. Biol. Hung.* 39, 39–46.

Simpson, C. L., Chyb, S., and Simpson, S. J. (1990b). Changes in chemoreceptor sensitivity in relation to dietary selection by adult *Locusta migratoria* L. *Entomol. Exp. Appl.* 56, 259–268.

Simpson, S. J., and Abisgold, J. D. (1985). Compensation by locusts for changes in dietary nutrients: behavioural mechanisms. *Physiol. Entomol.* 10, 443–452.

Simpson, S. J., and Ludlow, A. R. (1986). Why locusts start to feed, a comparison of causal factors. *Anim. Behav.* 34, 480–496.

Simpson, S. J., and Raubenheimer, D. (1993a). The central role of the haemolymph in the regulation of nutrient intake in insects. *Physiol. Entomol.* 18, 395–403.

Simpson, S. J., and Raubenheimer, D. (1993b). A multi-level analysis of feeding behaviour: the geometry of nutritional decisions. *Philos. Trans. R. Soc. London Ser B* 342, 381–402.

Simpson, S. J., and Simpson, C. L. (1990). The mechanisms of nutritional compensation by phytophagous insects. In: Bernays, E. A. (ed.), *Insect–Plant Interactions,* vol. 2. CRC Press, Boca Raton, FL, pp. 111–160.

Simpson, S. J., and Simpson, C. L. (1992). Mechanisms controlling modulation by amino acids of gustatory responsiveness in the locust. *J. Exp. Biol.* 168, 269–287.

Simpson, S. J., and White, P. R. (1990). Associative learning and locust feeding: evidence for a "learned hunger" for protein. *Anim. Behav.* 40, 506–513.

Simpson, S. J., Simmonds, M. S. J., and Blaney, W. M. (1988). A comparison of dietary selection behaviour in larval *Locusta migratoria* and *Spodoptera littoralis.* *Physiol. Entomol.* 13, 225–238.

Simpson, S. J., Simmonds, M. S. J., Blaney, W. M., and Jones, J. P. (1990). Compensatory dietary selection occurs in larval *Locusta migratoria* but not *Spodoptera littoralis* after a single deficient meal during *ad libitum* feeding. *Physiol. Entomol.* 15, 235–242.

Simpson, S. J., Simmonds, M. S. J., and Blaney, W. M. (1991). Variation in chemosensitivity and the control of dietary selection behaviour in the locust. *Appetite* 17, 141–154.

Slansky, F., Jr. (1992). Allelochemical–nutrient interactions in herbivore nutritional ecology. In: Rosenthal, G. A., and Berenbaum, M. R. (eds.), *Herbivores: Their Interactions with Secondary Plant Metabolites,* vol. 2. Academic Press, New York, pp. 135–174.

Slansky, F., Jr. (1993). Nutritional ecology: the fundamental quest for nutrients. In: Stamp, N. E., and Casey, T. M. (eds.), *Caterpillars. Ecological and Evolutionary Constraints on Foraging.* Chapman and Hall, New York, pp. 29–91.

Slansky, F., Jr., and Wheeler, G. S. (1991). Food consumption and utilization responses to dietary dilution with cellulose and water by velvetbean caterpillars, *Anticarsia gemmatalis.* *Physiol. Entomol.* 16, 99–116.

Slansky, F., Jr., and Wheeler, G. S. (1992). Caterpillars' compensatory feeding response to diluted nutrients leads to toxic allelochemical dose. *Entomol. Exp. Appl.* 65, 171–186.

Taylor, M. F. J. (1989). Compensation for variable dietary nitrogen by larvae of the salvinia moth. *Funct. Ecol.* 3, 407–416.

Timmins, W. A., and Reynolds, S. E. (1992). Physiological mechanisms underlying the control of meal size in *Manduca sexta* larvae. *Physiol. Entomol.* 17, 81–89.

Timmins, W. A., Bellward, K., Stamp, A. J., and Reynolds, S. E. (1988). Food intake, conversion efficiency, and feeding behaviour of tobacco hornworm caterpillars given artificial diet of varying nutrient and water content. *Physiol. Entomol.* 13, 303–314.

Trumper, S., and Simpson, S. J. (1993). Regulation and salt intake by nymphs of *Locusta migratoria*. *J. Insect Physiol.* 39, 857–864.

Trumper, S., and Simpson, S. J. (1994). Mechanisms regulating salt intake in nymphs of *Locusta migratoria*. *Physiol. Entomol.* 19, 203–215.

Turunen, S. (1985). Absorption. In: Kerkut, G. A., and Gilbert, L. I. (eds.), *Comprehensive Insect Physiology, Biochemistry and Pharmacology*, vol. 4. Pergamon Press, Oxford, pp. 241–277.

van Loon, J. J. A. (1988). Sensory and Nutritional Effects of Amino Acids and Phenolic Plant Compounds on the Caterpillars of Two Pieris Species. Doctoral thesis, Agricultural University, Wageningen.

van Loon, J. J. A. (1991). Measuring food utilization in plant feeding insects—towards a metabolic and dynamic approach. In: Bernays, E. A. *Insect–Plant Interactions*, vol. 3. CRC Press, Boca Raton, FL, pp. 79–124.

van Loon, J. J. A. (1993). Gravimetric vs respirometric determination of metabolic efficiency in caterpillars of *Pieris brassicae*. *Entomol. Exp. Appl.* 67, 135–142.

Waldbauer, G. P. (1968). The consumption and utilization of foods by insects. *Adv. Insect Physiol.* 5, 229–288.

Waldbauer, G. P., and Battacharya, A. K. (1973). Self-selection of an optimum diet from a mixture of wheat fractions by the larvae of *Tribolium confusum*. *J. Insect Physiol.* 19, 407–418.

Waldbauer, G. P., and Friedman, S. (1988). Dietary self-selection by insects. In: Sehnal, F., Zabza, A., and Denlinger, D. L. (eds.), *Endocrinological Frontiers in Physiological Insect Ecology*. Wroclaw Technical University Press, Wroclaw, pp. 403–442.

Waldbauer, G. P., and Friedman, S. (1991). Self-selection of optimal diets by insects. *Annu. Rev. Entomol.* 36, 43–63.

Waldbauer, G. P., Cohen, R. W., and Friedman, S. (1984). Self-selection of an optimal nutrient mix from defined diets by larvae of the corn earworm, *Heliothis zea* (Boddie). *Physiol. Zool.* 57, 590–597.

Wheeler, G. S., and Slansky, F., Jr. (1991). Compensatory responses of the fall armyworm (*Spodoptera frugipterda*) when fed water- and cellulose-diluted diets. *Physiol. Entomol.* 16, 361–374.

Yang, Y., and Joern, A. (1994a). Influence of diet quality, developmental stage and temperature on food residence time in *Melanoplus differentialis*. *Physiol. Zool.* 67, 598–616.

Yang, Y., and Joern, A. (1994b). Gut size changes in relation to variable food quality and body size in grasshoppers. *Funct. Ecol.* 8, 36–45.

Zanotto, F. P., Simpson, S. J., and Raubenheimer, D. (1993). The regulation of growth by locusts through post-ingestive compensation for variation in the levels of dietary protein and carbohydrate. *Physiol. Entomol.* 18, 425–434.

Zanotto, F. P., Raubenheimer, D., and Simpson, S. J. (1994). Selective egestion of lysine by locusts fed nutritionally unbalanced diets. *J. Insect Physiol.* 40, 259–265.

10

Effects of Experience on Feeding

E. A. Bernays

10.1. Introduction

When animals contact, and feed on, particular foods, the experience of the foods may influence future food choices. In phytophagous insects, plant chemicals are often involved, and the experiences can profoundly influence food intake behavior, determining whether or not a particular plant will be ingested and how much of it will be eaten. The effects range from very-short-term reversible changes in sensory thresholds to long-term effects on the nervous system.

This chapter discusses the effects of experience on the behavioral plasticity of feeding and food choice behavior. Neural changes underlying the behaviors are discussed, although knowledge of what these are is quite limited. In some cases, mechanisms of neural change are well known, although not necessarily in relation to a particular experiential effect; in other cases the mechanisms can be partially deduced. Numerous papers have been published that indicate an effect of experience, but specific controls to address the possible different mechanisms are lacking. For example, a food may become more acceptable as a result of experience, but there are several different mechanisms that could each alone explain the effect. Often experiential effects are likely to be the result of more than one kind of neural change. For this reason, the chapter is divided into sections that group together either mechanisms that are known to be involved or types of behavioral experiments with their possible underlying mechanisms.

10.2. Nonassociative Changes

10.2.1. Changes in Peripheral Receptors

Very-short-term changes in sensory input result from adaptation of the sensory neurons. Adaptation is usually ascribed to "fatigue"—that is, the reduced ability to continue production of electrical impulses. Sensory adaptation has sometimes been considered a basis for discontinued feeding [i.e., ending a meal (Barton Browne et al., 1975)], but it is probably not a major influence because of the intermittent contact of mouthpart sensilla with food, and because of the pauses between feeding bouts within a meal which give time for disadaptation (Blaney and Duckett, 1975). However, if different classes of chemoreceptor neurons adapt at different rates, the relative acceptability of different feeding substrates may alter. Such differential rates of adaptation have been described in caterpillars (Schoonhoven, 1977) and provide a simple explanation of why feeding bouts are shorter in the presence of mildly deterrent chemicals. The cells responsive to feeding stimulants have only to adapt more quickly than do the cells responsive to deterrents, to alter the balance of positive and negative inputs, and, thus, to alter the code received by the central nervous system. These experiential changes are very short-lived, however, with disadaptation (return to the original sensitivity) occurring within seconds or minutes. The effects would therefore influence meal length rather than initiation of feeding on the different substrates.

Longer-term changes in sensitivity in chemoreceptors may also occur in response to the nutritional status of an insect. This has been studied in most detail with locusts. Here, relatively high levels of amino acids in the hemolymph result in a reduced sensitivity of receptors on the tips of the maxillary palps to amino acids applied to them. Similarly, high concentrations of blood sugar cause a reduction in sensitivity of the receptors to sucrose. With normal feeding patterns, in which there are discrete meals separated by long intermeal intervals, the sensitivity of receptors rises and declines in relation to the changing composition of the hemolymph. The time scale is on the order of minutes to hours. These changes and their potential role in food choice behavior on artificial diets are covered in detail elsewhere (see Chapter 9).

Changes in chemoreceptor sensitivity that last on the order of days have also been described in insects. The first data on such changes were published by Schoonhoven (1969). *Manduca sexta* larvae were reared on artificial diet or on leaves of their host plants, and then their chemoreceptors were tested electrophysiologically with a variety of compounds. Responses to some compounds were significantly different in caterpillars from the two treatments, with a tendency for many chemicals to elicit the stronger responses from individuals reared on the artificial diet. For example, when the insects had been fed on artificial diet, cells responding to sodium chloride, salicin, inositol, and solanine (a host plant glycoalkaloid) showed up to twice the firing rate compared with the sensitivity of the same receptors of insects fed on their host plants. In some cases the changes in the receptors were correlated with changes in behavioral sensitivity to these compounds: higher firing rates of the inositol cell were correlated with increased responsiveness to inositol in the food.

It is difficult to interpret how such changes in chemoreceptors might normally influence feeding behavior, but some attempts have been made. By feeding insects on artificial diets with or without added chemicals, long-term effects on various chemoreceptors were studied in *M. sexta*. For example, when the phagostimulant inositol was added to the diet, the firing rate of the inositol receptor was subsequently reduced. It is not known if such a change would be due to repeated stimulation of the receptors directly, or whether it may result for postingestive feedbacks of the kind described in Chapter 9. Additional experiments have shown that different types of deterrent chemical incorporated into the diet cause a reduced firing of the deterrent receptor cells in response to the same as well as other deterrents. In one example, larvae were exposed to salicin in the diet for a period of days. When the sensilla were tested with salicin or with caffeine, the deterrent cell gave relatively low firing rates in response to both (Fig. 10.1) (Blaney et al., 1986; Schoonhoven, 1969). However, when given additional sugar in the diet there was no effect of the experience on the sugar receptor. Similarly, with *Spodoptera littoralis,* given experience of nicotine hydrogen tartrate and subsequently tested with nicotine hydrogen tartrate, there was a reduction in firing rates of the chemoreceptors when stimulated with it (Blaney and Simmonds, 1987).

Insects so pretreated with deterrents were also less deterred by the compounds at the behavioral level; they ate relatively more of foods containing these deterrents. Furthermore, the individuals of *M. sexta* that were most tolerant of salicin behaviorally showed the lowest firing rates to salicin in their maxillary sensilla

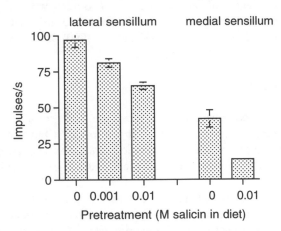

Figure 10.1. Changes in sensitivity of deterrent receptors in the larva of *Manduca sexta* as a result of previous exposure to a deterrent. Caterpillars were reared on plain diet, diet with the deterrent salicin added at a concentration of 0.001 M, and diet with salicin added at a concentration of 0.01 M. Shown here are the responses of both the lateral and the medial maxillary styloconic taste sensilla, each of which contain a deterrent receptor, when stimulated with salicin. In both sensilla, a marked reduction in the firing rate in response to salicin occurred in insects with previous experience of salicin. (After Schoonhoven, 1969.)

(Schoonhoven, 1976). These data strongly suggest that experience-induced changes in the sensitivity of chemoreceptors to particular compounds could influence the choice of foods thereafter. It is possible that, in some insects at least, habituation to deterrents (see Section 10.2.2) is simply mediated by changes in peripheral sensilla. In addition, the process of "induction of preference," in which some caterpillars tend to prefer the food they have been reared on (see Section 10.2.4), may be explained at least in part by these kinds of peripheral receptor changes.

The examples given by Simpson and his co-workers involve primary nutrients, and he proposes (see Chapter 9) that simple feedbacks from the hemolymph can regulate behavior by acting locally on the peripheral chemoreceptors (Simpson et al., 1991). He has direct evidence of this from experiments which involve injections into the hemolymph space immediately beneath a group of sensilla. Local increase in the concentration of amino acids causes a decrease in the firing rates of neurons to amino acids applied externally. The examples given by Schoonhoven tend to emphasize non-nutrients, and he reviews additional possible mechanisms for the changes found. In either case, the changing balance of inputs from the peripheral sensilla, received by the central nervous system, probably does influence behavior and is increasingly thought to be an important short-term experiential effect on the acceptability of particular foods.

Experience can also alter the sensitivity of odor receptors. Changes in odor receptors on the antennae of parasitic wasps that correlate with changes in behavioral preference have been described, and they have been implicated in behavioral changes that resulted from experience (Vet et al., 1990). It is possible that similar changes occur in relation to feeding.

10.2.2. Habituation

Habituation is defined behaviorally as the waning of a response to a stimulus with repeated exposure to that stimulus. It is a loss of habitual responses rather than the acquisition of new ones. The classic example is that of an animal hearing a loud noise, which initially elicits an escape response, but with repetition is ignored.

Among insects, studies with honeybees have been most detailed. For example, if one antenna of a bee is touched with sugar water, the bee extends its proboscis. If no sugar solution is encountered by the proboscis, the bee will, on subsequent stimulations of the same antenna with sugar water, gradually cease to extend its proboscis. If chemoreceptors on any of the mouthparts of such a habituated bee are stimulated with sugar, it will again extend its proboscis. Also, if the other antenna is stimulated, the proboscis is again extended. Such experiments suggested that habituation is a result of a local change in the antennal lobe on one side of the brain (Menzel et al., 1991).

A few experiments have explicitly examined habituation to feeding deterrents while ensuring that the feeding status of the insects was not involved. A detailed study was carried out with the polyphagous noctuid *Pseudaletia unipuncta,* for which caffeine is a feeding deterrent. Individual caterpillars were exposed to caffeine-treated maize leaves or untreated control leaves overnight. Then each caterpillar was allowed to take two meals on untreated corn to equalize their feeding states. After this, individuals of both groups were presented with caffeine-treated leaves: those that had previously experienced caffeine ate nearly twice as much as those that were naive (Fig. 10.2). (Usher et al., 1988). Clearly, they had habituated to this deterrent.

In a study with locusts, the deterrent, nicotine hydrogen tartrate, was painted onto leaves of sorghum and presented to individual insects on consecutive days for 18 hr each day. In the remaining hours each day they received uncontaminated food so that they could retain the status of well-fed insects and grow at a normal rate. On the first day of the experiment the treated leaves were deterrent to all individuals. By the third day, the nicotine-treated leaves were relatively acceptable to individuals of the desert locust, *Schistocerca gregaria,* although naive insects that were similar in age and state of feeding were still deterred by them (Fig. 10.3). The nicotine-experienced insects had habituated to this deterrent.

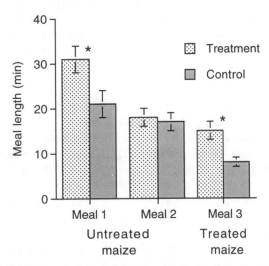

Figure 10.2. Habituation to a deterrent measured behaviorally in *Pseudaletia unipuncta.* Insects were allowed to feed on leaves of *Zea mays* coated with caffeine at a deterrent concentration, or on untreated leaves. After 24 hr, individual caterpillars were observed. Meals 1 and 2 were on untreated corn leaves to equalize the state of feeding of insects from the two treatments. Meal 3 was on corn leaves coated with caffeine. The caterpillars which had previously experienced caffeine ate markedly more than those which had not. (Based on data in Usher et al., 1988.)

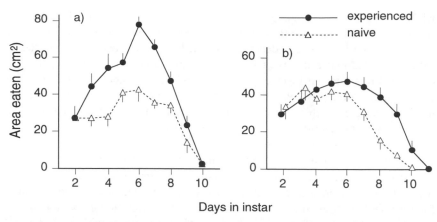

Figure 10.3. Habituation to deterrents in two species of locusts, showing the quantities of sorghum leaf treated with the deterrent nicotine hydrogen tartrate (NHT) consumed on different days of the final instar by (**a**) *Schistocerca gregaria* and (**b**) *Locusta migratoria*. Insects were either "experienced," having previously been exposed to sorghum treated with NHT, or "naive," having fed only on untreated sorghum. Experienced insects were the same individuals tested daily; naive insects were a new set of insects each day. The degree of habituation is marked in the polyphagous *S. gregaria* (**a**), even on the day after the start of the experiment. In the grass-feeding *L. migratoria* (**b**), no significant differences between the two treatments are seen until day 6, and these differences may be explained by the difference in instar length, because insects having a longer instar will have greater food intake for longer. Vertical lines represent the standard errors (After Jermy et al., 1982.)

Interestingly, individuals of the oligophagous migratory locust, *Locusta migratoria,* showed little habituation (Jermy et al., 1982).

Considerable variation in habituation may occur among individuals. An example of this variation was demonstrated in experiments on the polyphagous noctuid *Mamestra brassicae*. The caterpillars were tested with deterrent plants at intervals, while being allowed to feed on acceptable food between tests. Some individuals did not alter in their behavior over time, some habituated to the deterrent food and ate more, and others apparently became sensitized; that is, they showed increased deterrence (see Section 10.2.3) (Jermy et al., 1987).

It seems increasingly likely that habituation is involved in the process termed "induction" whereby insects such as caterpillars sometimes show an increased acceptance of the food they have experienced previously and a greater likelihood of rejecting an alternative food. Induction probably also involves other processes, however, and is discussed further in Section 10.2.4.

Habituation to deterrents may be very common, especially because many plant secondary compounds are deterrent but not toxic (Bernays, 1990; Bernays and

Chapman, 1987). There may be biological reasons for selecting a particular subset of plants in an environment, but this bias may be overcome, with no immediate costs, if the preferred plants are absent at any time. Field observations are required to determine how frequently such changes might occur in nature.

Other types of habituation may occur in relation to feeding. For example, insects may develop patterns of behavior based on previous experience of mechanical factors. In an example with locusts, individuals were fed daily in one of two different regimes. Some were allowed to take large meals at 12-hr intervals, while others were allowed to take only small meals at 4-hr intervals, but the total intake over a 24-hr period was the same. After 2 days they were allowed to empty their guts and then offered food *ad libitum*. The first meal size ingested by each insect was measured by weighing the foregut immediately after the end of the meal. Those insects that had been preconditioned on small meals ate markedly less than those preconditioned on large meals when they were given the test meals after the same periods of food deprivation (Bernays and Chapman, 1972).

The mechanisms underlying habituation to simple harmless sensory inputs have been studied in vertebrates and in molluscs, but less has been done with insects. Habituation to chemical deterrents probably involves changes in the chemoreceptors as described above, at least in some cases. It is possible that there is a reduction of acceptor sites on the chemoreceptor neurons, or that there is a change in the number of acceptor sites involved in the generation of a response by the cell (Blaney et al., 1986). Habituation to chemical deterrents is potentially a means of altering host selection behavior, enabling an insect to eat a previously unacceptable food. Habituation is not, however, a matter of just eating more of a deterrent food as a result of prior deprivation. Such an effect is driven by internal factors related to nutrient deficiency, rather than by a reduced sensitivity of deterrent receptors.

Further studies with *S. gregaria* showed that peripheral stimulation produced changes in the central nervous system and also that postingestive factors played a role (Szentesi and Bernays, 1984). The separation of these different factors was carried out by a series of different experiments.

Insects were fitted with small sections of microcapillaries over the tips of the maxillary palps. Some insects had a weak solution of nicotine hydrogen tartrate placed into the capillaries for a period of time each day, after which it was removed with absorbent paper. Control insects had just water placed in the capillaries for similar periods. Insects that were made to taste the nicotine hydrogen tartrate for periods of time without being able to ingest it subsequently accepted nicotine food more readily than did those treated with water. Because the capillaries extended beyond the tips of the palps, they also prevented the use of the maxillary palps in feeding decisions, so that sensilla other than the treated ones were providing the information to the central nervous system. The design of this experiment therefore precluded an explanation based on changing input from peripheral sensilla, and the change must have been somewhere more central in the nervous system.

In a further experiment, insects were dosed with encapsulated nicotine hydrogen tartrate to prevent it from being tasted, or with empty capsules. These capsules, made of gelatin, dispensed their contents into the gut some minutes after being swallowed. Subsequently, nicotine-treated food was accepted more readily by the nicotine-dosed insects than by insects dosed with empty capsules. Because the altered responses of the insects were caused by postingestive effects, it was assumed that there must have been some feedback to the central nervous system other than from the peripheral chemoreceptors.

In studies of habituation in other animals, synaptic changes in specific neural pathways to the central nervous system have been found: short-term habituation involving a decrease in neurotransmitter release at particular synapses and longer-term habituation involving also a decrease in productivity of neurotransmitters at synapses. While further studies are needed on insects, it is clear that the behavioral phenomenon of habituation may also depend on other types of neural change, including changes in peripheral sensilla.

In some cases, apparent habituation to feeding deterrents is brought about without any input from taste receptors. Thus *Manduca sexta* larvae will eat increasing amounts over time, of a diet laced with nicotine. In this case, the larvae do not taste the chemical and the diet is ingested by naive individuals. However, the noxious effects of the ingested chemical causes feeding to cease. With experience, larvae produce higher levels of detoxification enzymes so that the postingestive effects are less serious, and this in turn means that an individual encountering the nicotine diet will then feed for longer (J. Glendinning, *unpublished data*).

10.2.3. Sensitization

Sensitization is a fairly short-term experiential effect, usually lasting only for minutes. It is a process that results from the experience of biologically significant or intense sensory responses. It involves changes in the nervous system beyond the peripheral sensilla, and it causes an increased responsiveness to a stimulus on repeated exposure without any learned associations. For example, a deterrent that allows a potential food to be ingested for only a few seconds before it is rejected may, on subsequent encounter, prevent feeding at all. In some situations the taste of a strongly deterrent compound not only increases the responsiveness to that deterrent, but, in addition, leads temporarily to a reduced responsiveness to any food. Various physiological mechanisms have been proposed for sensitization, including presynaptic facilitation in interneurons and changes in the brain resulting in a heightened general state of excitation. One of the functional values of sensitization may be an increase in foraging efficiency; the decision not to eat may be made immediately, so the inspection and selection of potential foods can be relatively rapid.

Sensitization to stimuli producing positive effects always seems to be associated with increased activity. If contact is lost with a highly acceptable food, an insect may move around extremely rapidly, giving the appearance of being "excited." This was first described in flies by Dethier (1957), who found that when a fly encountered a small drop of sugar solution and tasted it, then, after losing contact with the stimulus, its speed of movement and its turning rate increased. This increasing propensity to search and find food after a very positive experience, such as tasting a highly acceptable food, may increase the chances of obtaining more of the same without any associative learning. This is especially likely if the food resource occurs in discrete patches. An example of the effect of such excitation is shown for locusts in Fig. 10.4 (Bernays and Chapman, 1974), but it is widespread in insects and other animals (Bell, 1991, p. 85).

The level of overall excitation or arousal may have a role in selectivity also. For example, insects artificially aroused by presentation with a highly stimulating sugar have been shown to ingest water which they otherwise would have rejected. This has been demonstrated in both flies and grasshoppers (reviewed by Barton Browne, 1975). This type of arousal is probably very short-lived. Extensive studies with flies showed that the decay time was between 1 and 5 min (Dethier et al., 1965). Similarly, honeybees sensitized by touching the antennae with a sugar solution are immediately more responsive to floral odors, but this effect is lost after about 3 min (Menzel et al., 1991).

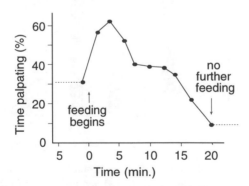

Figure 10.4. Excitation after contact with food in the migratory locust, *Locusta migratoria*. Individuals which had been deprived of food for 2 hr were given very small fragments of seedling wheat scattered over the substrate. After eating one piece, the insects actively turned and palpated in an apparent search behavior. The figure gives percentage of time spent palpating following loss of contact with the food at different stages in a meal. Early in the meal, very active palpation (up to 60% of time) occurs in these periods of loss of contact, illustrating excitation that probably represents sensitization. (After Bernays and Chapman, 1974.)

10.2.4. Induced Preferences

The term *induction* is used almost exclusively in studies with phytophagous insects in which individuals in choice tests tend to prefer the plant they have already experienced, over one they have not experienced, whether or not this plant is most appropriate for development. It has been demonstrated in different insect groups. Some authors have drawn parallels between induction and imprinting (Szentesi and Jermy, 1990).

Induction has been extensively studied among larvae of Lepidoptera where individuals of over 24 species have been shown to develop an altered preference in favor of the plant already experienced. Induction has also been demonstrated in Phasmatodea, Heteroptera, Homoptera, and Coleoptera (Hanson, 1976; Jermy, 1987; Szentesi and Jermy, 1990). In some cases the effect is quite extreme. For example, one study with the saturniid *Callosamia promethia* showed that experience on one host plant almost completely precluded acceptance of an alternative one (Fig. 10.5) (Hanson, 1976). Similarly, it was shown that the cabbage butterfly, *Pieris brassicae,* could be reared on *Brassica* or on *Tropaeolum,* yet if larvae were first induced on *Brassica,* they refused *Tropaeolum* and died of starvation (Ma, 1972). In most published cases, however, induction of preference is much less extreme. Often, the normal host plant will remain favored,

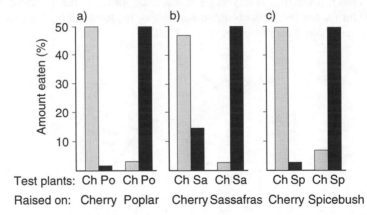

Figure 10.5. Induction of feeding preference in the larvae of *Callosamia promethia.* Results of three experiments with larvae reared and tested on different pairs of plants. (a) Insects were raised on wild cherry (*Prunus serotina*) or tulip poplar (*Liriodendron tulipifera*) and then tested for preference between these two. (b) Insects were raised on wild cherry or sassafras (*Sassafras albidum*) and then tested for preference between the two. (c) Insects were raised on wild cherry or spicebush (*Lindera benzoin*) and then tested for preference between the two. Height of columns represents the percentage of each test plant eaten. Experiments were conducted in petri dishes using leaf discs of alternating plant species. The test was terminated when 50% of one plant species had been eaten. (After Hanson, 1976.)

but if larvae are forced onto an alternative plant that is accepted although not so readily, the new plant may become relatively more acceptable thereafter, even though not as acceptable as the normal host.

Induction in Lepidoptera occurs in species with different host ranges, but it seems that the more taxonomically different the plants in any comparison, the greater the likelihood of finding induction. It has been suggested that the experience of very different plants leads to a greater difference in their relative acceptability in a choice test thereafter (Fig. 10.6) (de Boer and Hanson, 1984).

Most experiments on induction have been performed using a choice test, where relative amounts of the two foods eaten are measured, so it is not easy to distinguish increased acceptability of the plant experienced from decreased acceptability of the alternative plant, and such experiments tend to enhance very minor differences in acceptability that may not always have ecological relevance. In recent extensive studies (M. Weiss, *unpublished data*) it has become clear that even among species in which induction has been repeatedly demonstrated, the effects are extremely variable and there are always some individuals in which no changes occur.

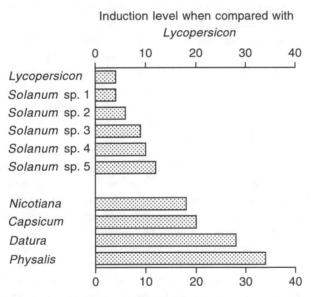

Figure 10.6. Induction of feeding preference in *Manduca sexta* in relation to plant taxon. Larvae were reared on one of ten different plants in the family Solanaceae (the host family). There were five species of *Solanum*, one in the closely related *Lycopersicon*, and species in four other genera. All insects were tested with a choice of the same plant versus *Lycopersicon*, to test for the level of induction on the different plants. The figure represents the level of induction (in arbitrary units) and illustrates the fact that within the two closely related genera there was little induction but when tests were between other, different genera, the induction was greater. (Based on data in de Boer and Hanson, 1984.)

Recently it has been shown that in the oligophagous larvae of *Manduca sexta*, induction could be explained mainly by decreases in deterrence of certain secondary metabolites (de Boer, 1992). This could thus be considered habituation. In some cases, changed thresholds in chemoreceptors are correlated with induction; these are described above under peripheral changes.

There may be elements of sensitization in cases where the adopted food has become increasingly acceptable. There may be some long-term facilitation effect in some central nervous pathways resulting from an early experience, such as occurs in the process of imprinting in vertebrates. This is suggested by the fact that in a variety of insects, responsiveness to particular chemicals depends on exposure to those chemicals at or before the time of hatching from the egg or eclosion from the pupa (Corbet, 1985) or soon afterwards (Jaenike, 1988). There may also be elements of associative learning, as in the example of *M. sexta* showing an increased attraction to an odor associated with its food (see below).

The process needs to be examined under natural conditions in the field to determine its prevalence and possible significance, but in any case there has been considerable speculation as to its biological meaning. In some field settings, one can envisage a benefit of induction; for example, when a larva falls off its host and must refind it, a heightened sensitivity to host odors may be useful. There may be some benefit of induction during normal feeding activities on the host plant if it heightens arousal and minimizes interruptions to feeding. In the laboratory a few experiments have demonstrated that larvae will sometimes die of starvation rather than eat the new food which, if it had been provided since hatching, would have been accepted. Translating this into a natural environment, the larva would presumably move away from the new plant after encounter, and perhaps at some later time refind the original species.

It would be an advantage to have an induced preference if larvae became particularly adept at digesting or detoxifying the preferred plant, but evidence in this direction is contradictory so far. However, in one case there is evidence for this (Karowe, 1989), and it may be that sufficiently detailed studies have not yet been done. This is mainly because it is difficult to separate the effects of simply not feeding from those of reduced digestive performance.

Because ecological or physiological benefits are not very clear, it may be that the phenomenon has its origin in terms of a more general requirement to narrow or simplify neural pathways, allowing the individual to make decisions more rapidly. Speed of decisions can, at least in theory, improve efficiency and decrease danger, allowing a relatively polyphagous animal to attain the benefits of genetically determined narrow host range (Bernays and Wcislo, 1994). It is also possible that many alterations in food acceptance behavior have no particular significance. Experiments with grasshoppers showed that prior experience of particular food plants could totally alter the acceptability hierarchy of a range of novel plants, although there were no obvious patterns and no explicable benefits to be derived from the change (Howard and Bernays, 1991).

Jermy (1987) noted that induction is a "statistical" phenomenon. In other words, there is a lot of variation, with only a proportion of individuals in any population showing the effect. A reexamination of data collected in "induction experiments" demonstrates that differences in effects of experience in different experiments reflect differences in the proportion of individuals that change.

10.2.5. Stimulating Effect of Novelty

In contrast to the phenomenon of induction, there are cases in which there appears to be an absolute requirement for change, or novelty. The highly polyphagous grasshopper *Taeniopoda eques* refuse to eat the same food plant in the laboratory over a long period, and it cannot be reared without having a mixture of plants available. Every one of 10 foods tested on isolated individuals showed a declining acceptability over time. Whatever the initial level of acceptability, the food was rejected by about the fifth encounter. Further study was undertaken with good-quality artificial diets that were all identical but had different added harmless flavors. After two meals on food with one flavor, individuals were likely to eat large meals on food with a novel flavor and small meals on the food with the flavor already experienced (Bernays et al., 1992). After experience with one of two different odors in the absence of food, *T. eques* ate sooner and for longer on an artificial diet laced with the novel odor than on the diet laced with the one already experienced (Fig. 10.7) (Bernays, 1992). This may be an exaggerated example of the well-known general phenomenon of an increased state of arousal

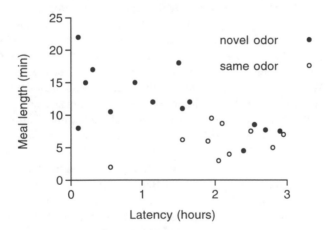

Figure 10.7. Effect of novel odors on the feeding behavior of *Taeniopoda eques*. Insects were placed in one of two cages without any food. One cage had the odor of citral and the other cage had the odor of coumarin. After 1 hr in the cage, each insect was observed in the presence of artificial diet laced with one of the two odors. Those that experienced the novel odor ate sooner and for longer than those experiencing the same odor. (After Bernays, 1992.)

when environmental stimuli change. It is perhaps also related to the process of sensitization. It is, of course, unlikely to occur in any but polyphagous species, and, of those, only in species that are mobile and individually polyphagous.

10.2.6. Indirect Nutritional Feedbacks

Experience of particular foods that are unbalanced with respect to certain nutrients can lead to short-term changes in behavior, and the changes result in an increased probability of selecting some complementary food or feeding for longer than usual on it. These changes in behavior are often due to associative learning (see Section 10.3). They may also be due to changes in the peripheral sensilla (see above). In some cases there is a suggestion that other mechanisms are at work.

Some insects have been shown to select wet or dry foods according to previous experience and/or current needs, and the changes in preference may be so rapid that individuals may switch among foods with different water content within a meal. The locust, *Schistocerca gregaria,* for example, was offered (under experimental conditions) fresh wheat seedlings and lyophilized wheat seedlings from the same batch. These were assumed to all be similar in their levels of basic nutrients. All individuals alternated between the wet and dry foods during the course of one meal, perhaps indicating short-term postingestive feedbacks driving preference first for the food that had excess water and then for the food with inadequate water (Lewis and Bernays, 1985).

In individuals that have been relatively deprived of an important nutrient, and then allowed access to it, rapid positive feedback from the gut can influence the amounts of the food then eaten. In Chapter 9, the case is made that concentrations of nutrients in the blood may alter during a meal and feed back directly on the peripheral chemoreceptors. Other, as yet unknown, feedbacks may also be possible. For example, experiments with the grasshopper *Schistocerca americana* showed within-meal feedbacks in relation to satisfying a sterol requirement. Individuals that had been pretreated with a food containing unutilizable sterols were allowed access to sugar-impregnated filter paper with either a utilizable sterol or a nonutilizable sterol. In the latter case, meal lengths were approximately 10 min, whereas with the satisfactory sterol, meal lengths were significantly longer, often 20 min (Champagne and Bernays, 1991). Evidence for an ability to taste the sterols under these circumstances is weak, but perhaps a short-term effect of sterol uptake that could result in the longer meals is via one of the several chemicals released into the blood from the gut during feeding (e.g., Endo and Nishiitsutsuji-Uwo, 1982; Montuenga et al., 1989). By analogy with molluscs, there may be such endogenous chemicals that modulate feeding and that play a part in the regulation of feeding rate and meal size (Susswein et al., 1984).

10.3. Associative Learning

In associative learning, experience results in an animal being able to associate a stimulus having no specific meaning (i.e., it is neutral), with some meaningful stimulus producing either positive or negative effects. As a result, on subsequent encounter, the response elicited previously only by the meaningful stimulus is then elicited by the neutral stimulus. In the literature on learning, the meaningful stimulus is termed the *unconditioned stimulus* (US), whereas the neutral one that comes to be associated with it is termed the *conditioned stimulus* (CS). Associative learning is a process that usually occurs in the brain, and a number of different models for potentiation of pathways as a result of experience have been developed (e.g., Carew and Sahley, 1986; Dudai, 1989; Marler and Terrace, 1984).

10.3.1. Association of Stimuli with Positive Effects

Positive associative learning in relation to feeding has been demonstrated with several insect species. Where the unconditioned stimulus is the taste of ingested food, both visual and chemical stimuli have been shown to be relevant conditioned stimuli. Most studies have been with honeybees. They rapidly learn to associate particular odors with sugar rewards, and some types of odors are more readily learned than others. Colors visible to bees, as well as different floral shapes, are also associated with food rewards, though the visual learning process may not be quite as rapid as odor learning. Some of the extensive literature on honeybee learning may be found in reviews (Bitterman, 1988; Gould, 1993; Menzel, 1985; Menzel et al., 1993). Rapid learning of colors and shapes of flowers in relation to food rewards has been demonstrated in other pollinators including bumblebees (Laverty, 1980) and, more recently, butterflies (Lewis and Lipani, 1990).

Among flies, learning has been demonstrated in the laboratory. A blowfly can be trained to extend its proboscis in response to salt or water on its tarsi if it is rewarded with sugar (Nelson, 1971). Similar experiments with the housefly, *Musca domestica*, demonstrated extensive learning ability (Yuval and Galun, 1987). Houseflies have also been shown to learn an association between odors and food. Flies were given sugar solution to drink in the presence of a variety of different single odors, and then tested 24 hr later to see if they responded to these odors. If presented with the odor previously accompanying the sugar, most flies extended their proboscises to drink. The response was highly specific to the precise odor experienced by each individual fly (Fukushi, 1973). Such learning ability is likely to be of considerable value in terms of successful foraging, but more realistic studies have not been undertaken except with species of *Rhagoletis*. Prokopy and his co-workers have demonstrated that these flies have an ability to learn colors, shapes, and odors of different host fruit in which they oviposit (reviewed in Papaj and Prokopy, 1989), so that it is likely that they can also learn cues associated with sources of food for themselves.

Among foliage-feeding species, there are a few examples of learning. For example, the grasshopper, *Melanoplus sanguinipes,* learned to forage preferentially in a place associated with a particular color/light intensity if that visual stimulus was associated with good-quality food (Fig. 10.8). Individuals were observed in an arena where two different-colored boxes were provided (dark green or yellow to mimic foliage and flowers of sunflower, which are both fed upon by this species in nature). The ambient temperature was suboptimal, and a light provided warmth on the roost or resting place, so that between foraging bouts insects returned to the resting place. Naive individuals left the roost to forage and would eventually stumble on the box with the food. After feeding they would return to the roost. The next feed was preceded by a very short search time, even if the positions of the colored boxes were reversed, indicating an ability to learn features of the environment related to color or light intensity (Bernays and Wrubel, 1985). The conditioned stimulus in this case was spectral reflectance, and the unconditioned stimulus was the food.

In a laboratory study with the desert locust, *Schistocerca gregaria,* individuals were trained to respond to specific odors associated with major essential nutrients. The locusts were trained for 2 days in boxes with free access to two artificial foods

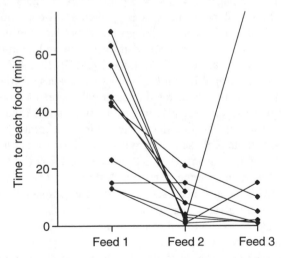

Figure 10.8. Visual learning in the grasshopper *Melanoplus sanguinipes.* Insects were observed in an arena in which there were two boxes of different color/light intensity. Food was placed in only one of the boxes and was not visible to the insects. Naive insects ready to feed took a relatively long time to locate the food when they left their basking site. However, after a single experience of finding the food in a particular colored box, they located the same-colored box very quickly on the second foraging bout. (Based on data in Bernays and Wrubel, 1985.)

that were similar in all respects except that one lacked digestible carbohydrate and the other lacked protein. Each was paired with one of two distinctive odors, carvone or citral (Fig. 10.9a). After the training period, insects were put in clean boxes having no added odors and were given only one of the two diets so that they became deficient in either carbohydrate or protein. On retesting, insects deprived of carbohydrate did not respond differentially to the odors, but those deprived of protein responded very differently to the two odors, approaching the source of the odor previously paired with the protein diet significantly more often than expected (Fig. 10.9b) (Simpson and White, 1990). The experiments demonstrated that the insects had associated an odor with the availability of a

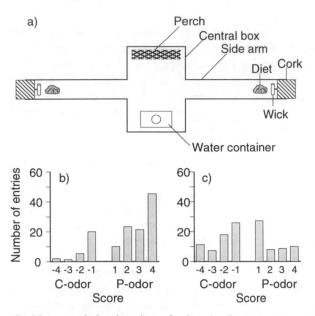

Figure 10.9. Positive associative learning of odors by *Locusta migratoria*. (**a**) The experimental apparatus. A small quantity of artificial diet was used, containing either (i) protein but no carbohydrate or (ii) carbohydrate but no protein. The wick close to the food was dosed with a volatile compound, either carvone or citral, so that insects placed in the central box could associate a specific odor with one of the diets. During the test, the same odors were given, but both the side arms contained artificial diet lacking both carbohydrate and protein. Each insect was scored according to the distance it moved along the arms: 1 indicates that it moved into the arm, but did not move far from the central box; 4 indicates that it moved to the vicinity of the diet. (**b**) Graph showing the results of tests with insects deprived of protein for 4 hr following the training period. They showed a strong tendency to move into the arm containing the odor previously associated with the protein-containing diet. (**c**) Graph showing the results of tests with insects deprived of carbohydrate for 4 hr following the training period. They showed no tendency to move preferentially into either arm. (After Simpson and White, 1990.)

key nutrient (i.e., protein) that they were lacking. The conditioned stimulus was the odor of the chemical; and the unconditioned stimulus, while unidentified, was presumably some nutrient feedback associated with ingestion of protein. In a field situation, it may be possible for individual grasshoppers to learn to feed on a particular food plant that best satisfies a regular deficit of such a major nutrient.

A possible example of associative learning was reported in studies of induction in the caterpillar M. sexta. The caterpillars showed increased orientation responses toward a source of the odor citral after being fed on artificial diet with added citral. Insects having had plain diet, on the other hand, turned significantly more frequently toward the plain diet and away from the odor of citral (Saxena and Schoonhoven, 1982).

Mechanisms of associative learning have been studied in honeybees. There are three phases. First there is a short-term memory of up to about 3 min which is dominated by the nonassociative process of sensitization, and in the case of stimuli from the antennae it is restricted to effects in the antennal lobes. Then there is an intermediate-term memory of about 3–30 min which consolidates this effect. Finally there is a long-term memory. These last two result from changes that occur in the mushroom bodies (Menzel, 1990).

10.3.2. Association of Stimuli with Negative Effects

Although the phenomenon of learning to reject food as a result of negative experience is obviously associative, it has often been given special treatment in earlier literature because of the characteristic time delay that occurs between ingestion of a food and the consequent deleterious effects. However, it has now become obvious that there is a continuum between avoidance learning unrelated to food chemistry (such as to an unacceptable physical stimulus), avoidance learning of bad taste, and avoidance learning of food that causes deleterious consequences upon ingestion [see Bernays (1993) for a review].

Clearly, an ability to learn to avoid a plant due to a noxious effect following ingestion would be of considerable potential value. Field behavioral studies would not allow a distinction between aversion learning, sensitization, or changes in behavior brought about by other, unknown variables. Its potential importance in food selection must initially be demonstrated in laboratory experiments, where relevant controls can ensure that the results are unambiguous.

In order to demonstrate an ability to associate a taste with a noxious effect following ingestion, nymphs of the grasshopper Schistocerca americana were injected with nicotine hydrogen tartrate after feeding on a particular plant, and then they were tested on the same or a different plant. It was found that individuals given the different plants, known to be acceptable, after this experience ate normal-sized meals. On the other hand, individuals given the plants that had been eaten immediately before the injection ate little or none (Bernays and Lee,

1988). This demonstrates that the insects were not too sick to eat, but rejected the type of plant eaten just before they suffered the symptoms. Additional controls showed that the chemical injected was critical, and not the injection process itself. The conditioned stimulus, the taste of the first plant, and the unconditioned stimulus, a factor associated with the chemically induced sickness, became associated. The results were clear-cut when test plants were species of moderate or low acceptability, but plants that were very highly acceptable in the first instance never became unacceptable. In a series of experiments, a variety of different secondary metabolites were found to be effective; the grasshoppers learned to reject foods that were eaten immediately before an experience of toxicity (Fig. 10.10) (Lee and Bernays, 1990).

Learned aversions to noxious chemicals may be widespread, even though there has been so little study of the process in insects. Studies of two predator types provide some evidence of aversion learning, although the tests did not involve specific controls. Mantids refused milkweed bugs after experiencing them (Berenbaum and Militczky, 1984; Gelperin, 1968; Paradise and Stamp, 1991), and

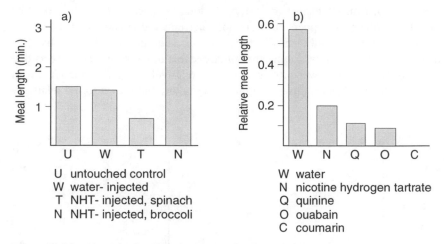

Figure 10.10. Aversion learning in the grasshopper *Schistocerca americana*. Individual grasshoppers were given one meal on the rearing food and one on spinach. Following this, they were either injected with a mild toxin, injected with water, or left untouched. Amounts injected were not usually enough to cause any obvious symptoms, although in some cases regurgitation occurred about 15 min after injection. (a) The lengths of tests meals on spinach or a novel food (broccoli) after injection. NHT-injected insects took only very short meals on spinach, but long meals on broccoli. This shows that the injection did not cause a general reduction in feeding, and the small meals on spinach indicate a specific association. (b) Effects of injection of different toxins. After the injection, the insects were given a test meal on spinach. The duration of this meal is expressed as a proportion of the length of the spinach meal before injection. All four compounds tested caused a reduction in meal length. (Based on data in Lee and Bernays, 1990).

vespid wasps avoided aposematic species of caterpillars after initial attempts to feed on them (Bernays, 1993).

The probable occurrence of aversion learning in relation to poor nutrient availability has been shown with grasshoppers feeding on spinach leaves. Successive meals on this plant became smaller until, after about four meals, it was rejected (Lee and Bernays, 1988). Spinach contains phytosterols which are totally unsuitable for this grasshopper; and because dietary sterols are essential nutrients, it appeared that rejection might have been associated with the absence of usable sterols or the presence of unusable sterols. Later experiments demonstrated that when appropriate sterols were added to the spinach leaves, the decline in acceptability did not occur (Fig. 10.11) (Champagne and Bernays, 1991). The sterols themselves appeared not to be tasted, so that the results indicate that the flavor of spinach, which was initially acceptable, became unacceptable as the insect obtained feedback concerning its nutritional unsuitability.

Aversions may also be induced by an inadequate protein concentration in the food. In a series of experiments with *S. americana,* artificial diets were prepared that were either very low in protein (2% wet weight) or higher in protein (4% wet weight). Either tomatine or rutin was added at concentrations that could be detected by the grasshoppers but were not deterrent. Individuals were fed on one of the diets for 4 hr and then offered the low-protein diet with the familiar or a different flavor (tomatine, rutin, or nontoxic levels of nicotine). The insects that had experienced the lower-protein foods fed relatively longer on the diets with different flavors than on those with the same flavor, and they also fed relatively

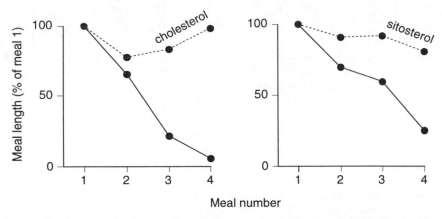

Figure 10.11. Aversion learning by grasshoppers of a food with unsuitable sterols. The solid lines show successive meal lengths by nymphs of *Schistocerca americana* on leaves of spinach. Successive meals decline in length, and the fourth encounter is usually followed by rejection. When either cholesterol or sitosterol, both utilizable sterols, are added (dashed lines), the learned aversion does not develop. (After Champagne and Bernays, 1991.)

longer than did the insects which had experienced the higher protein (Bernays and Raubenheimer, 1991). The conclusion was that insects fed protein-deficient foods subsequently showed an aversion for the flavor associated with the poor food as well as a preference for novel foods (neophilia). Similar kinds of results have often been demonstrated in studies on vertebrates, whereby learned aversions of one food go hand in hand with an increased acceptability of a novel food. In fact, in choice tests, the two cannot be separated. In any case the learned aversion for flavor coupled with low protein indicates that the neutral flavor had acquired significance (i.e., became the conditioned stimulus) related to the poor protein content of the food.

These data indicate that aversion learning, in grasshoppers at least, may have an important role in dietary mixing to obtain a suitable nutrient balance. Field studies have shown considerable individual polyphagy in grasshoppers, and the dietary mixing may have some basis in improving nutrient mixes ingested. It is possible, for example, that successive aversion learning experiences on a series of plants that are each imperfect would lead to a better diet than would remaining on one plant only. A laboratory experiment to test this idea was carried out also with *S. americana*. Two unbalanced but complementary artificial diets were offered to grasshoppers, with or without distinctive flavors. The behavior of the grasshoppers was then monitored over 3 days. In all cases the grasshoppers were mobile enough to encounter and eat both diets, but the added flavors enhanced the amount of mixing (Bernays and Bright, 1991). It may be that the distinctive secondary chemistry of plants is of considerable value to phytophagous insects that are individually polyphagous, because they would provide distinctive signatures and aid in the process of learning (Bernays and Bright, 1993).

The evidence so far indicates that aversion learning relating to nutrient profiles causes altered preferences among different foods, and that relative acceptability of a food will vary according to its nutrient content and, in addition, the insect's nutrient status/experience. *S. americana* is highly polyphagous; and it is perhaps to be expected that an ability to learn from experience in this manner would be highly valuable, just as has been proposed in vertebrates. In the experiments showing learned aversion to poisons in this grasshopper, the memory of the experience was extinguished after 4 days. This period would probably allow plenty of time to move into different areas with different foods to choose from. With nutrient deficiency, the length of the memory has not been tested, but short-term effects would be adequate to improve the efficiency of foraging.

There is one report of apparent aversion learning in caterpillars. Experiments were carried out with two species of polyphagous arctiids, *Diacrisia virginica* and *Estigmene congrua*. These species initially responded positively to petunia leaves. After ingestion of this plant, the caterpillars regurgitated the leaf material; and when subsequently given a choice of petunia and another plant, they selected the alternative (Dethier, 1980). The effect was not produced in the oligophagous caterpillar *Manduca sexta*, even though it, too, regurgitated. There is the suggestion that polyphagous species are better able to learn such negative associations (Dethier and Yost, 1979). Gelperin and Forsythe (1975) also suggest that polyphagous species would learn more readily than oligophagous or monophagous ones. Species which are hardwired to accept a narrow range of plants and reject all others are less likely to be able to learn to avoid plants through experience. However, while extreme specialists may never ingest food that is toxic to them, some at least move from low- to high-quality plants. This could be due to increasing deterrence of the low-quality plant but could also result from aversion learning. For example, Wang (1990) showed that the creosote bush grasshopper, *Ligurotettix coquilletti,* moved away from bushes they started to develop on and accumulated on bushes known (by the observer) to be of higher quality for development. Similarly, Parker (1984) found that the grasshopper *Hesperotettix viridis* had a shorter tenure time on damaged host plants than on undamaged ones.

Because there is often a time delay between sensory patterns associated with food intake and the (negative) consequences of ingestion, it is to be expected that certain patterns of feeding would enhance the likelihood of aversion learning. For example, discrete meals on single food items will allow associations to be made more readily than would grazing on a mixture of foods within a meal (Zahorik and Houpt, 1981). Another feeding habit that may allow learned aversions to form is that of short-term fidelity to a particular resource such that a learned aversion can develop over a series of meals on a single food type, whereupon rejection and movement away may follow if the food is unsuitable. Species that tend to rest on or near their food, as do most plant-feeding insects, are in this category. It may be that many such insects have extensive capabilities to learn associations between food characteristics and unsuitability, but establishing this with certainty requires long-term continuous observations.

Aversion learning of successively encountered plant species could theoretically improve the nutrient mix ingested over any random selection. Field studies were conducted to investigate the likelihood of this possibility by making continuous 6- to 10-hr observations on individual grasshoppers, recording the behavior on hand-held computers. Contrasts were encountered among species. For example, three species engaged in such rapid switching between plant species that learning based on any postingestive feedback would have been impossible. On the other hand, two species retained host fidelity for long enough that some short-term aversion learning could be invoked for movement between food types.

Inability to move readily from one food resource to another is a constraint in many insect species; and, in others, distances between potential alternative foods may be prohibitive for any learning to be really useful. For example, many holometabolous larvae have no alternative but to remain on or in their food source, and aversion learning has no relevance for them. This is clearly the case with many fly larvae, such as leaf miners or carcass dwellers, and for many beetle larvae, such as wood-boring species. Certain homopterans such as scale insects show the most extreme restriction; they are totally immobile and must feed in the position first selected.

Blaney and others (e.g., Blaney and Simmonds, 1985, 1987; Blaney et al., 1985; Chapman and Sword, 1993) have demonstrated with both grasshoppers and caterpillars that individuals presented with a relatively unfavorable plant will bite and reject the plant at first encounter. On successive contacts a greater proportion of individuals reject at palpation—that is, before biting (Fig. 10.12). The authors suggested that insects learn to associate the superficial taste or smell with the internal constituents of the plant. Avoidance learning appeared to be occurring, even though there had been no ingestion.

10.4. Conclusions

Experience, especially of different chemicals, has a variety of effects on food selection behavior and is one of the factors leading to the variability of response seen in nature. Various different neural phenomena are involved, including changes in the input from peripheral sensilla. These chemoreceptor changes vary in their extent and in how long they last, from adaptation lasting seconds or minutes, to longer-term changes with unknown mechanisms, lasting hours or days. An apparent difference from vertebrates is this plasticity of the chemoreceptors, which may then govern some of the changes in food selection behavior.

Other neural changes involve synaptic changes in particular neural pathways,

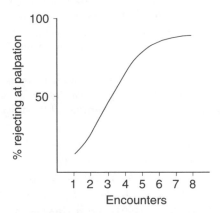

Figure 10.12. The effect of experience on rejection of deterrent plants by *Locusta migratoria*. Insects were offered a plant that was not an acceptable host. On first encounter they commonly palpated a leaf and then bit into it before rejection. On subsequent encounters they were more likely to reject the leaf at palpation—that is, before they bit into it. The insects apparently learned to associate factors at the surface of the leaf (volatiles or wax components) with the unacceptable constituents of the leaf interior. (Based on data in Blaney and Simmonds, 1985.)

such as those invoked for habituation. In addition, there may be a more generalized arousal such as is believed to occur in the process of sensitization. Other nonlearned factors may involved feedbacks via the blood, from nutrients or hormones. These could alter the peripheral receptors, central nervous system receptors, or neural pathways, and thus alter the final integrated information.

Finally, there are clear demonstrations of associative learning or conditioning, in which stimuli which are usually neutral or insignificant become cues associated with good food, poor food, specific nutrients, or toxins. However, in many cases where effects of experience that alter feeding behavior have been described, it is likely that both associative and nonassociative effects contributed to the effects. This is particularly the case with "induction"; but even where something as apparently simple as habituation has been demonstrated, there are probably effects at both peripheral and central neural levels.

References

Barton Browne, L. (1975). Regulatory mechanisms in insect feeding. *Adv. Insect Physiol.* 11, 1–116.

Barton Browne, L., Moorhouse, J. E., and van Gerwen, A. C. M. (1975). Sensory adaptation and the regulation of meal size in the Australian plague locust, *Chortoicetes terminifera*. *J. Insect Physiol.* 21, 1633–1639.

Bell, W. J. (1991). *Searching Behavior*. Chapman and Hall, New York.

Berenbaum, M. R., and Militczky, E. (1984). Mantids and milkweed bugs. Efficacy of aposematic coloration against invertebrate predators. *Am. Midl. Nat.* 111, 64–68.

Bernays, E. A. (1990). Plant secondary compounds deterrent but not toxic to the grass specialist *Locusta migratoria:* implications for the evolution of graminivory. *Entomol. Exp. Appl.* 54, 53–56.

Bernays, E. A. (1992). Dietary mixing in generalist grasshoppers. In: Menken, S. B. J., Visser, J. H., and Harrewijn, P. (eds.), *Proceedings of the 8th International Symposium on Insect–Plant Interactions*. Centre for Agricultural Publishing and Documentation, Wageningen, pp. 146–148.

Bernays, E. A. (1993). Food aversion learning. In: Lewis, A. C., and Papaj, D. (eds.), *Insect Learning*. Chapman and Hall, New York, pp. 1–17.

Bernays, E. A., and Bright, K. L. (1991). Dietary mixing in grasshoppers: switching induced by nutritional imbalances in foods. *Entomol. Exp. Appl.* 61, 247–254.

Bernays, E. A., and Bright, K. L. (1993). Dietary mixing in grasshoppers: a review. *Comp. Biochem. Physiol.* 104A, 125–131.

Bernays, E. A., and Chapman, R. F. (1972). Meal size in nymphs of *Locusta migratoria*. *Entomol. Exp. Appl.* 15, 399–410.

Bernays, E. A., and Chapman, R. F. (1974). The regulation of food intake by acridids. In: Barton Browne, L. (ed.), *Experimental Analysis of Insect Behaviour*. Springer-Verlag, Berlin, pp. 48–59.

Bernays, E. A., and Chapman, R. F. (1987). Evolution of deterrent responses in plant-feeding insects. In: Chapman, R. F., Bernays, E. A., and Stoffolano, J. G. (eds.), *Perspectives in Chemoreception and Behavior*. Springer-Verlag, New York, pp. 1–16.

Bernays, E. A., and Lee, J. C. (1988). Food aversion learning in the polyphagous grasshopper *Schistocerca americana*. *Physiol. Entomol.* 13, 131–137.

Bernays, E. A., and Raubenheimer, D. (1991). Dietary mixing in grasshoppers: changes in acceptability of different plant secondary compounds associated with low levels of dietary protein. *J. Insect Behav.* 4, 545–556.

Bernays, E. A., and Wcislo, W. T. (1994). Sensory capabilities, information processing, and resource specialization. *Q. Rev. Biol.* 69, 187–204.

Bernays, E. A., and Wrubel, R. P. (1985). Learning by grasshoppers: association of colour/light intensity with food. *Physiol. Entomol.* 10, 359–369.

Bernays, E. A., Bright, K., Raubenheimer, D., and Champagne, D. (1992). Variety is the spice of life: frequent switching between foods in the polyphagous grasshopper *Taeniopoda eques*. *Anim. Behav.* 44, 721–731.

Bitterman, M. E. (1988). Vertebrate–invertebrate comparisons. In: Jerison, H. J., and Jerison, I. (eds.), *Intelligence and Evolutionary Biology*. Springer-Verlag, Berlin, pp. 251–276.

Blaney, W. M., and Duckett, A. M. (1975). The significance of palpation by the maxillary palps of *Locusta migratoria*: an electrophysiological and behavioural study. *J. Exp. Biol.* 63, 701–712.

Blaney, W. M., and Simmonds, M. S. J. (1985). Food selection by locusts: the role of learning in rejection behaviour. *Entomol. Exp. Appl.* 39, 273–278.

Blaney, W. M., and Simmonds, M. S. J. (1987). Experience: a modifier of neural and behavioral sensitivity. In: Labeyrie, V., Fabres, G., and Lachaise, D. (eds.), *Insects–Plants. Proceedings of the 6th Int Symposium on Insect–Plant Relationships*. Junk, Dordrecht, pp. 237–241.

Blaney, W. M., Winstanley, C., and Simmonds, M. S. J. (1985). Food selection by locusts: an analysis of rejection behaviour. *Entomol. Exp. Appl.* 38, 35–40.

Blaney, W. M., Schoonhoven, L. M., and Simmonds, M. S. J. (1986). Sensitivity variations in insect chemoreceptors: a review. *Experientia* 42, 13–19.

Carew, T. J., and Sahley, C. L. (1986). Invertebrate learning and memory: from behavior to molecules. *Annu. Rev. Neurosci.* 9, 435–487.

Champagne, D. E., and Bernays, E. A. (1991). Phytosterol unsuitability as a factor mediating food aversion learning in the grasshopper *Schistocerca americana*. *Physiol. Entomol.* 16, 391–400.

Chapman, R. F., and Sword, G. (1993). The importance of palpation in food selection by a polyphagous grasshopper (Orthoptera: Acrididae). *J. Insect Behav.* 6, 79–92.

Corbet, S. A. (1985). Insect chemosensory responses: a chemical legacy hypothesis. *Ecol. Entmol.* 10, 143–151.

de Boer, G. (1992). Diet-induced food preference by *Manduca sexta* larvae: acceptable non-host plants elicit a stronger induction than host plants. *Entomol. Exp. Appl.* 63, 3–12.

de Boer, G., and Hanson, F. (1984). Food plant selection and induction of feeding preferences among host and non-host plants in larvae of the tobacco hornworm *Manduca sexta. Entomol. Exp. Appl.* 35, 177–194.

Dethier, V. G. (1957). Communication by insects: physiology of dancing. *Science* 125, 331–336.

Dethier, V. G. (1980). Food aversion learning in two polyphagous caterpillars, *Diacrisia virginica* and *Estigmene congrua. Physiol. Entomol.* 5, 321–325.

Dethier, V. G., and Yost, M. T. (1979). Oligophagy and the absence of food-aversion learning in tobacco hornworms, *Manduca sexta. Physiol. Entomol.* 4, 125–130.

Dethier, V. G., Soloman, R. L., and Turner, L. H. (1965). Sensory input and central excitation and inhibition in the blowfly. *J. Comp. Physiol. Psychol.* 60, 303–313.

Dudai, Y. (1989). *The Neurobiology of Memory.* Oxford University Press, Oxford.

Endo, Y., and Nishiitsutsuji-Uwo, J. (1982). Exocytic release of secretory granules from endocrine cells in the midgut of insects. *Cell Tissue Res.* 222, 515–522.

Fukushi, T. (1973). Olfactory conditioning in the housefly *Musca domestica. Ann. Zool. Jpn.* 46, 135–143.

Gelperin, A. (1968). Feeding behavior of the praying mantis: a learned modification. *Nature* 286, 149–150.

Gelperin, A., and Forsythe, D. (1975). Neuroethological studies of learning of mollusks. In: Fentress, J. C. (ed.), *Simpler Networks and Behavior.* Sinauer, New York, pp. 239–250.

Gould, J. L. (1993). Ethological and comparative perspectives on honey bee learning. In: Papaj, D., and Lewis, A. C. (eds.), *Insect Learning,* Chapman and Hall, NY, pp. 18–50.

Hanson, F. E. (1976). Comparative studies on induction of food choice preferences in lepidopterous larvae. *Symp. Biol. Hung.* 16, 71–77.

Howard, J. J., and Bernays, E. A. (1991). Effects of experience on palatability hierarchies of novel plants in the polyphagous grasshopper *Schistocerca americana. Oecologia* 87, 424–428.

Jaenike, J. (1988). Effects of early adult experience on host selection in insects: some experimental and theoretical results. *J. Insect Behav.* 1, 3–16.

Jermy, T. (1987). The role of experience in the host selection of phytophagous insects. In: Chapman, R. F., Bernays, E. A., and Stoffolano, J. G. (eds.), *Perspectives in Chemoreception and Behavior.* Springer-Verlag, New York, pp. 143–157.

Jermy, T., Bernays, E. A., and Szentesi, A. (1982). The effect of repeated exposure to feeding deterrents on their acceptability to phytophagous insects. In: Visser, J. H., and Minks, A. K. (eds.), *Insect–Plant Relationships.* Pudoc, Wageningen, pp. 25–30.

Jermy, T., Horvath, J., and Szentesi, A. (1987). The role of habituation in food selection of lepidopterous larvae: the example of *Mamestra brassicae.* In: Labeyrie, V., Fabres, G., and Lachaise, D. (eds.), *Insects–Plants.* Junk, Dordrecht, pp. 231–236.

Karowe, D. (1989). Facultative monophagy as a consequence of prior feeding experience: behavioral and physiological specialization in *Colias philodice* larvae. *Oecologia* 78, 106–111.

Laverty, T. M. (1980). The flower-visiting behavior of bumblebees: floral complexity and learning. *Can. J. Zool.* 58, 1324–1335.

Lee, J. C., and Bernays, E. A. (1988). Declining acceptability of a food plant for the polyphagous grasshopper, *Schistocerca americana:* the role of food aversion learning. *Physiol. Entomol.* 13, 291–301.

Lee, J. C., and Bernays, E. A. (1990). Food tastes and toxic effects: associative learning by the polyphagous grasshopper *Schistocerca americana. Anim. Behav.* 39, 163–173.

Lewis, A. C., and Bernays, E. A. (1985). Feeding behaviour: selection of both wet and dry food for increased growth in *Schistocerca gregaria* nymphs. *Entomol. Exp. Appl.* 37, 105–112.

Lewis, A. C., and Lipani, G. (1990). Learning and flower use in butterflies. In: Bernays, E. A. (ed.), *Insect–Plant Interactions, vol. II.* CRC Press, Boca Raton, FL, pp. 95–110.

Ma, W. C. (1972). Dynamics of feeding responses in *Pieris brassicae* Linn. as a function of chemosensory input: a behavioural, ultrastructural and electrophysiological study. *Meded. Landbouwhogesch. Wageningen* 11, 1–162.

Marler, R., and Terrace, H. S. (eds.) (1984). *The Biology of Learning.* Springer-Verlag, Berlin.

Menzel, R. (1985). Learning in honeybees in an ecological and behavioral context. In: Holldobler, B., and Lindauer, M. (eds.), *Experimental Behavioral Ecology and Sociobiology.* Fischer Verlag, Stuttgart, pp. 55–74.

Menzel, R. (1990). Learning, memory and "cognition" in honeybees. In: Kesner, R. P., and Olten, D. S. (eds.), *Neurobiology of Comparative Cognition.* Lawrence Erlbaum Associates, Hillsdale, NJ, pp. 237–292.

Menzel, R., Hammer, M., Braun, G., Mauelshagen, J., and Sugawa, M. (1991). Neurobiology of learning and memory in honeybees. In: Goodman, L. J., and Fisher, R. C. (eds.), *The Behaviour and Physiology of Bees.* C.A.B. International, Wallingford, Oxford, pp. 323–353.

Menzel, R., Greggers, U., and Hammer, M. (1993). Functional organization of appetitive learning and memory in a generalist pollinator, the honey bee. In: Papaj, D., and Lewis, A. C. (eds.), *Insect Learning.* Chapman and Hall, NY, pp. 79–125.

Montuenga, L. M., Barrenechia, M. A., Sesma, P., Lopez, J., and Vazquez, J. J. (1989). Ultrastructure and immunocytochemistry of endocrine cells in the midgut of the desert locust, *Schistocerca gregaria. Cell Tissue Res.* 258, 577–583.

Nelson, M. C. (1971). Classical conditioning in the blowfly (*Phormia regina*). *J. Comp. Physiol. Psychol.* 77, 353–368.

Papaj, D. R. and Prokopy, R. J. (1989). Ecological and evolutionary aspects of learning in phytophagous insects. *Annu. Rev. Entomol.* 34, 315–350.

Paradise, C. J., and Stamp, N. E. (1991). Prey recognition times of praying mantids and consequent survivorship of unpalatable prey. *J. Insect Behav.* 4, 265–273.

Parker, M. A. (1984). Local food depletion and the foraging behavior of a specialist grasshopper, *Hesperotettix viridis*. *Ecology* 65, 824–835.

Saxena, K. N., and Schoonhoven, L. M. (1982). Induction of orientational and feeding preferences in *Manduca sexta* larvae for different food sources. *Entomol. Exp. Appl.* 32, 172–180.

Schoonhoven, L. M. (1969). Sensitivity changes in some insect chemoreceptors and their effect on food selection behaviour. *Proc. Kon. Ned. Akad. Wet. C* 72, 491–498.

Schoonhoven, L. M. (1976). On the variability of chemosensory information. *Symp. Biol. Hung.* 16, 261–266.

Schoonhoven, L. M. (1977). On the individuality of insect feeding behavior. *Proc. Kon. Ned. Akad. Wtsch. C* 80, 341–350.

Simpson, S. J., and White, P. R. (1990). Associative learning and locust feeding: evidence for a "learned hunger" for protein. *Anim. Behav.* 40, 506–513.

Simpson, S. J., James, S., Simmonds, M. S. J., and Blaney, W. M. (1991). Variation in chemosensitivity and the control of dietary selection behaviour in the locust. *Appetite* 17, 141–154.

Susswein, A. J., Klaudiusz, F., Weiss, R., and Kupfermann, I. (1984). Internal stimuli enhance feeding behavior in the mollusc *Aplysia*. *Behav. Neural Biol.* 41, 90–95.

Szentesi, A., and Bernays, E. A. (1984). A study of behavioural habituation to a feeding deterrent in nymphs of *Schistocerca gregaria*. *Physiol. Entomol.* 9, 329–340.

Szentesi, A., and Jermy, T. (1990). The role of experience in host plant choice by phytophagous insects. In: Bernays, E. A. (ed.), *Insect–Plant Interactions,* vol. 2. CRC Press, Boca Raton, FL, pp. 39–74.

Usher, B. F., Bernays, E. A., and Barbehenn, R. V. (1988). Antifeedant tests with larvae of *Pseudaletia unipuncta:* variability of behavioral response. *Entomol. Exp. Appl.* 48, 203–212.

Vet, L. E. M., De Jong, R., Giessen, W. A. van, and Visser, J. H. (1990). A learning-related variation in electroantennogram responses of a parasitic wasp. *Physiol. Entomol.* 15, 243–247.

Wang, G. Y. (1990). Dominance in territorial grasshoppers: studies of causation and development. Ph.D. thesis, University of California, Los Angeles, 139pp.

Yuval, B, and Galun, R. (1987). Aspects of compound conditioning to gustatory stimuli in the housefly *Musca domestica* L. *J. Insect Physiol.* 33, 159–165.

Zahorik, D. M., and Houpt, K. A. (1981). Species differences in feeding strategies, food hazards, and the ability to learn aversions. In: Kamil, A., and Sargent, T. (eds.), *Foraging Behavior: Ecological, Ethological and Psychological Approaches.* Garland Press, New York, pp. 289–310.

11

Ontogenic Changes in Feeding Behavior

L. Barton Browne

11.1. Introduction

The feeding behavior of insects changes, often radically, according to their stage of development, with individuals in different developmental stages differing as regards amounts and, often, the kinds of food they eat. It is well known, for example, that the larvae and adults of most holometabolous insects have very different feeding behaviors and diets. Perhaps less widely recognized are the less striking, but nonetheless still substantial, changes which can occur during larval development, both within and between instars, and during adult life in relation to stages of somatic and reproductive development and to the level of an insect's reproductive activity.

The overall development of an insect can usefully be considered to consist of a series of distinct major developmental programs, such as development within each larval instar, pupal development, any posteclosion development of somatic tissues, and the development of the reproductive organs and products. Each program has its distinct characteristics as regards the nature of the growing tissues, the temporal pattern of their growth, and the nature and source of the required nutrients. The nutritional requirements for tissue growth vary according to which tissues are growing and the stage to which their growth has proceeded. As might be expected, there is considerable evidence that such variation in the nutritional requirements of the developing tissues is reflected in more or less contemporaneous variation in the amounts and kinds of foods ingested, except in those instances where nutrients for the growth of the tissues are derived wholly or largely from endogenous sources.

The concept of growth, nutritional, and intake targets proposed by Raubenheimer and Simpson (1993) and by Simpson and Raubenheimer (1993a) (see also Chapter 10) provides a framework against which developmental stage-related changes in food intake can usefully be considered. According to this concept, the growth target is the optimal amount and balance of nutrients that can be allocated to developing tissues (somatic, storage, and reproductive) within a given time period. The nutritional target is the optimal amount and balance of nutrients required for all biological functions, including growth and maintenance. The intake target is the level of intake of nutrients that enables the insect to achieve the nutritional target with the optimal level of postingestive processing. Thus, when considered in terms of the target concept, the issue of developmental stage-related changes in food intake becomes that of the relationship between the position of the intake target and the attempts by the insect to reach this by selecting among foods and by regulating the amounts of each eaten. It is the proximate mechanisms underlying the tracking of the intake target in relation to stage of development on which I will be focusing in this chapter.

Developmental stage-related changes in feeding behavior can manifest themselves in two ways. First, there are changes according to stage of development in the rate of total food intake. In this context, I will use the term *rate of intake* to refer to the amount of food consumed over some substantial unit of time. Many insects take their food in the form of discrete meals, and, in the case of insects which do so, the unit of time will be one during which the insect has consumed at least several meals. This is usually a day or a substantial part thereof. This terminology is used to distinguish rate of intake over a period from the rate of intake within a meal, which is referred to as *rate of ingestion*. Second, insects at different stages of development may consume different relative amounts of two or more foods when these differ as regards their nutrient and/or water content. When such changes occur, the insect is often said to have changed its food preference (Simpson and Simpson, 1990) or to exhibit self-selection behavior (Waldbauer and Friedman, 1991).

11.2. Categories of Physiological Mechanisms

Physiological mechanisms responsible for developmental stage related changes in feeding behavior and responsible for there being, in many instances, a general correspondence between the amounts and kinds of food eaten and an insect's nutritional requirements at any particular stage of its development fall into two categories, which I will refer to as *demand-mediated* and *non-demand-mediated*. Interplay between an insect's genetic constitution and environmental factors determines which one or more developmental programs are active at any particular stage in its overall development. The two categories differ as regards the kind of causal pathway by which the activation and progress of developmental programs influence the insect's feeding behavior. When a mechanism is in the *demand-mediated* category, developmental stage-related changes in feeding behavior are caused in some direct manner by variation in the rate of uptake of nutrients by the tissues (Fig. 11.1). Mechanisms in this category represent direct ways in which the amounts and kinds of food ingested by an insect can be matched to the changing nutritional requirements of its growing tissues, or, in the terminology of Raubenheimer and Simpson (1993) and of Simpson and Raubenheimer (1993a), matched to the nutritional target. Mechanisms in the *non-demand-mediated* category are those in which developmental stage-related changes in feeding behavior are not caused by changes in rate of nutrient uptake (Fig. 11.2). In such mechanisms, changes in feeding behavior are mediated by a variety of neural or hormonal signals acting directly on central or peripheral neural or possibly muscular elements involved in the location, recognition, or ingestion of food, or are the result of developmental changes in such elements. It can be expected, however, that the separate mechanisms regulating feeding behavior and the rate of uptake of nutrients by the tissues have evolved such that there is at least general correspondence between intake and the nutritional target. The two categories of mechanisms, although distinct, are not mutually exclusive. As will become apparent in my later discussions, there is evidence to suggest that

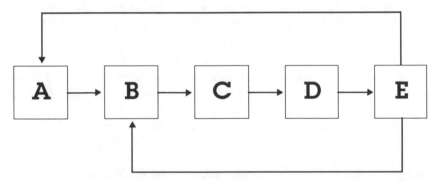

Figure 11.1. Causal relationships involved in the *demand-mediated* category of mechanisms. **A** represents the activation of a developmental program; **B,** the requirements of the tissues for nutrients during the course of the program; **C,** the mechanisms controlling feeding; **D,** feeding behavior; **E,** tissue growth. The main feature of mechanisms in this category is that the demand for nutrients by the tissues (**B**) directly affects the mechanisms controlling feeding (**C**) and, thus, feeding behavior (**D**). The demand for nutrients (**B**) at any particular stage of a developmental program is determined both by the nature of that program (**A**) and the stage to which growth and development have proceeded (**E**). The arrow from **E** to **A** is included to indicate the existence of feedback mechanisms which result in the deactivation of developmental programs when the development of the relevant tissues has been completed. For sake of simplicity, this diagram assumes that only one developmental program is active. The possibility exists, however, that different programs might be active contemporaneously at some stages of an insect's overall development.

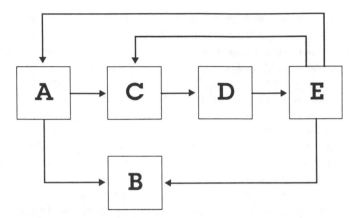

Figure 11.2. Causal relationships in the *non-demand-mediated* category of mechanisms. A to E are the same as for Fig. 11.1. As in that figure, it is assumed that only one developmental program is active. The main feature of mechanisms in this category is that the mechanisms controlling feeding (**C**) are influenced directly by the nature of the developmental program activated (**A**) and by feedback concerning the state of development of growing tissues (**E**). Independently of this, the tissue requirements (**B**) are determined by the nature of the developmental program activated and the stage to which it has progressed.

a number of different mechanisms, which are not necessarily restricted to either one of the categories, may act sequentially and possibly even simultaneously during the course of one developmental program.

Mechanisms in the *demand-mediated* category are based on nutrient feedback with changes in hemolymph nutrient concentrations directly affecting relevant neural and possibly muscular elements, or acting via an effect on the rate of passage of nutrients from the gut and hence on the rate of gut emptying (see Simpson and Simpson, 1990; Simpson and Raubenheimer, 1993b). In contrast, hemolymph nutrient concentrations are not involved in the range of mechanisms categorized as being *non-demand-mediated*. A major consequence of this difference between the two categories is that an insect in which developmental stage-related changes in feeding behavior are largely or wholly the result of *demand-mediated* mechanisms would be expected to display various kinds of nutritional compensation, whereas those in which *non-demand-mediated* mechanisms were of prime importance would not do so. Important manifestations of compensatory feeding are the ability of the insect to (i) increase its rate of intake in response to reductions in the concentrations of all or some nutrients in its food, (ii) select from between two or more foods of different nutrient contents such that its growth rate is optimal, and (iii) change its relative rates of intake of two foods of different nutrient contents after being deprived of particular nutrients or classes of nutrients (Simpson and Simpson, 1990; Waldbauer and Friedman, 1991;

Raubenheimer and Simpson, 1993; Simpson and Raubenheimer, 1993b). Another feature which would be expected to be apparent in an insect in which *demand-mediated* mechanisms are playing a major role, but which would not be expected to be displayed if mechanisms in the *non-demand-mediated* category were dominant, is that the rate of intake would be reduced by a range of treatments that prevent the growth of tissues programmed to grow at the time in question.

A variety of mechanisms which can be considered to be in the *non-demand-mediated* category have been identified or can be postulated. As will become apparent in ensuing discussions, some may be recognized by the effects on feeding behavior of treatments which alter the titers of relevant hormones and/or neural inputs to the neural elements involved in the regulation of feeding. The involvement of such mechanisms can also be inferred from temporal correlations between changes in feeding behavior and developmental stage-related changes in hormone titers or neural inputs in unmanipulated insects.

11.3. Changes During Larval Development

Larval feeding must provide the nutrients for somatic growth during larval development and, in some species, for postlarval growth of somatic and reproductive tissues. In addition, it must provide the energy required to support metabolic processes both while the insect is feeding and during any periods when it is not doing so (e.g., before and after ecdysis), during pupal development, and for the remainder of life in species which do not feed as adults. The larval period, itself, is a time in which growth is wholly or largely restricted to the somatic and storage tissues. The absolute amount and rate of growth increases from instar to instar, and rate changes in a more or less similar way over the course of each instar. As would be expected on the basis of the target concept, changes in rate of intake of food reflect such changes in growth rate. There is evidence also that the proportions of different kinds of food ingested by insects with access to more than one kind of food can change both between and within instars.

11.3.1. Changes Within Larval Instars

The rate of intake of larvae with access to one kind of food, which is nutritionally adequate in the sense that it supports growth and development, varies in a highly characteristic way within each instar. First there is usually a period, following ecdysis or hatching from the egg during which a larva does not feed (Adler and Adler, 1988; Beck et al., 1958; Blaney et al., 1973; Stoffolano and Bernays, 1980). There then follows a period in which intake per day is either constant (Woodring et al., 1977) or more usually increases progressively (Chapman and Beerling, 1990; Horie and Watanabe, 1983; Reynolds et al., 1986; Schoonhoven et al., 1991; Simmonds et al., 1991; Simpson, 1982a) over at least the first half of the instar. This period of relatively high and usually increasing rate of intake is followed by one during which the rate of intake declines more or less abruptly (Beck et al., 1958; Clark, 1980; Schoonhoven et al., 1991; Woodring et al., 1977), with feeding ceasing at some time (typically 24 hr or more) before ecdysis.

Two studies in which larvae were given continuous access to two foods, one containing protein but no utilizable carbohydrate and the other carbohydrate but no protein, have indicated that changes over the course of an instar in the rates of intake of the two diets follow different time courses. In the first study, intake patterns of such diets were examined in the second and final instar larvae of the cockroach *Supella longipalpa* (Cohen et al., 1987a). Intake of the carbohydrate-containing diet is high for the first half of both instars but decreases markedly over the second half. Intake of the protein-containing diet, on other hand, remains at a constant relatively low level virtually throughout the instar, with the result that the ratio of the amounts of carbohydrate-containing to protein-containing diet ingested tends to decrease. The second study examined changes in food-related behavior over the final instar of *Helicoverpa zea* (Cohen et al., 1987b). It was shown that the period for which larvae are in contact with, and therefore probably feeding on, a carbohydrate-containing diet increases progressively over the instar, whereas the period for which they are in contact with protein-containing-diet does not change significantly. This suggests that the trend, in terms of the ratio of carbohydrate-containing diet to protein-containing diet ingested, is the opposite of that recorded for *S. longipalpa*.

11.3.1.1. BEHAVIORAL MECHANISMS

The behavioral basis for the increase in the rate of intake over the days following ecdysis has been examined in the final-instar larvae of *Locusta migratoria* (Blaney et al., 1973; Simpson, 1982b) and *Manduca sexta* (Reynolds et al., 1986) and the first instar of the *Schistocerca americana* (Chapman and Beerling, 1990). There is some evidence that the behavioral basis for the increase in larvae of *L. migratoria* varies according to the type of food. When larvae are feeding on seedling wheat, the increase is the result of their taking larger meals at a more or less constant frequency (Simpson, 1982b), whereas there is a tendency for frequency of feeding to increase between days 1 and 5 of the instar when larvae are feeding on mature *Agropyron* (Blaney et al., 1973). The durations of meals taken by larvae feeding on this grass do not change over the period of increasing intake rate, but for reasons discussed in detail by Simpson (1990), this does not necessarily mean that meal size is constant. In final-instar larvae of *M. sexta*, the increasing rate of intake of either tobacco foliage or artificial diet between days 1 and 3 of the instar is largely the result of their taking progressively larger meals (Reynolds et al., 1986). First-instar larvae of *S. americana* feed little on the day of hatching, but on the following day they take larger meals with higher frequency. Over the following 3 days, intake rate increases gradually, largely because of an increase in the number of meals consumed.

Less information is available about the behavioral basis for the decline in intake over the later parts of the instars. In most instances it cannot even be said for certain whether intake by individuals declines progressively or abruptly, because the data presented are means for a number of insects. With data of this kind, a decline would appear to be graded if individuals were to cease feeding abruptly but at different times. This is illustrated by the available data for the final instar larva of *L. migratoria*. Data for groups of larvae show a decline in intake over 2 days, with that on the penultimate day of the instar being of the order of 50% that of the peak value (Hill and Goldsworthy, 1968; Hoekstra and Beenakkers, 1976; Simpson, 1982a). However, the only available data derived from detailed observations on the pattern of intake of individual larvae indicate that this changes only about 3 hr before cessation of feeding, with only the last three meals taken being of shorter duration than preceding ones and only the last interfeed period being notably long (Blaney et al., 1973). Although data derived from the observation on feeding patterns of individuals are not available, it appears that intake by individual larvae of *Acheta domesticus* might, in fact, decline progressively over the last half of the instar because the variance for daily consumption on the days over which pooled data indicate a progressive decline is no higher than that during the preceding period of constant intake (Woodring et al., 1977).

The behavioral basis for the changes over the course of the final instar of *H. zea* in the relative proportions of time spent on diets lacking protein or carbohydrate have been examined using time-lapse photography (Cohen et al., 1987b). This study showed that the duration of individual periods for which larvae remain in contact with the diet containing sugar but lacking protein increases significantly over the instar, whereas the duration of bouts of contact with diet containing protein but no sugar does not.

11.3.1.2. PHYSIOLOGICAL MECHANISMS

The physiological bases for the increase in intake rate over the first part of an instar and its subsequent decline have not been fully elucidated in any species. However, larvae of all species for which relevant information is available appear to display characteristics which indicate that mechanisms in the *demand-mediated* category play a significant role in bringing about the developmental stage-related changes in their rate of intake. The findings relevant to the criteria suggested earlier as indicating the significant involvement of mechanisms in this category are as follows:

1. The ability to compensate for dilution of diet with non-nutrient material (see Barton Browne, 1975; Simpson and Simpson, 1990) or nutrient imbalance (Raubenheimer, 1992; Raubenheimer and Simpson, 1993) by increasing intake rate has been demonstrated in larvae of a number of species, including

several lepidopteran and acridid species which are known to display typical changes in rate of intake over the course of an instar (Simpson et al., 1988).

2. Dietary selection behavior, in which insects feed from two foods of different nutrient contents such that their growth rate is maximal, has been demonstrated in larvae of *Spodoptera littoralis* and *Locusta migratoria* (Simpson et al., 1988), *Helicoverpa zea* (Cohen et al., 1987b) and *Supella longipalpa* Cohen et al., 1987a).

3. Larvae of *S. littoralis* and *L. migratoria* which have fed on a food which is deficient in either protein or carbohydrate show a preference for a food containing the nutrient in which their previous food was deficient over that on which they have previously fed.

Also of relevance to the question of the causation of intra-instar changes in intake rate is the finding by Simmonds et al. (1991) and Schoonhoven et al. (1991) that the responsiveness of gustatory receptors on the mouthparts of final-instar larvae of leafworm *Spodoptera littoralis* changes with age such that responsiveness is generally correlated with the rates of intake, especially over that part of the instar when feeding rates are increasing. These authors have proposed that changes in intake rate in the early part of the instar might be the direct result of the changes in receptor sensitivity, because it would be expected that the amount of excitation generated in the central nervous system (CNS) by a particular feeding stimulant would be correlated with the level of excitatory input. The extent to which the changes in receptor responsiveness are due to developmental changes in the receptor neurons or to effects on receptor sensitivity of changes in hemolymph composition as the duration of the postecdysis fast increases is uncertain. If the former, the proposed mechanism would fall into the *non-demand-mediated* category and, if the latter, into the *demand-mediated* category.

There is evidence that one or more hormones is implicated in bringing about the cessation of feeding prior to ecdysis. It is commonly accepted that the reduction in rate of intake is associated with apolysis (e.g., Chapman, 1990), but the evidence that the two actually coincide appears to be lacking. A close temporal association between reduction in feeding and apolysis would suggest, however, that release of ecdysone and the resulting increase in hemolymph ecdysteroid titer, or the release of prothoracicotropic hormone (PTTH) which triggers ecdysone release, might be involved. No studies appear to have been made of the relationship between hormonal events during an instar and rate of intake per se, but there are data for *Manduca sexta* and *L. migratoria* which are consistent with the view that releases of ecdysone and/or PTTH might be involved. It has been shown that final-instar larvae of *M. sexta* leave the food and start wandering soon after a small, brief increase in hemolymph ecdysteroid titer which takes place more than a day before apolysis, which occurs coincidentally with a second larger increase in ecdysteroid titer. Each of the releases of ecdysone is preceded by one or more releases of PTTH (Bollenbacher and Gilbert, 1982). When the moult is from one larval instar to another, however, there is only a single release of ecdysone and of PTTH, and it is likely that one or other (or both) of these events causes feeding to be terminated just before apolysis. Dominick and Truman (1985) obtained evidence which suggests that ecdysteroid is itself responsible for the cessation of feeding and initiation of wandering in final-instar larvae of *M. sexta*. They found that larvae from which the prothoracic gland, the source of ecdysone, had been removed continue to increase in weight for 6 days after control larvae begin wandering, that wandering could be induced by perfusion of such larvae with 20-hydroxyecdysone, and that perfusion with the same compound would induce premature wandering in intact larvae. In final-instar *L. migratoria*, the ecdysteroid titer of the hemolymph is very low until about 60% of the way through the instar at which time it increases sharply, reaching a peak at about the 75% mark (Baehr et al., 1979). The timing of this peak corresponds reasonably well with the reduction in rates of food intake and growth (Baehr et al., 1979; Simpson, 1982a).

There are data which indicate that feeding in the final instar of the codling moth *Laspeyresia pomonella* is correlated with the continuing presence of juvenile hormone (JH) in the hemolymph. Sieber and Benz (1978) showed that "short-day" larvae, which are destined to enter diapause, feed for several days longer and attain a greater weight than do non-diapause "long-day" larvae. They showed further that the period of feeding and final weight of long-day larvae is increased by the topical application of the juvenoid, methoprene, early in the instar. This is in accord with an earlier finding by these authors (Sieber and Benz, 1977) that the titer of JH in the hemolymph falls to zero within the first 24 hr following the moult into the final instar of non-diapause larvae but remains at levels readily detected by bioassay in larvae destined to diapause. Further evidence indicating that feeding continues if the titer of JH remains elevated is provided by the finding of Hintze-Podufal and Fricke (1971) that topical application of either of two juvenoids substantially prolongs the feeding period in the final instar of another lepidopteran, *Cerura vinula*. The finding that "long-day" larvae of *L. pomonella* do not resume feeding when treated with methoprene after their feeding has ceased suggests that the titer of JH in the hemolymph does not itself control feeding (Sieber and Benz, 1978).

Generally, a change in the titer of a hormone might influence feeding behavior either by acting directly on mechanisms regulating feeding or via an effect on the rate of tissue growth. If the former, a mechanism responsible for bringing about a change in feeding behavior would be considered to fall into the *non-demand-mediated* category and, if the latter, into the *demand-mediated* category. In the case of the hormone-mediated cessation of feeding before molting, it seems likely that the mechanism is in the *non-demand-mediated* category, with ecdysteroid or possibly JH or PTTH acting directly on the mechanism regulating feeding. This is suggested by the facts that feeding actually ceases before each moult, rather than continuing at a reduced rate, and that cessation of feeding is associated with the initiation of wandering, a specific locomotor behavior, late in the final instar. If feeding rate were being determined by tissue demand, it would be expected that feeding would continue at a reduced rate reflecting the ongoing nutritional requirements for tissue maintenance.

The physiological basis for the failure of final-instar larvae of *L. migratoria* to feed for a number of hours after ecdysis has been examined by Stoffolano and Bernays (1980), who concluded that this was due largely to a general low level of arousal in the CNS. This conclusion was based mainly on the findings that the duration of the postecdysial fast is reduced by mechanical disturbance of the larva and that undisturbed insects which had shorter fasts showed higher levels of locomotor activity than those which fasted for longer period. The investigation ruled out any major involvement of air in the gut or of blood-borne factors. The mechanism responsible for the low level of arousal following ecdysis is unknown, but it appears likely that this reflects the state of the CNS and, thus, that the mechanism responsible falls into the *non-demand-mediated* category.

The physiological basis for within-instar changes in the relative and absolute changes in rates of intake of foods with different nutrient contents, as has been documented in larvae of *S. longipalpa* (Cohen et al., 1987a) and *H. zea* (Cohen et al., 1987b), has not been established with certainty. The nature of the physiological mechanisms underlying dietary selection behavior has been extensively discussed in recent articles (Simpson and Simpson, 1990; Waldbauer and Friedman, 1991). However, in view of evidence for the involvement of nutrient feedback in dietary selection behavior by larvae of *H. zea* (Friedman et al., 1991), it seems likely that mechanisms in the *demand-mediated* category play a major part in bringing about the observed developmental stage-related changes in dietary selection in this species. Evidence that dietary selection behavior involves nutrient feedback-based mechanisms in larvae of another lepidopteran, *S. littoralis,* and of the locust *L. migratoria* (Simpson et al., 1988) suggests that this is probably so for other species, including *S. longipalpa* (Cohen et al., 1987b).

11.3.2. Changes Between Larval Instars

It is well known that insects generally consume larger total amounts of food in each successive instar with some, especially some lepidopterans, taking over half of their total larval intake in the last instar. This increase in intake over successive instars has been the subject of surprisingly few detailed studies. However, the results of these have been similar in principle, all having shown that the increased intake in later instars results largely from larvae in later instars having higher rates of daily intake on those days within the instar for which they are feeding and, to a lesser degree, from there being a tendency for larvae in later instars to feed for a greater number of days, because of increased instar duration (Capinera, 1979; Hill and Goldsworthy, 1968; Horie and Watanabe, 1983; McAvoy and Smith, 1979; Retnakaran, 1983; Woodring et al., 1977).

Different larval instars may feed on different foods. There are documented examples in which larvae of different instars of phytophagous insects feed on different parts of their host plant (Cohen et al., 1988; McAvoy and Kok, 1992; Rogers, 1978), show different degrees of selectivity [with later instars usually becoming less selective (Parrott et al., 1983; Johnson, 1984)], and/or change their method of feeding (Gaston et al., 1991). In predators, there is a tendency for the size of prey taken to increase over successive instars (Blois and Cloarec, 1983; Thompson, 1978).

11.3.2.1. BEHAVIORAL MECHANISMS

It appears that there have been very few investigations of the behavioral basis for differences in rates of intake between instars. Adler and Adler (1988) obtained time budgets for three 30-min periods—one near the beginning of the feeding period, one near its midpoint, and one a few hours before the cessation of feeding—in all instars of the corn earworm, *Helicoverpa zea*. Data resulting from observations made early and late in the instars reveal no distinct trends in either the duration or frequency of meals. The only obvious effect of instar is that the mean duration of meals taken by first- and second-instar larvae near the midpoints of the feeding periods is about half that of those taken in the three later instars. Because the proportion of time spent feeding differs relatively little between instars (Adler and Adler, 1988), it follows that the very marked increases in intake rate in successive instars (Huffman and Smith, 1979) must be largely due to increases in the rate of ingestion. Observations on meal duration and frequency over the four last larval instars of the pentatomid bug *Acrosternum hilare* have shown that meal durations differ little between instars but that there is a progressive instar-to-instar increase in both the peak and the mean number of meals taken per day (Simmons and Yeargan, 1988). Information is not available on the total amount of food ingested or on the peak daily rate of intake in the four instars, but these data indicate that increased frequency of feeding is at least partly responsible for the increase in intake which undoubtedly occurs between successive instars. There is an obvious need for more extensive and detailed observations on the behavioral bases for changes in intake rate between instars, especially in insects for which some of the mechanisms regulating the intake of food have been elucidated.

11.3.2.2. PHYSIOLOGICAL MECHANISMS

Little can be said about the physiological basis for instar-dependent changes in rate of intake. As might be expected, amounts of food ingested and increases in body weight over the course of different instars are correlated (Capinera, 1979; Hill and Goldsworthy, 1968; Horie and Watanabe, 1983; Woodring et al.,

1977). However, the ratio of the two can differ somewhat between instars because of changes in the efficiency of conversion of ingested food to body substance (ECI) (Waldbauer 1968). In view of the evidence favoring the involvement of *demand-mediated* mechanisms in changes in intake rate within an instar, it seems reasonable to believe that differences in rates of intake between instars can also be explained at least largely in these terms.

It is likely that many of the reported changes between instar as regards the kind of food they consume are direct results of physical constraints related to the size of the larva, but in one instance there is evidence that changes in the feeding behavior reflect changes in nutritional requirements. Cohen et al. (1988) showed that first- and second-instar larvae of the corn earworm, *Helicoverpa zea,* tend to prefer maize silk to kernels but that the fourth- and the final-instar larvae feed almost exclusively on kernels. First-instar larvae moult after 3 days irrespective of whether they are feeding on silk or kernel, but the total duration of instars 2–4 is 12 days when larvae were given only silk and 5 days when given kernels. These findings suggest that the switch from silk to kernels has a nutritional basis and is consistent with the involvement of mechanisms in *demand-mediated* category in the bringing about of changes in feeding behavior between instars.

11.4. Changes During Adult Development

Adult insects need nutrients for energy for any posteclosion somatic growth and for reproductive development, with the nutritional requirements for reproduction being, in general, greater in the female than in the male. The extent to which these nutrients are acquired through feeding during the adult stage varies between species and, in some instances, between subspecies or races. Some species (e.g., *Lymantria dispar, Hyalophora cecropia*) do not feed at all as adults but use endogenous reserves acquired through larval feeding to provide all their posteclosion nutritional requirements. Most, however, obtain at least some of their posteclosion requirement through feeding as adults.

Some adult insects normally obtain all the nutrients they require from only one kind of food, whereas others ingest foods of more than one kind which differ, to different degrees, in nutrient content. Such differences can be very great, as in the case of females of most anautogenous dipterans which obtain their carbohydrate requirements at least largely from nectar and their protein needs for reproductive development from foods of high protein and low carbohydrate content. In other instances, all kinds of food eaten contain somewhat different but still substantial amounts of the major categories of nutrients. The foods eaten may also differ markedly in water content.

11.4.1. Changes in Relation to Somatic Growth

In some insects, somatic tissues undergo considerable growth over a number of days after adult eclosion, with increases in the weight of somatic tissues ranging up to 100% or more. Posteclosion somatic development typically involves growth of muscle and cuticle (Kitching and Roberts, 1975; Mordue and Hill, 1970), with the result that there is an increase in both the total protein and carbohydrate contents of the somatic tissues. In some insects, such as the blowfly *Lucilia cuprina,* a substantial proportion of nutrients for somatic growth are derived from endogenous reserves resulting from larval feeding (Williams, 1972), with the result that the whole-body protein and/or carbohydrate content changes less than that of the somatic tissues. However, it is probable that the nutrients necessary for somatic growth of many insects are obtained largely or wholly from food ingested after eclosion, in which case increases in both the protein and carbohydrate content of the whole body are of the same order as those in the somatic tissues, as has been demonstrated in *Schistocerca gregaria* (Mordue and Hill, 1970). The posteclosion food intake of insects can, during any period of somatic growth, be expected to reflect an intake target congruent with the changes in the composition of the whole body.

Changes in feeding behavior in relation to posteclosion somatic development are more clearly seen in males than in females because there is less tendency for these to be obscured by contemporaneous or temporally contiguous changes associated with reproductive development. I will, therefore, consider here only results obtained with males. Available evidence indicates, however, that the feeding behavior of females during somatic development is generally similar to that of conspecific males (Chyb and Simpson, 1990; Hill et al., 1968; Walker et al., 1970). I will make some reference to feeding behavior of teneral females when comparing general characteristics of this with feeding behavior by females during ovarian development (see Section 11.4.3).

A number of studies with adult males have shown that the mean rate of intake in the period following eclosion, during which somatic growth is taking place, is several times that in the post-teneral period. Intake rate typically increases over the first 1 or 2 days following eclosion and declines over the last day or two of somatic growth. This has been shown for *S. gregaria* (Norris, 1960; Walker et al., 1970), *Locusta migratoria* (Strong, 1967), and the milkweed bug *Oncopeltus fasciatus* (Slansky, 1980). Increases in the weight of the somatic tissue of 100%, 70%, and 40% have been recorded in males of *S. gregaria* (Walker et al., 1970), *L. migratoria* (Chyb and Simpson, 1990), and *O. fasciatus* (Slansky, 1980), respectively, over a period of 8–10 days following eclosion.

When adult males of *L. migratoria* have continuous access, from the time of eclosion, to two artificial diets, one containing protein but no utilizable carbohydrate and the other carbohydrate but no protein, their rates of intake of both change with age of the insect, but in different ways. Calculations based on the data presented by Chyb and Simpson (1990) indicate that the intake of the protein-containing diet is maximal during days 2–6 after eclosion, whereas that of the carbohydrate-containing diet peaks between days 5 and 8. The changes in the rates of intake of the two diets are such that the amount of protein relative to carbohydrate ingested declines progressively over the 15 days following eclosion. The total intake rate is greatest on days 5 and 6, this being about twice the daily rate in the post-teneral period.

The rate of intake of a solution of yeast extract by males of the blowfly *Phormia regina* which have access to both this protein-rich solution and sucrose solution is maximal between 15 and 24 hr after eclosion, whereas that of sucrose solution remains approximately constant up to 60 hr and, thereafter, tends to increase (Belzer 1978a). As with *L. migratoria,* therefore, the age-related changes in rates of intake of the two nutritionally contrasting solutions result in there being a progressive decrease in the ratio of protein to carbohydrate ingested after this reaches a maximum soon after eclosion. While it is likely that the peak in the rate of intake of the yeast extract by males of *P. regina* is associated with the provision of nutrients needed for somatic development, this cannot be stated with any certainty, because it has been shown that the accessory reproductive glands of males of this species develop fully only after the ingestion of protein (Stoffolano, 1974), which suggests that these glands have a requirement for protein. Furthermore, in males of another blowfly, *Lucilia cuprina,* protein intake is not clearly higher during the 2 days following eclosion (Roberts and Kitching, 1974), when substantial somatic growth takes place (Kitching and Roberts, 1975), than over the subsequent 4 days. In this species, however, there is evidence that the nutrients for somatic development are derived from reserves acquired through larval feeding (Williams, 1972), but there is no certainty that this is so in *P. regina.*

11.4.1.1. BEHAVIORAL MECHANISMS

The behavioral basis for changes in rate of intake during the posteclosion period of somatic development has been investigated in one insect, the locust *L. migratoria.* In this species, the stage-related changes in rate of total intake outlined above were shown to be due mainly to changes in meal frequency (Chyb and Simpson, 1990).

11.4.1.2. PHYSIOLOGICAL MECHANISMS

Some insights into the physiological mechanisms involved in changes in feeding behavior in relation to somatic growth are provided by a study of the effects of allatectomy on the rate of intake by adult males of *S. gregaria* (Walker and Bailey, 1971). This showed that intake by males which had been allatectomized 4 days after eclosion is higher from day 6 onwards than in sham-operated controls. The weight of the somatic tissue of allatectomized males increases by about 30% between days 8 and 14 following eclosion, whereas that of controls shows no significant increase. Furthermore, the weight of fat body of allatectomized males increases fourfold over this period, whereas that of controls remains constant. These results suggest that hormone released by the corpus allatum is involved in bringing about the substantial reduction in rate of intake which normally occurs at 5–6 days after eclosion (Walker et al., 1970). Norris (1960) observed a positive correlation between the age at which rate of food intake declines from its initial high rate and the age at which individuals become sexually mature, as evidenced by their developing yellow coloration. Because it has been established that both sexual maturation and the development of yellow coloration in *S. gregaria* depend on the juvenile hormone (Pener and Lazarovici, 1979), it seems likely that it is the lack of this hormone which is responsible for the prolongation of the period for which intake rate remains high in allatectomized males.

As pointed out in Section 11.3.1.2, the mechanisms by which hormones bring about developmental stage-related changes may be either *demand-mediated* or *non-demand-mediated*. One line of evidence suggests that the mechanism responsible for the decline in food intake in *S. gregaria* might be in the *non-demand-mediated* category. Data for males of this species indicate that the rate of intake declines somewhat in advance of the reduction in the rate of somatic growth; rate of intake declines by a factor of about two approximately 5 days after eclosion, whereas rate of somatic growth does not show any marked reduction until about the eighth day, at which time it falls abruptly from a high rate to zero (Walker et al., 1970). However, evidence favoring the involvement of *demand-mediated* mechanisms in acridids is provided by the finding by Chyb and Simpson (1990) that adults of *L. migratoria* compensate for reductions in the protein and carbohydrate contents of their diets by increasing their rate of intake during the later part of the period of posteclosion somatic growth. This conclusion is supported by electrophysiological studies by Simpson et al. (1990) which showed that there are age-related changes in the responsiveness of gustatory chemosensilla on the maxillary palp of adults of *L. migratoria,* which can plausibly be explained in terms of the effects on the chemoreceptors of demand-induced changes in hemolymph composition, as has been demonstrated by Abisgold and Simpson (1988). It is significant, however, that no compensation for reduction in either protein or carbohydrate content is apparent over the first 4–5 days following eclosion. Chyb and Simpson (1990) suggest that this may be because the insects are, during this period, using reserves acquired through larval feeding.

11.4.2. Changes in Relation to the Acquisition of Energy Reserves

In some insects, reserves of carbohydrate and/or lipid in the fat body increase after eclosion. The triglyceride and glycogen content of females of the mosquito *Aedes taeniorhynchus* increases about fourfold over the first 9–12 days of adult life and remain at a fairly constant level thereafter. This posteclosion period of accumulation of carbohydrate reserves is reflected in changes in the insect's rate of intake of sucrose solutions, with intake rates over the first 5 or so days being higher than subsequent ones (Nayar and Sauerman, 1974). The finding that these insects compensate strongly (by increasing their rate of intake) for dilution of the sucrose solution suggests that the mechanism responsible for the relatively high rates of sugar intake in the period following eclosion is in the *demand-mediated* category.

Females of the mosquito *Culex pipiens* overwinter as adults in reproductive diapause. After eclosion, diapausing females accumulate about twice as much of mainly lipid reserves in their fat body as do nondiapausing individuals (Mitchell and Briegel, 1989). Bowen (1992) has shown that there are associated differences in posteclosion feeding behavior. The incidence of feeding on fruit by nondiapausing females peaks on the third day after eclosion, falling to a relatively low level thereafter. In contrast, the incidence of fruit feeding by diapausing individuals remains elevated until at least 15 days after eclosion. That ovarian development in diapausing mosquitoes ceases in a developmental stage similar to that seen in females which have been allatectomized shortly after eclosion (Wilton and Smith, 1985) suggests that diapause is associated with a low titer of JH in the hemolymph. In view of the previously mentioned indication of the involvement of mechanisms in the *demand-mediated* category in *A. taeniorhynchus,* it seems likely that the prolongation of the period during which the incidence of fruit feeding is high in diapausing females of *C. pipiens* can be attributed to a continuing demand by the fat body for carbohydrates. Such demand may be terminated by the release of JH in nondiapausing females.

11.4.3. Changes in Relation to Reproduction

Most insects acquire at least some of the nutrients required for reproduction through feeding as adults. There are a considerable number of studies of the relationship between reproduction and feeding behavior in female insects. There are, in contrast, few in males despite the fact that those of some species make considerable nutritional investment in reproduction through the passing, at mating, of large spermatophores representing, in some species, over 10% of the male's body weight.

11.4.3.1. Changes in Females

The temporal pattern of change in rate of uptake of nutrients by the ovaries varies according to several aspects of the female's reproductive physiology. As might be expected, on the basis of the target concept (Raubenheimer and Simpson, 1993; Simpson and Raubenheimer, 1993a), changes in rate of uptake are reflected, at least in general terms, by changes in feeding behavior, except in the case of autogeny where the nutrients for ovarian development are substantially or wholly derived from endogenous material acquired through larval feeding. Because mature oocytes have high protein content, ovarian development in anautogenous females represents an increase mainly in the insect's protein content (Mordue and Hill, 1970; Williams et al., 1979). It can be expected, therefore, that the feeding behavior of such females during ovarian development would reflect an intake target which emphasizes the intake of protein-rich food. In some species, females display discrete ovarian cycles in which batches of oocytes develop synchronously, whereas in other species the oocytes develop more or less continuously.

Rate of intake in relation to age has been examined in females of three species which do not display discrete ovarian cycles: the milkweed bug *Oncopeltus fasciatus* (Slansky, 1980), the cricket *Acheta domesticus* (Clifford and Woodring, 1986), and the fruit fly *Rhagoletis pomonella* (Webster et al., 1979). Rate of intake by mated females of *O. fasciatus* and *A. domesticus* of the only food provided, and by females of *R. pomonella* of yeast hydrolysate, remains fairly constant throughout the preoviposition period and over a substantial period during which oviposition is taking place. On the evidence discussed earlier in relation to males (Slansky, 1980), it appears that the rate of food intake by females of *O. fasciatus* during the reproductive period is similar to that in the preceding period of posteclosion somatic growth.

Females of a range of species which display discrete ovarian cycles, including some cockroaches, flies, and locusts, show their highest rates of intake of foods containing nutrients needed for reproductive development during the early part of each cycle.

Rate of food intake falls to a very low level 1 or 2 days prior to ovulation in the oviparous cockroaches *Periplaneta americana* (Bell, 1969; Verrett and Mills, 1973), *Parcoblatta fulvensis*, *Parcoblatta pennsylvanicus* (Cochran, 1986), *Supella longipalpa* (Hamilton et al., 1990), and *Blattella germanica* (Hamilton and Schal, 1988) and remains so until deposition of the ootheca. In the ovoviviparous species, *Leucophaea maderae*, rate of intake shows a similar decline at the time of ovulation and remains at a relatively low level until parturition (Englemann and Rau, 1965).

Food intake throughout the posteclosion period of somatic growth and two or more ovarian cycles has been examined in two species of locusts, *Schistocerca gregaria* and *Locusta migratoria*. Females of *S. gregaria* with continuous access to both bran and lettuce show greater age-related changes in the rate of intake of bran than of lettuce (Hill et al., 1968; Mordue and Hill, 1970). The rate of intake of both foods on the day before most locusts lay each of their first two egg pods is less than that in the preceding part of the ovarian cycle, but the reduction in intake of bran is somewhat more marked. The daily rate of intake of bran during the period of somatic development is about twice the maximum recorded during the first three ovarian cycles, whereas the corresponding rates of intake of lettuce are of the same order, with the result that bran constitutes a smaller proportion of the female's diet during reproductive development than during the period of somatic growth. Changes in the rates of intake of the two foods are such that the maximum total rate of intake of dry matter during the ovarian cycles is about half that during the period of somatic growth and about double that on the day prior to each oviposition. Estimates of total intake based on daily fecal production indicate that the pattern in *L. migratoria* is generally similar except that the maximum rate of intake during ovarian cycles is of the same order as, or rather less than, that during somatic development (Strong, 1967).

When females of the anautogenous blowflies *Calliphora erythrocephala* (Strangways-Dixon, 1961), *Phormia regina* (Belzer, 1978a; Dethier, 1961), *Lucilia cuprina* (Roberts and Kitching, 1974), and *Sarcophaga bullata* (de Clerck and de Loof, 1983) are given continuous access to solutions of a protein-rich material and sugar, the rate of intake of protein-rich solution declines markedly in each ovarian cycle as the synchronously developing oocytes enter the later stages of vitellogenesis and increases again after each oviposition or larviposition. The rate of intake of the sugar solution varies more or less inversely to that of protein or remains fairly constant. In a fifth cyclorrhaphous species, the face fly, *Musca autumnalis,* a similar pattern of intake is indicated by changes in the protein and sucrose contents of the crop (van Geem and Broce, 1986).

The *ad libitum* rates of leucine solution and phosphate buffer ingested by gravid females of *Musca domestica* are substantially less than those of protein-deprived females (Barton Browne and Kerr, 1986). This finding shows that intake of these materials, which apparently elicit sensory input recognized by the fly as being specifically indicative of protein-rich materials, is influenced by stage of ovarian development.

In the blowflies *P. regina* and *L. cuprina,* the size of meals of protein-rich solutions declines as females approach ovarian maturity and, thus, reflect changes seen in the same species in *ad libitum* feeding. In *P. regina,* this decline in meal size was found to occur fairly early in vitellogenesis (Rachman, 1980), whereas in *L. cuprina* it did not occur until the oocytes were approaching maturity (Barton Browne and van Gerwen, 1992). It is uncertain, however, as to whether the two species really differ as regards the timing of the reduction in meal size in relation to ovarian development, because the treatment given to the flies prior to test differed in some important respects [see Barton Browne and van Gerwen (1992) for discussion]. Gravid females of *P. regina* take smaller meals of a range of inorganic salt solutions than do protein-deprived females (Rachman et al., 1982; Busse and Barth, 1985), a result which is in accord with the results obtained with *ad libitum* feeding houseflies (Barton Browne and Kerr, 1986).

Females of tsetse flies (*Glossina* spp.) are viviparous. The proportion of females of *Glossina morsitans* (Denlinger and Ma, 1974) and *G. pallidipes* (Leegwater-van der Linden, 1981) which feed when given access to a mammalian host for a limited period each day is high over the first 6 days of the 9-day pregnancy cycle but then declines, reaching a low level on the day of parturition. Females of *G. morsitans* show a somewhat different relationship between feeding and stage in the pregnancy cycle, with intake being greatest on days 1 and 2 after parturition and on days 5–8 (Langley and Pimley, 1974). The mean size of blood meal taken by those females of *G. austeni* which do feed on the last 3 days of the reproductive cycle is about half that of meals taken during the first 6 days (Tobe and Davey, 1972).

Mosquitoes are another group of dipterans which display discrete ovarian cycles. Females of many mosquitoes require at least one blood meal in every ovarian cycle. A number of studies with such anautogenous mosquitoes have shown that gravid females are much less responsive to host volatiles than are nulliparous females that have not fed on blood or parous females after they have oviposited (Klowden, 1990; Bowen, 1991; also see Chapter 7). The findings of Meola and Readio (1987) indicate, however, that readiness to bite and ingest blood is less affected by stage of ovarian development than is responsiveness to host odors. Only a small proportion ($\sim$ 5%) of gravid females of *Culex pipiens* ingest blood when given access to a host in a cage of such a size that it is necessary for them to engage in host-seeking behavior, whereas about 50% do so when they are confined virtually in contact with the host. In both circumstances, a high proportion of postoviposition females take blood meals. The only information on the effect of stage of ovarian development on size of blood meal is a statement to the effect that meals taken by gravid females of *Aedes aegypti, A. taeniorrhynchus,* and *A. sollicitans* are "not very large," presumably in comparison with those taken by nongravid individuals (Meola and Lea, 1972).

In considering the ontogeny of behavior directed toward the acquisition of protein-rich material by anautogenous females, I have considered, thus far, only the decline in rate of intake or responsiveness to host stimuli which occurs when the synchronously developing oocytes enter an advanced state of development. In some species, behavior directed toward the acquisition of protein-rich material does not develop fully for one or more days after eclosion. This has been shown to be so in relation to host-seeking behavior of virgin females of *A. aegypti* (Davis, 1984) and *Aedes triseriatus* (Porter et al., 1986) and also in relation to protein feeding by the bushfly *Musca vetustissima* (Jones and Walker, 1974). Females of *A. triseriatus* also show a delay on about 1.6 days before they reestablish their responsiveness to the mammalian host following oviposition (Porter et al., 1986).

As is the case with their food intake, the rate of water intake varies in relation to reproductive cycles in several species of cockroaches. Females of *P. fulvescens* and *P. pennsylvanica* show a high rate of water consumption immediately following deposition of the ootheca, with the rate of intake on the day of deposition of the ootheca being several times that on any other day on the reproductive cycle (Cochran, 1986). Similarly, the rate of water intake by *L. maderae* is high on the day following parturition (Scheurer and Leuthold, 1969). The pattern of water intake by *P. americana* appears to differ somewhat from that of these species in that it is high early in a reproductive cycle, then declines briefly to increase again at about the time of ootheca formation (Verrett and Mills, 1973). The rate of water intake by ootheca-carrying females of *B. germanica* is about half that in the period between deposition of one ootheca and the formation of the next (Cochran, 1983).

Some females which display discrete ovarian cycles can mature a single batch of eggs without feeding on protein-rich material as adults, but must ingest such material to support subsequent ovarian cycles. Comparisons of feeding behavior have been made between nulliparous autogenous and anautogenous females of the same species and between nulliparous and parous autogenous females. Spradbery and Schweizer (1979) produced anautogenous females of the normally autogenous screw-worm fly, *Calliphora bezziana*, by prematurely terminating their larval feeding. Such females with continuous access to both meat exudate and sucrose solution ingest a greater amount of exudate during their first ovarian cycle than do normal autogenous individuals, despite their smaller size (Spradbery and Sands, 1981) and reduced number of ovarioles (Spradbery and Schweizer, 1981). In autogenous females of *C. bezziana*, exudate makes up 21% of total intake in the second cycle as compared with only 9% in the first (Spradbery and Schweizer, 1979). In the mosquito *Aedes taeniorhynchus*, the proportion of nulliparous females taking blood, when given access to a host twice a day, is five times higher in an anautogenous strain than in an autogenous strain (Nayar and Sauerman, 1975). Field studies have shown that nulliparous autogenous females of the midges *Culicoides melleus* and *C. hollensis* (Magnarelli, 1981), the mosquito *Culex tarsalis* (Nelson and Milby, 1982), and tabanids *Chrysops atlanticus* (Anderson, 1971) and *Hybomitra frontalis* (Leprince, 1989) are never or only rarely taken at mammalian hosts or carbon-dioxide-baited traps, whereas nullipars make up a substantial proportion of the catch of females of anautogenous forms (e.g., Leprince, 1989).

The effect of reproductive diapause on responsiveness to host volatiles and/ or the likelihood that an individual will ingest blood once in contact with a host have been examined in females of several mosquito species. Diapausing females of *Culex pipiens* are essentially unresponsive to olfactory host stimuli (Bowen et al., 1988). The results of tests in which hosts have been provided for diapausing females in cages of different sizes confirm that diapausing females are unresponsive to host volatiles but indicate that a substantial proportion of females of *Culex tarsalis* will take a blood meal if in close proximity with the host (Mitchell, 1981). In *C. pipiens*, also, blood feeding by diapausing females is rare when provided with a host in a cage of a size which requires them to engage in host-seeking behavior in order to establish contact with the host (Eldridge and Bailey, 1979; Wilton and Smith, 1985).

11.4.3.2. BEHAVIORAL MECHANISMS IN FEMALES

The behavioral basis for changes in rates of intake of food and water in relation to reproductive events has been examined directly only in females of *B. germanica*. Ootheca-carrying females feed only about one-fifth (Cochran, 1983) to one-third (Silverman, 1986) as frequently as do females without oothecae. In addition, the mean duration of each visit to food is briefer by a factor of about two in ootheca-carrying females (Silverman, 1986). Frequency of drinking in ootheca-carrying females is about half that in females without oothecae (Cochran, 1983; Silverman, 1986), but mean duration of visits to water sources is similar in the two categories of female (Silverman, 1986). Although no detailed observations have been made on their *ad libitum* feeding behavior, it appears, on the basis of results discussed earlier, that both the frequency and size of meals taken by female tsetse flies change in relation to stage in the pregnancy cycle and that decline in meal size plays a role in the reduction in intake rate in the later part of ovarian cycles in blowflies *P. regina* and *L. cuprina*.

11.4.3.3. PHYSIOLOGICAL MECHANISMS IN FEMALES

Physiological mechanisms responsible for changes in feeding behavior in relation to ovarian development have been investigated in a range of species. Evidence is somewhat conflicting as to whether or not mechanisms in the *demand-mediated* category play a significant role, but the results of a number of studies have indicated the involvement of mechanisms in the *non-demand-mediated* category, especially in relation to the factors responsible for the reduced rate of overall intake or of protein-rich foods by females carrying numbers of mature or nearly mature oocytes or by pregnant viviparous females.

Evidence for a Demand-mediated Mechanism

The two main expectations if mechanisms in the *demand-mediated* category were playing an important role in bringing about changes in development in relation to stage of ovarian development are that the insects would display some form of compensatory feeding behavior and that the rate of intake of food containing relevant nutrients would be reduced by treatments which prevent the uptake of nutrients by the ovaries.

An important manifestation of compensatory feeding is that an insect will increase its rate of intake of all food or of foods rich in particular nutrients in response to a reduction in nutrient concentration. Because mature oocytes have high protein content (e.g., Mordue and Hill, 1970; Williams et al., 1979), it would be expected that if *demand-mediated* mechanisms were playing a major role, females would compensate particularly for reductions in the protein content of their diet. There have been only few studies of the effects of nutrient concentration on rate of intake, and these have given somewhat equivocal results. Rates of intake of diets of different protein contents have been measured in females of the cockroaches *Blattella germanica* (Hamilton and Schal, 1988) and *Supella longipalpa* (Hamilton et al., 1990). In the diets used in these particular studies, however, reduction in protein content was achieved by replacing protein with utilizable carbohydrate, with the result that the concentations of the two were inversely related. Under these circumstances, if an insect were capable of compensation in relation to both protein and carbohydrate, it would be expected that the mechanisms regulating intake of the two nutrients would oppose one another with the result that clear-cut compensation for either might not be discernible (see Raubenheimer, 1992). Some evidence was found for compensation for reduced nitrogen content of the diet in *B. germanica* but not in *S. longipalpa*. The rate of intake by females of *B. germanica* of diet containing 5% protein is about 30% higher than that of a diet containing 25% protein, and the length of ovarian cycles is independent of the protein content of the diet. Females of *S. longipalpa* show no increase in intake in response to reduced protein content, and the duration of ovarian cycles is longer in females feeding on diets of low protein content. Further evidence that females of *S. longipalpa* lack the ability to exhibit nutritional compensation has been provided by the result of experiments in which rat food, an adequate diet, was diluted with nonutilizable material (Hamilton and Schal, 1991).

Another characteristic expected of insects in which mechanisms in the *demand-mediated* category play a major role in bringing about developmental stage-related changes in feeding is an ability to select optimally when given access to two or more diets of complementary nutrient contents. This appears not to have been formally demonstrated in relation to the intake of nutrients for the support of ovarian development. However, it has been shown that females of the cockroach *Parcoblatta fulvescens* with access to one high-protein diet and two low-protein diets neither stores nor excretes urates, which suggests that nitrogen intake is in balance with demand (Lembke and Cochran, 1990). Females given access to only high-protein diets both excrete and store urates (Cochran, 1986) and have ovarian cycles of similar length to those of the self-selecting individuals (Lembke and Cochran, 1990).

Ovariectomy, chemosterilization, and allatectomy are treatments which permanently eliminate the uptake of nutrients by developing oocytes. The effects on feeding of ovariectomy have been examined in the locusts *Locusta migratoria* (Strong, 1967) and *Schistocerca gregaria* (Hill et al., 1968), those of chemosterilization have been studied in the blowfly, *Phormia regina* (Belzer, 1978b, 1979), and those of allatectomy have been reported in *L. migratoria* (Strong, 1967), *S. gregaria* (Hill and Izatt, 1974), the cockroach *Leucophaea maderae* (Englemann and Rau, 1975), and *P. regina* (Belzer, 1978b; Belzer, 1979), all of which display discrete ovarian cycles. In addition, the effects of ovariectomy have been examined in the cricket *Acheta domesticus* which does not display discrete cycles (Clifford and Woodring, 1986). Neither ovariectomy nor allatectomy results in a reduction in rate of intake in either of the locusts or in the cockroach throughout at least most of the period for which the controls are developing their first batch of oocytes. The situation with *P. regina* is less clear-cut. Both allatectomized and chemosterilized females with access to yeast extract from the day of eclosion consume somewhat less of this material than do untreated females at the time when their oocytes are undergoing vitellogenesis (Belzer, 1978b). However, neither treatment affects yeast intake in females which are given access to this material after having been maintained on only sugar solution for some days after eclosion (Belzer, 1979). Ovariectomy has no effect on food consumption by females of *A. domesticus* over the first 3 days after eclosion, which corresponds to the period before ovulation begins. Thereafter, however, the rate of intake of the ovariectomized females is substantially lower than that of intact, mated and therefore egg-laying females.

The results of these investigations suggest that the changes in the rate of ingestion of foods containing nutrients needed for ovarian development are not the direct result of changes in rate of uptake of nutrient by the ovaries themselves. However, interpretation of the results of these treatment in relation to the more general question of the involvement of regulatory mechanisms in the *demand-mediated* category is uncertain because the nutrients required for the synthesis of yolk protein, the major constituent of mature oocytes, are generally not taken up directly by the oocytes, but are taken up by the fat body, where the yolk protein precursor, vitellogenin, is synthesized. Thus, the possibility exists that changes in the rate of uptake of nutrients by the fat body, rather than by the ovaries, might form the basis for a mechanism in the *demand-mediated* category. The lack of effect on feeding of treatments which prevent the development of oocytes does not rule out this possibility because there is evidence that such treatments do not prevent uptake of nutrients by the fat body. It has been shown, in fact, that allatectomy results in hypertrophy of the fat body in several species including *L. migratoria* (Strong, 1967), *S. gregaria* (Hill and Izatt, 1974), and *P. regina* (Orr, 1964) and that ovariectomy has a similar effect in *L. migratoria* (Strong, 1967). The findings with *A. domesticus* (Clifford and Woodring, 1986) are compatible with the view that the rate of uptake of nutrients by the fat body of ovariectomised females is similar to that of intact females up until the time that ovulation begins in the latter but that it declines thereafter.

Some temporal relationships between readiness to ingest protein-rich material and the course of development of batches of oocytes by some anautogenous dipterans which display discrete ovarian cycles provide further evidence suggesting that feeding behavior is not directly influenced by changes in rate of uptake of nutrients by the developing oocytes. It has been shown that oocyte development does not proceed beyond a previtellogenic or very early vitellogenic stage in a number of species of mosquitoes (Clements, 1992) and in several cyclorrhaphous species including *P. regina* (Stoffolano, 1974), *L. cuprina* (Barton Browne et al., 1976), and *M. domestica* (Adams and Nelson, 1990) unless females ingest protein-rich materials. In such species, the ingestion of protein-rich material initiates a cascade of endocrinological events which result in vitellogenin synthesis by the fat body and possibly the ovaries, and which render the ovaries competent to take up vitellogenin from the hemolymph (Clements, 1992; Yin et al., 1990); Adams and Nelson, 1990). Thus, such species readily ingest protein-rich materials at a time when there is essentially no demand for nutrients by their ovaries. In the case of those mosquitoes which normally take a single blood meal per ovarian cycle, all nitrogen-containing nutrients required for maturation of a batch of oocytes are ingested at a time when the demand of the ovaries is negligible. Females of *L. cuprina* require more than one meal of protein-rich material per ovarian cycle (Barton Browne et al., 1986) but can ingest a significant proportion of their requirements in a single meal (Barton Browne et al., 1976). These findings with mosquitoes and blowflies again suggest that if a mechanism in the *demand-mediated* category does in fact play a significant role in relation to stage of ovarian development, the demand is being exerted by some tissue other than the ovaries. The fat body is a likely candidate, in view of its role in nutrient storage and vitellogenin synthesis. There is, however, the alternative possibility that mechanisms in the *non-demand-mediated* category are dominant, with the readiness to ingest protein-rich material representing a default setting in the insect, possibly after posteclosion maturation of the necessary neural and muscular elements involved.

There are some correlations which suggest, in general terms, that mechanisms in the *demand-mediated* category may be involved. Females of the ovoviviparous cockroach *L. maderae* normally do not initiate development of their next batch of oocytes while they have an ootheca in the brood sac. If, however, four pairs of actively secreting corpora allata are transplated into such females, some will do so. Those females which do develop their oocytes have greater rates of food intake than do those which fail to do so after receiving the same treatment (Englemann and Rau, 1965). This result suggests that intake rate might be determined by demand in females which have depleted their reserves by developing their previous batch of oocytes. The rate of oviposition of virgin females is usually less than that of mated individuals, and in several studies it has been found that the rate of intake of food by mated females is greater than that by virgins. One such study (Slansky, 1980) has demonstrated that the feeding rate of virgin females of the milkweed bug *Oncopeltus fasciatus,* which lay eggs at about one-third the rate of mated females, is only one-third that of mated individuals from the time at which the latter commence oviposition. Similarly, virgin females of the housefly, *Musca domestica,* which lay at about 38% the rate of mated females (Riemann and Thorson, 1969), consume the protein-rich material which provides the nutrients for ovarian development at about one-third the rate (Greenberg, 1959). Virgin females of the cricket *A. domesticus,* which retain eggs in the oviduct for life, feed at a lower rate than do mated individuals (Clifford and Woodring, 1986), but, as discussed below, it is likely that the mechanism responsible for this difference is in the *non-demand-mediated* category.

Evidence for a Non-Demand-Mediated Mechanism

Several studies have provided evidence indicative of the involvement of mechanisms in the *non-demand-mediated* category. One or more mechanisms in this category apparently play a major role in causing the reduced rate of intake by females of the ovoviviparous cockroach *Leucophaea maderae* while they have an ootheca in the brood sac and probably during the few days prior to ovulation. The rate of food intake by females allatectomized soon after eclosion is substantially greater than that of control females from about 5 days before ovulation until after parturition by the latter. In addition, removal of the oothecae from the uterus of pregnant females results in an increase in rate of intake even when further ovarian development is prevented by allatectomy at the time of oothecae removal. Transection of the ventral nerve cord results in a similar increase in rate of intake (Englemann and Rau, 1965) These data suggest that neural input from receptors monitoring distention of the brood sac directly inhibits feeding. Evidence has been obtained that neural inputs from receptors monitoring abdominal distention play an important part in bringing about the reduction in rate of intake of protein-rich material by females of the blowfly *P. regina* as they approach ovarian maturity. Belzer (1979) showed that the injection of a volume of methyl cellulose approximately equal to that of the mature ovaries into the abdominal hemocoal of protein-deprived, 7-day posteclosion females causes a reduction in the amount of protein-rich material they consume over 1 hr. No such reduction is apparent in similarly injected females with transected median abdominal nerves. Noninjected females with transected abdominal nerves consume larger amounts of protein-rich material than do intact females. Reference was made above to the low rate of intake by virgin females of *A. domesticus*. In this species, oocytes are ovulated as they are matured, and up to 650 eggs accumulate in the oviduct of virgins. Clifford and Woodring (1986) interpret their findings as indicating that distention of the oviduct is responsible for reduced intake. This interpretation is supported by their observation that females which oviposit following being mated after accumulating the maximum number of eggs have a large amount of food in the gut within 1 day.

The physiological basis for the lower intake of protein-rich material and/or reduced responsiveness to mammalian hosts of nulliparous autogenous females as compared with parous autogenous or nulliparous anautogenous conspecifics has not been investigated in detail. However, it has been established that autogenous females of some mosquitoes and calliphorids have, at the time of their eclosion, larger reserves of stored nutrients relative to their body weight than do their anautogenous counterparts (Clements, 1992; Spradbery and Sands, 1981; Williams et al., 1977). It seems likely that the lower levels of food intake and responsiveness to host cues of autogenous nullipars are related to this in some way which is yet to be elucidated.

A considerable amount of information is available about the physiological bases for changes in the responsiveness of females mosquitoes to host volatiles in relation to ovarian development and adult reproductive diapause. Because the available data have been discussed in detail in two recent reviews (Klowden, 1990; Bowen, 1991; also see Chapter 7), I will give only a brief summary here.

Development of host-seeking behavior following emergence is dependent upon

JH in *Culex pipiens* and *C. quinquefasciatus* (Meola and Petralia, 1980), but not in *Aedes aegypti* (Bowen and Davis, 1989). The responsiveness of females of *A. aegypti* to host volatiles increases over the 4 days following eclosion. Over this period, there is an increase in the proportion of olfactory receptors showing a high level of response to the important host volatile lactic acid, which suggests that increasing receptor sensitivity is at least partly responsible for the increasing behavioral responsiveness (Davis, 1984).

The physiological basis for the reduced responsiveness of gravid females of *A. aegypti* to host volatiles has been investigated in considerable detail. It has been shown that a humoral factor produced by mature ovaries (Klowden, 1981) causes the fat body (Klowden et al., 1987) to produce another humoral factor which is responsible for inhibiting host-seeking behavior. Neither factor has been identified, but it has been suggested that the one produced by the ovary might be an ecdysteroid. The responsiveness of lactic-acid-sensitive chemoreceptors is reduced in gravid females; and it appears that lack of sensitivity of chemoreceptors, as in young females, is at least partly responsible for the reduced behavioral responsiveness. The lactic acid receptors of diapausing females of *C. pipiens* which, like gravid females of *A. aegypti* also have a low level of responsive to host stimuli, also show reduced sensitivity (Bowen et al., 1988).

The mechanisms involved in the modulation of host-seeking behavior would appear to fall into the *non-demand-mediated* category, with humoral factors acting directly on peripheral and possibly central neural elements and, in the case of development of host-seeking behavior in *Aedes aegypti,* possibly on developmental changes in neural elements in the peripheral nervous system. However, *demand-mediated* mechanisms based on effects of changes in the concentrations of nutrients in the hemolymph cannot be entirely ruled out. For example, JH might have its effect on the posteclosion development of host-seeking behavior in the two *Culex* species by making the fat body competent to take up nutrients. This would have the effect of causing a reduction in their concentrations in the hemolymph which could then affect the sensitivity of peripheral receptors.

11.4.3.4. CHANGES IN MALES

I am aware of only one study in which the food intake of frequently mating males has been compared with that of nonmating males or males mating less frequently. In this, the daily intake rate of males of the cockroach *Blattella germanica* which mated twice per week was found to be about one-third greater than that of males which were allowed to mate only once (Hamilton and Schal, 1988). This increase was similar irrespective of which of two contrasting diets were fed upon. One contained 5% protein and 70% utilizable carbohydrate, whereas the other contained 65% protein and 11% carbohydrate. The interpretation of these results suggested by the authors is that the greater rate of intake of the low-protein diet by the males that mated frequently is to accrue sufficient nitrogen for the increased production of the spermatophores and of uric acid (see Roth and Dateo, 1964), which are both transferred to the female at mating, and that the increase in intake of the carbohydrate-rich diet is to service an increased requirement for carbohydrate. This interpretation implies that the mechanism responsible for there being a direct relationship between rate of intake of food by males of *B. germanica* and their frequency of mating is in the *demand-mediated* category. If this were so, it would be expected that males of this species would exhibit compensatory feeding in relation to nutrient concentrations. The finding in this study that males consumed the high-protein, low-carbohydrate diet at a higher rate than they consumed the low-protein, high-carbohydrate diet suggests compensation in relation to carbohydrate intake. However, as indicated earlier in this section, there are problems in the interpretation of experiments in which concentrations of protein and carbohydrate in different diets are inversely related.

Some orthopterans produce very large spermatophores, and it has been shown in one of these, the katydid *Requena verticalis,* that males feeding on a poor diet mate less frequently than those receiving better nutrition (Gwynne, 1990). Examination of the obverse effect—that of frequency of mating on food intake in species such as this which produce large spermatophores—would be considerable interest.

11.5. Concluding Remarks

The material I have discussed in this chapter clearly demonstrates that an insect's feeding behavior may change, often quite substantially, according to its stage of development, and it indicates that some progress has been made toward the elucidation of the physiological basis for some of the observed behavioral changes. Current knowledge suggests that mechanisms in the *demand-mediated* category play an important role in causing changes in intake which are compatible with changes in the nutritional targets during the growth phases of each larval instar and that mechanisms in the *non-demand-mediated* category play major roles in bringing about reduction in feeding as developmental programs approach or reach completion. The mechanisms responsible for changes in intake during the phase of reproductive development of females during which oocytes are growing have yet to be satisfactorily elucidated. There is, for example, an urgent need for experiments to determine whether insects display compensatory feeding (Simpson and Simpson, 1990) during oocyte growth in relation particularly to concentration of nitrogen-rich nutrients.

It is notable that, apart from the elucidation of the physiological basis for ontogenetic changes in the host-seeking behavior of mosquitoes (Klowden, 1990; Bowen, 1991), the past decade has seen relatively little increase in our understanding of the physiological mechanisms involved. Perhaps the most significant recent advance has been the recognition of the fact that developmental stage-related changes in the sensitivity of gustatory (Simmonds et al., 1991; Schoonhoven et al., 1991; Simpson et al., 1990) and olfactory (Davis, 1984; Bowen et al., 1988; Bowen and Davis, 1989) receptors may be at least partly responsible for the observed behavioral changes.

Despite the only moderate rate of progress in recent years toward the elucidation of the physiological mechanisms underlying stage of development-related changes in feeding behavior, I believe that the time is now ripe for some fruitful research in this area. This belief is based on two factors. The first of these is the increasing realization that great insight can be gained through detailed observations on feeding behavior (Simpson, 1990). Results of such observations can form the basis for fine-grained analyses of causal factors in that they allow consideration of the effects of stage of development on the physiological basis for the timing of the initiation and termination of meals separately from the recognition of changes in rate of ingestion. The second factor is that the functional concept of growth, nutritional, and intake targets (Raubenheimer and Simpson, 1993; Simpson and Raubenheimer, 1993a) and the proximate concept of the distinction between *demand-mediated* and *non-demand-mediated* categories of mechanisms together provide a framework for the consideration of the physiological mechanisms involved in changes in feeding behavior in relation to change of development.

References

Abisgold, J. D., and Simpson, S. J. (1988). The effect of dietary protein levels and haemolymph composition on the sensitivity of the maxillary palp chemoreceptors of locusts. *J. Exp. Biol.* 135, 215–229.

Adams, T. S., and Nelson, D. R. (1990). The influence of diet on ovarian maturation, mating, and pheromone production in the housefly, *Musca domestica*. *Invert. Reprod. Dev.* 17, 193–201.

Adler, P. H., and Adler, C. R. L. (1988). Behavioral time budgeted for larvae of *Heliothis zea* (Lepidoptera: Noctuidae) on artificial diet. *Ann. Entomol. Soc. Am.* 81, 682–688.

Anderson, J. F. (1971). Autogeny and mating and their relationship to biting in the saltmarsh deer fly, *Chrysops atlanticus* (Diptera: Tabanidae). *Ann. Entomol. Soc. Am.* 64, 1421–1424.

Baehr, J.-C., Porcheron, P., Papillon, M., and Dray, F. (1979). Haemolymph levels of juvenile hormone, ecdysteroids and protein during the last two larval instars of *Locusta migratoria*. *J. Insect Physiol.* 25, 415–421.

Barton Browne, L. (1975). Regulatory mechanisms in insect feeding. *Adv. Insect. Physiol.* 11, 1–116.

Barton Browne, L., and Kerr, R. W. (1986). Influence of sex and prior protein feeding on preferences by the housefly, *Musca domestica*, between sucrose solutions and solutions of *l*-leucine of sodium phosphate buffer. *Entomol. Exp. Appl.* 41, 135–138.

Barton Browne, L., and van Gerwen, A. C. M. (1992). Volume of protein meals taken by females of the blowfly, *Lucilia cuprina:* ovarian development related and direct effects of protein ingestion. *Physiol. Entomol.* 17, 9–18.

Barton Browne, L., Bartell, R. J., van Gerwen, A. C. M., and Lawrence, L. J. (1976). Relationship between protein ingestion and sexual receptivity in females of the Australian sheep blowfly *Lucilia cuprina. Physiol. Entomol.* 1, 235–240.

Barton Browne, L., van Gerwen, A. C. M., and Roberts, J. A. (1986). Ovarian development in females of the Australian sheep blowfly, *Lucilia cuprina*, given limited opportunity to feed on protein-rich material at different ages. *Intern. J. Invert. Reprod. Dev.* 10, 179–186.

Beck, S. D., Edwards, C. A., and Medler, J. T. (1958). Feeding and nutrition of the milkweed bug, *Oncopeltus fasciatus* (Dallas). *Ann. Entomol. Soc. Am.* 51, 283–288.

Bell, W. J. (1969). Continuous and rhythmic reproductive cycle observed in *Periplaneta americana. Biol. Bull. Mar. Biol. Lab. Woods Hole* 137, 239–249.

Belzer, W. R. (1978a). Patterns of selective protein ingestion by the blowfly *Phormia regina. Physiol. Entomol.* 3, 169–175.

Belzer, W. R. (1978b). Factors conducive to increased protein feeding by the blowfly *Phormia regina. Physiol. Entomol.* 3, 251–257.

Belzer, W. R. (1979). Abdominal stretch in the regulation of protein ingestion by the black blowfly, *Phormia regina. Physiol. Entomol.* 4, 7–13.

Blaney, W. M., Chapman, R. F., and Wilson, A. (1973). The pattern of feeding of *Locusta migratoria* (L.) (Orthoptera, Acrididae). *Acrida* 2, 119–137.

Blois, C., and Cloarec, A. (1983). Density-dependent prey selection in the water stick insect, *Ranatra linearis* (Heteroptera). *J. Anim. Ecol.* 52, 849–866.

Bollenbacher, W. E., and Gilbert, L. I. (1982). Neuroendocrine control of postembryonic development in insects: the prothoracicotropic hormone. In: Farner, D. S., and Lederis, K. (eds.), *Neurosecretion: Molecules, Cells, Systems*. Plenum Press, New York, pp. 361–370.

Bowen, M. F. (1991). The sensory physiology of host-seeking behavior in mosquitoes. *Annu. Rev. Entomol.* 36, 139–158.

Bowen, M. F. (1992). Patterns of sugar feeding in diapausing and nondiapausing *Culex pipiens* (Diptera: Culicidae) females. *J. Med. Entomol.* 29, 43–49.

Bowen, M. F., and Davis, E. E. (1989). The effect of allatectomy and juvenile hormone replacement on the development of host-seeking behaviour and lactic acid receptor sensitivity in the mosquito *Aedes aegypti Med. Vet. Entomol.* 3, 53–60.

Bowen, M. F., Davis, E. E., and Haggart, D. A. (1988). A behavioural and sensory analysis of host-seeking behaviour in the diapausing mosquito *Culex pipiens. J. Insect Physiol.* 34, 805–813.

Busse, F. K., Jr., and Barth, R. H., Jr. (1985). Physiology of feeding-preference patterns of female black blowflies (*Phormia regina* Meigen): modification in response to salts subsequent to salt feeding. *J. Insect Physiol.* 31, 23–26.

Capinera, J. L. (1979). Zebra caterpillar (Lepidoptera: Noctuidae): foliage consumption and development of larvae on sugarbeet. *Can. Entomol.* 111, 905–909.

Chapman, R. F. (1990). General anatomy and function. In: *Insects of Australia,* vol. 1. Melbourne University Press, Melbourne, pp. 33–67.

Chapman, R. F., and Beerling, E. A. M. (1990). The pattern of feeding of first instar nymphs of *Schistocerca americana. Physiol. Entomol.* 15, 1–11.

Chyb, S., and Simpson, S. J. (1990). Dietary selection in adult *Locusta migratoria. Entomol. Exp. Appl.* 56, 47–60.

Clark, J. V. (1980). Changes in the feeding rate and receptor sensitivity over the last instar of the African armyworm, *Spodoptera exempta. Entomol. Exp. Appl.* 27, 144–148.

Clements, A. N. (1992). *The Biology of Mosquitoes,* vol. 1. Chapman and Hall, London.

Clifford, C. W., and Woodring, J. P. (1986). The effects of virginity and ovariectomy on growth, food consumption, fat body mass and oxygen consumption in the house cricket, *Acheta domesticus. J. Insect Physiol.* 32, 425–431.

Cochran, D. G. (1983). Food and water consumption during the reproductive cycle of female German cockroaches. *Entomol. Exp. Appl.* 34, 51–57.

Cochran, D. G. (1986). Feeding, drinking and urate excretory cycles in reproducing female *Parcoblatta* cockroaches. *Comp. Biochem. Physiol.* 84A, 677–682.

Cohen, R. W., Heydon, S. L., Waldbauer, G. P., and Friedman, S. (1987a). Nutrient self-selection by the omnivorous cockroach, *Supella longipalpa. J. Insect Physiol.* 33, 77–82.

Cohen, R. W., Waldbauer, G. P., Friedman, S., and Schiff, N. M. (1987b). Nutrient self-selection by *Heliothis zea* larvae: a time-lapse film study. *Entomol. Exp. Appl.* 44, 65–73.

Cohen, R. W., Waldbauer, G. P., and Friedman, S. (1988). Natural diets and self-selection: *Heliothis zea* larvae and maize. *Entomol. Exp. Appl.* 46, 161–171.

Davis, E. E. (1984). Development of lactic acid-receptor sensitivity and host-seeking behaviour of newly emerged females *Aedes aegypti* mosquitoes. *J. Insect Physiol.* 30, 211–215.

de Clerck, D., and de Loof, A. (1983). Effect of dietary ecdysterone on protein ingestion and copulatory behaviour of the blowfly, *Sarcophaga bullata. Physiol. Entomol.* 8, 243–249.

Denlinger, D. L., and Ma, W.-C. (1974). Dynamics of the pregnancy cycle in the tsetse *Glossina morsitans. J. Insect Physiol.* 20, 1015–1026.

Dethier, V. G. (1961). Behavioral aspects of protein ingestion by the blowfly *Phormia regina* Meigen. *Biol. Bull. Mar. Biol. Lab. Woods Hole* 121, 456–470.

Dominick, O. S., and Truman, J. W. (1985). The physiology of wandering behaviour in *Manduca sexta.* II. The endocrine control of wandering behaviour. *J. Exp. Biol.* 117, 45–68.

Eldridge, B. F., and Bailey, C. L. (1979). Experimental hibernation studies in *Culex pipiens* (Diptera: Culicidae): reactivation of ovarian development and blood-feeding in prehibernating females. *J. Med. Entomol.* 15, 462–467.

Engelmann, F., and Rau, I. (1965). A correlation between feeding and the sexual cycle in *Leucophaea maderae* (Blattaria). *J. Insect Physiol.* 11, 53–64.

Friedman, S., Waldbauer, G. P., Ertmoed, J. E., Naeem, M., and Ghent, A. W. (1991). Blood trehalose levels have a role in the control of dietary self-selection by *Heliothis zea* larvae. *J. Insect Physiol.* 37, 919–928.

Gaston, K. J., Reavey, D., and Valladares, G. R. (1991). Changes in feeding habit as caterpillars grow. *Ecol. Entomol.* 16, 339–344.

Greenberg, B. (1959). House fly nutrition. I. Quantitative studies of the protein and sugar requirements of males and females. *J. Cell. Comp. Physiol.* 53, 169–177.

Gwynne, D. T. (1990). Testing parental investment and the control of sexual selection in katydids: the operational sex ratio. *Am. Nat.* 136, 474–484.

Hamilton, R. L., and Schal, C. (1988). Effects of diet protein levels on reproduction and food consumption in the German cockroach (Dictyoptera: Blattellidae). *Ann. Entomol. Soc. Am.* 81, 969–976.

Hamilton, R. L., and Schal, C. (1991). Effects of dextrin and cellulose on feeding and reproduction in female brown-banded cockroaches, *Supella longipalpa*. *Physiol. Entomol.* 16, 57–64.

Hamilton, R. L., Cooper, R. A., and Schal, C. (1990). The influence of nymphal and adult dietary protein on food intake and reproduction in female brown-banded cockroaches. *Entomol. Exp. Appl.* 55, 23–31.

Hill, L., and Goldsworthy, G. J. (1968). Growth, feeding activity, and the utilization of reserves in larvae of *Locusta*. *J. Insect Physiol.* 14, 1085–1098.

Hill, L., and Izatt, M. E. G. (1974). The relationship between corpora allata and fat body and haemolymph lipids in the adult female desert locust. *J. Insect Physiol.* 20, 2143–2156.

Hill, L., Luntz, A. J., and Steele, P. A. (1968). The relationships between somatic growth, ovarian growth, and feeding activity in the adult desert locust. *J. Insect Physiol.* 14, 1–20.

Hintze-Podufal, C., and Fricke, F. (1971). The effect of farnesol derivatives on the mature larva of *Cerura vinula* L. (Lepidoptera). *J. Insect Physiol.* 17, 1925–1932.

Hoekstra, A., and Beenakkers, A. M. Th. (1976). Consumption, digestion, and utilization of various grasses of fifth-instar larvae of the migratory locust. *Entomol. Exp. Appl.* 19, 130–138.

Horie, Y., and Watanabe, K. (1983). Daily utilization and consumption of dry matter in food by the silkworm, *Bombyx mori* (Lepidoptera: Bombycidae). *Appl. Entomol. Zool.* 18, 70–80.

Huffman, F. R., and Smith, J. W., Jr. (1979). Bollworm: peanut foliage consumption and larval development. *Environ. Entomol.* 8, 465–467.

Johnson, S. J. (1984). Larval development, consumption, and feeding behavior of the cotton leafworm, *Alabama argillacea* (Hubner). *Southwestern Entomol.* 9, 1–6.

Jones, R. E., and Walker, J. M. (1974). Some factors affecting protein feeding and egg development in the Australian bushfly *Musca vetustissima. Entomol. Exp. Appl.* 17, 177–25.

Kitching, R. L., and Roberts, J. A. (1975). Laboratory observations on the teneral period in sheep blowflies, *Lucilia cuprina* (Diptera: Calliphoridae). *Entomol. Exp. Appl.* 18, 220–225.

Klowden, M. J. (1981). Initiation and termination of host-seeking inhibition in *Aedes aegypti* during oocyte maturation. *J. Insect Physiol.* 27, 799–803.

Klowden, M. J. (1990). The endogenous regulation of mosquito reproductive behavior. *Experientia* 46, 660–670.

Klowden, M. J., Davis, E. E., and Bowen, M. F. (1987). Role of the fat body in the regulation of host-seeking behaviour in the mosquito, *Aedes aegypti. J. Insect Physiol.* 33, 643–646.

Langley, P. A., and Pimley, R. W. (1974). Utilization of U-^{14}C amino acids or U-^{14}C protein by adult *Glossina morsitans* during *in utero* development of larva. *J. Insect Physiol.* 20, 2157–2170.

Leegwater-van der Linden, M. E. (1981). Effect of a four-day-a-week feeding regimen versus daily feeding on the reproduction of *Glossina pallidipes. Entomol. Exp. Appl.* 29, 169–176.

Lembke, H. F., and Cochran, D. G. (1990). Diet selection by adult female *Parcoblatta fulvescens* cockroaches during the oothecal cycle. *Comp. Biochem. Physiol.* 95A, 195–199.

Leprince, D. J. (1989). Gonotrophic status, sperm presence and sugar feeding patterns in southwestern Quebec tabanid (Diptera) populations. *J. Am. Mosq. Contr. Assoc.* 5, 383–386.

Magnarelli, L. A. (1981). Parity, follicular development, and sugar feeding in *Culicoides melleus* and *C. hollensis. Environ. Entomol.* 10, 807–811.

McAvoy, T. J., and Kok, L. T. (1992). Development, oviposition, and feeding of the cabbage webworm (Lepidoptera: Pyralidae). *Environ. Entomol.* 21, 527–533.

McAvoy, T. J., and Smith, J. C. (1979). Feeding and developmental rates of the Mexican bean beetle on soybeans. *J. Econ. Entomol.* 72, 835–836.

Meola, R. W., and Lea, A. O. (1972). Humoral inhibition of egg development in mosquitoes. *J. Med. Entomol.* 9, 99–103.

Meola, R. W., and Petralia, R. S. (1980). Juvenile hormone induction of biting behavior in *Culex* mosquitoes. *Science* 209, 1548–1550.

Meola, R., and Readio, J. (1987). Juvenile hormone regulation of the second biting cycle in *Culex pipiens. J. Insect Physiol.* 33, 751–754.

Mitchell, C. J. (1981). Diapause termination, gonactivity, and differentiation of host-seeking and blood-feeding behavior in hibernating *Culex tarsalis. J. Med. Entomol.* 18, 386–394.

Mitchell, C. J., and Briegel, H. (1989). Inability of diapausing *Culex pipiens* to use blood for producing lipid reserves for overwintering survival. *J. Med. Entomol.* 26, 318–326.

Mordue (Luntz), A. J., and Hill, L. (1970). The utilisation of feed by the adult female desert locust, *Schistocerca gregaria. Entomologia Exp. Appl.* 13, 352–358.

Nayar, J. K., and Sauerman, D. M., Jr. (1974). Long-term regulation of sucrose intake by the female mosquito, *Aedes taeniorhynchus. J. Insect. Physiol.* 20, 1203–1208.

Nayar, J. K., and Sauerman, D. M., Jr. (1975). Flight and feeding behavior of autogenous and anautogenous strains of the mosquito *Aedes taeniorhynchus. Ann. Entomol. Soc. Am.* 68, 791–796.

Neosn, R. L., and Milby, M. M. (1982). Autogeny and blood-feeding by *Culex tarsalis* (Diptera: Culicidae) and the interval between oviposition and feeding. *Can. Entomol.* 114, 515–521.

Norris, M. J. (1960). Group effects on feeding in adult males of the desert locust, *Schistocerca gregaria* (Forsk.), in relation to sexual maturation. *Bull. Entomol. Res.* 51, 731–753.

Orr, C. W. M. (1964). The influence of nutritional and hormonal factors on egg development in the blowfly *Phormia regina* (Meig.). *J. Insect Physiol.* 10, 53–64.

Parrott, W. L., Jenkins, J. N., and McCarty, J. C., Jr. (1983). Feeding behavior of first-stage tobacco budworm (Lepidoptera: Noctuidae) on three cotton cultivars. *Ann. Entomol. Soc. Am.* 76, 167–170.

Pener, M. P., and Lazarovici, P. (1979). Effect of exogenous juvenile hormone on mating behaviour and yellow colour in allatectomized adult male desert locusts. *Physiol. Entomol.* 4, 251–261.

Porter, C. H., DeFoliart, G. R., Miller, B. R., and Nemenyi, P. B. (1986). Intervals to blood feeding following emergence and oviposition in *Aedes triseriatus* (Diptera: Culucidae). *J. Med. Entomol.* 23, 222–224.

Rachman, N. J. (1980). Physiology of feeding preference patterns of female black blowflies (*Phormia regina* Meigen). I. The role of carbohydrate reserves. *J. Comp. Physiol.* 139, 59–66.

Rachman, N. J., Busse, F. K., Jr., Barth, R. H., Jr. (1982). Physiology of feeding-preference patterns of female black blowflies (*Phormia regina* Meigen): alterations in responsiveness to salts. *J. Insect Physiol.* 28, 625–630.

Raubenheimer, D. (1992). Tannic acid, protein, and digestible carbohydrate: dietary imbalance and nutritional compensation in locusts. *Ecology* 73, 1012–1027.

Raubenheimer, D., and Simpson, S. J. (1993). The geometry of compensatory feeding in the locust. *Anim. Behav.* 45, 954–964.

Retnakaran, A. (1983). Spectrophotometric determination of larval ingestion rates in the spruce budworm (Lepidoptera; Tortricidae). *Can. Entomol.* 115, 31–40.

Reynolds, S. E., Yeomans, M. R., and Timmins, W. A. (1986). The feeding behaviour of caterpillars (*Manduca sexta*) on tobacco and artificial diet. *Physiol. Entomol.* 11, 39–51.

Riemann, J. G., and Thorson, B. J. (1969). Effect of male accessory material on oviposition and mating by female house flies. *Ann. Entomol. Soc. Am.* 62, 828–834.

Roberts, J. A., and Kitching, R. L. (1974). Ingestion of sugar, protein and water by adult *Lucilia cuprina* (Wied.) (Diptera, Calliphoridae). *Bull. Entomol. Res.* 64, 81–88.

Rogers, C. E. (1978). Sunflower moth: feeding behavior of the larva. *Environ. Entomol.* 7, 763–765.

Roth, L. M., and Dateo, G. P., Jr. (1964). Uric acid in the reproductive system of the cockroach, *Blattella germanica*. *Science* 146, 782–784.

Scheurer, R., and Leuthold, R. (1969). Haemolymph proteins and water uptake in female *Leucophaea maderae* during the sexual cycle. *J. Insect Physiol.* 15, 1067–1077.

Schoonhoven, L. J., Simmonds, M. S. J., and Blaney, W. M. (1991). Changes in the responsiveness of the maxillary styloconic sensilla of *Spodoptera littoralis* to inositol and sinigrin correlate with feeding behaviour during the final larval stadium. *J. Insect Physiol.* 37, 261–268.

Sieber, R., and Benz, G. (1977). Juvenile hormone in larval diapause of the codling moth, *Laspeyresia pomonella* L. (Lepidoptera: Tortricidae). *Experientia* 33, 1598–1599.

Sieber, R., and Benz, G. (1978). The influence of juvenile hormone on the feeding behaviour of last instar larvae of the codling moth, *Laspeyresia pomonella* (Lep., Tortricidae), reared under different photoperiods. *Experientia* 34, 1647–1650.

Silverman, J. (1986). Adult German cockroach (Orthoptera: Blattellidae) feeding and drinking behavior as a function of density and harborage-to-resource distance. *Environ. Entomol.* 15, 198–204.

Simmonds, M. S. J., Schoonhoven, L. M., and Blaney, W. M. (1991). Daily changes in the responsiveness of taste receptors correlate with feeding behaviour in larvae of *Spodoptera littoralis*. *Entomol. Exp. Appl.* 61, 73–88.

Simmons, A. M., and Yeargan, K. V. (1988). Feeding frequency and feeding duration of the green stink bug (Hemiptera: Pentatomidae) on soybean. *J. Econ. Entomol.* 81, 812–815.

Simpson, C. L., Chyb, S., and Simpson, S. J. (1990). Changes in chemoreceptor sensitivity in relation to dietary selection by adult *Locusta migratoria*. *Entomol. Exp. Appl.* 56, 259–268.

Simpson, S. J. (1982a). Changes in the efficiency of utilisation of food throughout the fifth-instar nymphs of *Locusta migratoria*. *Entomol. Exp. Appl.* 31, 265–275.

Simpson, S. J. (1982b). Patterns in feeding: a behavioural analysis using *Locusta migratoria* nymphs. *Physiol. Entomol.* 7, 325–336.

Simpson, S. J. (1990). The pattern of feeding. In: Chapman, R. F., and Joern, A. (eds.), *Biology of Grasshoppers*. John Wiley and Sons, New York, pp. 73–103.

Simpson, S. J., and Raubenheimer, D. (1993a). A multi-level analysis of feeding behaviour: the geometry of nutritional decisions. *Philos. Trans. R. Soc. London Ser B* 342, 381–402.

Simpson, S. J., and Raubenheimer, D. (1993b). The central role of haemolymph in the regulation of nutrient intake in insects. *Physiol. Entomol.* 18, 395–403.

Simpson, S. J., and Simpson, C. L. (1990). The mechanisms of nutritional compensation by phytophagous insects. *In* Bernays E. A. (ed.), *Insect–Plant Interaction,* vol. 2. CRC Press, Boca Raton, FL, pp. 111–160.

Simpson, S. J., Simmonds, M. S. J., and Blaney, W. M. (1988). A comparison of dietary selection behaviour in larval *Locusta migratoria* and *Spodoptera littoralis. Physiol. Entomol.* 13, 225–238.

Slansky, F., Jr. (1980). Quantitative food utilization and reproductive allocation by adult milkweed buds, *Oncopeltus fasciatus. Physiol. Entomol.* 5, 73–86.

Spradbery, J. P., and Sands, D. P. A. (1981). Larval fat body and its relationship to protein storage and ovarian development in adults of the screw-worm fly *Chrysoma bezziana. Entomol. Exp. Appl.* 30, 116–122.

Spradbery, J. P., and Schweizer, G. (1979). Ingestion of feed by the adult screw-worm fly, *Chrysomya bezziana* (Diptera, Calliphoridae). *Entomol. Exp. Appl.* 25, 75–85.

Spradbery, J. P., and Schweizer, G. (1981). Oosorption during ovarian development in the screw-worm fly, *Chrysomya bezziana. Entomol. Exp. Appl.* 30, 209–214.

Stoffolano, J. G., Jr. (1974). Influence of diapause and diet on the development of the gonads and accessory reproductive glands of the black blowfly, *Phormia regina* (Meigen). *Can. J. Zool.* 52, 981–988.

Stoffolano, J. G., Jr., and Bernays, E. A. (1980). The post-ecdysial fast of fifth-instar nymphs of *Locusta migratoria. Entomol. Exp. Appl.* 28, 213–221.

Strangways-Dixon, J. (1961). The relationship between nutrition, hormones and reproduction in the blowfly *Calliphora erythrocephala* (Meig.). I. Selective feeding in relation to the reproductive cycle, the corpus allatum volume and fertilization. *J. Exp. Biol.* 38, 225–235.

Strong, L. (1967). Feeding activity, sexual maturation, hormones, and water balance in the female African migratory locust. *J. Insect. Physiol.* 13, 495–507.

Thompson, D. J. (1978). Prey size selection by larvae of the damselfly, *Ischnura elegans* (Odonata). *J. Anim. Ecol.* 47, 769–785.

Tobe, S. S., and Davey, K. G. (1972). Volume relationships during the pregnancy cycle of the tsetse fly *Glossina austeni. Can. J. Zool.* 50, 999–1010.

van Geem, T. A., and Broce, A. B. (1986). Fluctuations in the protein and carbohydrate content of the crop correlated to periodicities in ovarian development of the female face fly (Diptera: Muscidae). *Ann. Entomol. Soc. Am.* 79, 1–6.

Verrett, J. M., and Mills, R. R. (1973). Water balance during vitellogenesis by the American cockroach: translocation of water during the cycle. *J. Insect Physiol.* 19, 1889–1901.

Waldbauer, G. P. (1968). The consumption and utilization of food by insects. *Adv. Insect Physiol.* 5, 229–288.

Waldbauer, G. P., and Friedman, S. (1991). Self-selection of optimal diets by insects. *A. Rev. Entomol.* 36, 43–63.

Walker, P. R., and Bailey, E. (1971). Effect of allatectomy on the growth of the male desert locust during adult development. *J. Insect Physiol.* 17, 1125–1137.

Walker, P. R., Hill, L., and Bailey, E. (1970). Feeding activity, respiration, and lipid and carbohydrate content of the male desert locust during adult development. *J. Insect Physiol.* 16, 1001–1015.

Webster, R. P., Stoffolano, J. G., Jr., and Prokopy, R. J. (1979). Long-term intake of protein and sucrose in relation to reproductive behavior of wild and laboratory cultured *Rhagoletis pomonella. Ann. Entomol. Soc. Am.* 72, 41–46.

Williams, K. L. (1972). Protein synthesis during the metamorphosis of *Lucilia cuprina.* Ph.D. thesis, Australian National University, Canberra, Australia.

Williams, K. L., Barton Browne, L., and van Gerwen, A. C. M. (1977). Ovarian development in autogenous and anautogenous *Lucilia cuprina* in relation to protein storage in the larval fat body. *J. Insect Physiol.* 23, 659–664.

Williams, K. L., Barton Browne, L., and van Gerwen, A. C. M. (1979). Quantitative relationships between the ingestion of protein-rich material and ovarian development in the Australian sheep blowfly, *Lucilia cuprina* (Wied.). *Int. J. Invert. Reprod.* 1, 75–88.

Wilson, D. P., and Smith, G. C. (1985). Ovarian diapause in three geographic strains of *Culex pipiens* (Diptera: Culicidae). *J. Med. Entomol.* 22, 524–528.

Woodring, J. P., Roe, R. M., and Clifford, C. W. (1977). Relation of feeding, growth and metabolism to age in the larval, female house cricket. *J. Insect. Physiol.* 23, 207–211.

Yin, C.-M., Zou, B.-X., Yi, S.-X., and Stoffolano, J. G., Jr. (1990). Ecdysteroid activity during oogenesis in the black blowfly, *Phormia regina* (Meigen). *J. Insect. Physiol.* 36, 375–382.

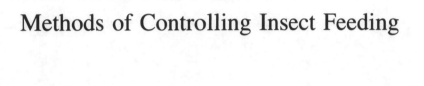

Methods of Controlling Insect Feeding

12

Stimulation of Feeding: Insect Control Agents

D. A. Avé

12.1. Introduction

There is an increased public and government interest in means of managing populations of pest insects with reduced amounts of insecticides in order to limit the impact on the environment and human health. In the past decade the search for commercial alternatives to the "classic chemical" insecticides has received new impetus. Established agricultural product companies, such as Ciba-Geigy, Sandoz and Abbott, and those formed during the 1980s, such as Ecogen (includes former Scentry), Mycogen, Entotech, Biosys (includes former AgriSense), are competing in this market. The "alternative" pest control materials include naturally occurring chemicals such as pheromones and plant chemicals and various insect disease organisms such as bacteria, nematodes, viruses, fungi, and protozoa (Ghassemi et al., 1983). All of these materials are used in what can be considered a biorational approach to pest control. Because some of the biopesticides such as bacteria and virusses do not act by contact but have to be ingested by the insect pest, the development of feeding stimulants to increase the uptake of the biopesticide by the pest has received new attention. Frequently, literature on materials that stimulate feeding on insecticides has not distinguished between food attractants and contact feeding stimulants, whereas nonpolar materials in principle can have both functions. The word "bait" is also used. This term includes either of the above mainly because the materials used are chemically not well-defined. In addition, a bait implies the presence of "a poisonous material." I will use baits and feeding stimulants where appropriate throughout the chapter in the context of the literature discussed.

The use of baits to aid in the control of insect pests is not a new concept. Palatable food substances were reportedly already in use in combination with insecticides for domestic cockroach control as early as 1858 (Appel, 1990). With these bait stations the dispersion of insecticides was limited, resulting in a reduced exposure to humans. On the agricultural front the use of baits has been in large area-wide insect eradication programs, which frequently included centers of human population. Again, a reduced insecticide exposure to humans, but increased insecticide contact with the pest insect, was the goal. An example can be found in successful research initiated in the 1950s for the control of the Mediterranean fruit fly. Insecticides mixed with protein hydrolysate as attractant/ food bait provided superior control of the Medfly in Hawaii. These tests provided the proving grounds for the use of Staley's insect bait SIB-2 [acid-hydrolyzed corn protein (gluten meal)] during the eradication program of the Medfly in Florida (1956–1957). Later this material was mixed with corn steep liquor and designated SIB-7 (Steiner et al., 1961). The flies were attracted to the food bait which enticed them to make contact and feed on the deposits. With the bait, less than half of the conventionally sprayed amount of insecticide was needed for control. Again, SIB-7 was used in the eradication of the Medfly in California (1979–1980).

For other agricultural pests, such as lepidopterous larvae and various species of the corn root worm and cucumber beetle complex, researchers have attempted in the past 20 years to develop successful feeding stimulants. Coax®, and similar products based on cottonseed flour, are available for caterpillars. Products containing cucurbitacins are potent feeding stimulants for corn rootworm beetles. The cucurbitacins are economically available as a root powder from the buffalo gourd *Cucurbita foetidissima,* and a few formulations are now available for agricultural use. An insecticide product currently on the market and baited with buffalo gourd root powder (Slam®, Microflo Company, Lakeland, FL) contains just 0.7 ounce of carbaryl insecticide per 0.5 pound of product applied per acre. Company brochures describe this amount of insecticide to be less than that in a pet flea collar. A recently marketed feeding stimulant product also based on cucurbitacins (Compel®, Ecogen) can be used with even lower rates of carbaryl than those otherwise used to control corn rootworm beetles.

It is evident from these examples that some inroads have been made in the development and commercialization of feeding stimulants for agricultural purposes, although their use is not yet widespread. Various reasons for the lack of success can be pointed out, and a quick comparison with insect bait development for household insect pests is useful at this point. Research with cockroaches has shown that not every insecticide is suitable for incorporation into a bait. In fact, the earlier roach baits with propoxur, diazinon, or chlordane were not consistent in their efficacy because of unwanted effects of the insecticide on cockroach feeding despite the presence of the bait. Baited traps became better with the discovery of the insecticide hydramethylnon, which elicited minimal avoidance (Appel, 1990). Obviously, the development of a successful insect feeding stimulant has to take into account all of the various steps of insect behavior involved in feeding and ingestion, and how the material is applied to the crop and presented to the insect pest.

In this chapter, important concepts of insect feeding stimulants will be discussed.

12.2. Effects of Insecticides on Feeding Behavior

Insecticides can have a pronounced effect on insect behavior, and, assuming an initial limited encounter with insecticide deposits, such an exposure can cause irritation and avoidance as behavioral reaction. Lockwood et al. (1984) considered these two phenomena as being stimulus-independent and stimulus-dependent, respectively. Avoidance, which is due to the insect's sensory evaluation of the insecticide deposit, is different from irritation, which is caused by physiological discomfort upon direct or vapor contact with the insecticide. Although irritation of the insect is described as being independent of concentration, a threshold level obviously exists. Irritation and avoidance can also occur at sublethal levels (Giles and Rothwell, 1983; Moriarty, 1969; Pluthero and Singh, 1984; Tan, 1981; for reviews see Gould, 1991; Haynes, 1988). Both behaviors prevent the insect from settling on the food substrate and reduce the amount of time the insect would have fed on the insecticide-treated area (Young and McMillian, 1979). In addition, even after the ingestion of a sublethal dose of insecticide, avoidance has been observed in the form of food-aversion learning as discussed in lepidopterous larvae by Dethier (Dethier, 1980; Dethier and Yost, 1979). The sickness resulting from having ingested the treated substrate caused the larvae to select other food sources after recovery (also see Chapter 10).

Irritation, avoidance, and postingestion avoidance behaviors have been difficult to separate during observations on the effects of pesticides on insect feeding. Polles and Vinson (1969) concluded that avoidance of malathion was a negative factor in the mortality of *Heliothis virescens* (tobacco budworm) larvae. Droplet density was such that it allowed enough spacing for the larvae to avoid touching the insecticide deposits. In another case, Ruscoe (1977) described the action of sublethal concentrations of permethrin as an antifeedant for neonate larvae of diamondback moth, *Plutella xylostella*. In addition, many other synthetic pyrethroids were found to be deterrent to *Spodoptera frugiperda* (fall armyworm) larvae at 5 ppm (Gist and Pless, 1985). The phenomenon of diamondback moth larvae avoiding leaves treated with lethal concentrations of permethrin was carefully reevaluated by Hall and co-workers (Hall et al., 1989; Hoy et al., 1990a) using a droplet generator producing a defined size and number of deposits. A choice of leaf disks sprayed with permethrin droplets of 110μm inflight diameter at 0, 5, 10, 20, and 40 drops/cm^2 was given to third-instar diamondback moth larvae from populations with different levels of susceptibility to permethrin.

A definite avoidance was observed among larvae from the most susceptible populations, but no effect was found on selection of a feeding site in the more resistant larvae. Their final suggestion was that irritation caused avoidance of the larvae to permethrin and that resistant larvae were apparently less irritated by the permethrin.

12.2.1. Biorational Insecticides

Effects on insect feeding have been reported for the biorational insecticide formulations based on *Bacillus thuringiensis* (Bt). Yendol et al. (1975) tested the feeding activity of *Lymantria dispar* (gypsy moth) larvae on leaves treated with a Bt formulation or purified active ingredient, delta-endotoxin, and found that the commercial material deterred feeding although the active ingredient did not. Other tests were conducted by Gould and co-workers (Gould and Anderson, 1991; Gould et al., 1991) in which tobacco budworm larvae were given a choice of control diet and diet with Bt, either Dipel® or delta-endotoxin. In both cases the larvae avoided diet containing Bt when they apparently were affected by the toxin, as evidenced by their reduction in growth. However, diets with Dipel® were avoided even at sublethal concentrations in the absence of growth reduction of the larvae (Gould et al., 1991). This example shows the complexity of analyzing the effects of insecticides on feeding behavior because postingestional avoidance (larvae fed from Bt substrate) may occur together with deterrence/repellence. Ramachandran et al. (1993) observed a preference of control diet over Bt-toxin-treated diet by *Choristoneura fumiferana* (spruce budworm) larvae but not by *Hyphantria cunea* (fall webworm). In the latter case, however, the toxin suspension was painted on the diet instead of being incorporated into it.

Deterrency and/or repellency of the active ingredient of Bt is not likely to occur because delta-endotoxin consists of large proteins which cannot be expected to stimulate insect sensory structures. Thus, we have to consider the by-products of the biopesticide manufacturing process (fermentation) and/or components of the commercial formulation such as wetting agents. Evidence for the deterrent/repellent effects of formulation ingredients has not been investigated thoroughly, and the available data show inconsistencies (Lockwood et al., 1984; Young and McMillian, 1979). However, it is conceivable that commercial insecticide formulations contain components which insects are capable of detecting on the plant surface. Analysis of feeding behavior in caterpillars at a resolution of individual bites has shown that deterrents added to acceptable food substrates will shorten the meals and make the intermeal intervals longer (Blaney, et al., 1984, 1988; Ma, 1972). For a discussion on insect feeding deterrents see Frazier (1986) and Chapter 13 by Frazier and Chyb.

Even after acceptance of the treated foliage, the insect has to maintain a good feeding rate in spite of the presence of the insecticide. This is especially important for the exclusive oral toxin Bt. Upon intake, Bt has a substantial effect on feeding behavior, especially of lepidopterous larvae. In fact, the correlation of ingestion and the resulting effects of Bt is so good that in the past the measurement of the potency of a Bt preparation was based on measuring the amount of feeding (see Rombach et al., 1989). The amount of Bt ingested will cause a disruption of feeding lasting until death or recovery in the case of sublethal dosages (Retnakaran et al., 1983; Rombach et al., 1989). Van Frankenhuyzen (1990) thoroughly investigated the dose acquisition of Bt by the spruce budworm. His conclusion was that ". . . dose acquisition is limited by feeding inhibition associated with the dose that is initially ingested." If feeding activity on the Bt deposits is not optimal, the total amount of Bt ingested when feeding inhibition takes effect could be sublethal.

In laboratory studies (Avé, *unpublished data*), the duration of the first meal after limited starvation was timed for several lepidopterous larvae exposed to substrates with and without Bt. Table 12.1 shows that the highest amount of Bt used caused an abrupt termination of feeding in second-instar *Helicoverpa zea* after as little as 1 to 1.5 min of feeding. In other tests on *Spodoptera exigua,* a feeding inhibition induced by 0.2% Bt (technical powder) did not differ from the 0.1% Bt treatment, the feeding times being 1:27 and 1:32 min, respectively. The lowest concentration at which an effect was still discernible appeared to be 0.002% (wt vol) Bt powder or just over 4 ppm of toxin protein per amount of agar substrate. Thus, the phenomenon of feeding inhibition in these caterpillars seems to occur over a 50-fold range of Bt, including sublethal concentrations. Coleopteran-active Bt's against Colorado potato beetle show less of an abrupt inhibition of feeding in a time course of several hours (Zehnder and Gelernter, 1989). In order to have insects spend more time within an area containing deposits of an insecticide (in other words, to enhance the toxin-dose transfer), the insect has to be encouraged to settle and feed at its optimal rate on the insecticide-treated substrate.

12.3. Status of Available Feeding Stimulants

In the mid-1960s, feeding stimulants for lepidopterous larvae were developed primarily as adjuvants to the early biopesticides based on Bt (Dipel®, Javelin®) and nuclear-polyhydrosis virus (NPV) (Elcar®) for the control of *Heliothis* and *Helicoverpa* species and for the control of *Pectinophora gossypiella* on cotton (Allen and Pate, 1966; Bell and Kanavel, 1975, 1977; Montoya et al., 1966). Some of these experimental feeding stimulants originated from larval host plants such as extracts of corn (maize) kernels, fresh corn, and cotton plant parts. Many other ingredients were commercially available, and the materials consisted of sugars as well as vegetable oils, torula yeast, yeast extract, and hydrolyzed protein (Table 12.2). For *Heliothis* and *Helicoverpa* larvae, the preferred materials were invert sugar and cottonseed oil (McLaughlin et al., 1971). Other preferred feeding

Table 12.1. Inhibition of feeding induced by Bacillus thuringiensis during the first meal of lepidopterous larvae on treated diet

Treatment BT source[a]	Concentration of Bt (% wt/vol)	Average feeding [min:sec (SEM)]	Number of larvae observed
Second-instar Helicoverpa zea			
Control	0	2:55 (± 17 sec)	8
EG2158	0.1	3:35 (± 26 sec)	10
EG2371	0.1	1:16 (± 3 sec)	6
	0.05	1:23 (± 5 sec)	10
	0.01	2:00 (± 8 sec)	10
	0.002	2:18 (± 6 sec)	15
	0.0004	3:06 (± 9 sec)	13
EG2424	0.1	0:59 (± 5 sec)	7
	0.05	1:17 (± 4 sec)	16
	0.01	1:52 (± 5 sec)	16
	0.002	2:10 (± 6 sec)	8
	0.0004	2:48 (± 8 sec)	8
Third-instar Spodoptera exigua			
Control	0	3:51 (±16 sec)	9
EG2158	0.05	4:03 (±19 sec)	15
EG2371	0.2	1:27 (±4 sec)	6
	0.1	1:32 (± 4 sec)	12
	0.05	1:51 (± 3 sec)	16
	0.01	2:31 (± 3 sec)	33
	0.002	3:14 (± 11 sec)	8
	0.0004	3:24 (± 15 sec)	8
EG2424	0.1	1:51 (± 7 sec)	16
	0.05	2:39 (± 6 sec)	16
	0.01	3:00 (± 8 sec)	16
	0.002	3:31 (± 7 sec)	16
	0.0004	3:55 (± 26 sec)	8

[a]Technical powders made from fermentations of EG2371 and EG2424 had approximately 18% lepidopteran-active Bt delta-endotoxin. EG2158 contained non-lepidopteran-active (coleopteran active) endotoxin at the same concentration. Powders were incorporated in warmed agar substrate which included various feeding stimulants.

stimulant mixtures included corn meal, wheat, and corn oil (Patti and Carner, 1974) or Pharmamedia® cottonseed flour, sucrose, and crude cottonseed oil (Bell and Kanavel, 1978). Also, bean-based materials such as soy flour, crude soy bean oil mixed with sucrose (Smith et al., 1982), or fresh and dried garbanzo beans were preferred in larval feeding tests (Potter and Watson, 1983b). As a result of this research, two types of commercial feeding stimulants became available: those based on cottonseed flour, such as Coax®, and those with soybean flour, as in Gustol® (see Table 12.3). Recently, a comparative laboratory study was performed on a series of commercial feeding stimulants using six lepidopteran species (Farrar and Ridgway, 1994).

Table 12.2. Materials tested for feeding stimulant activity in the development of baits for lepidopterous larvae

Ingredients	Species[a]	References[b]
Corn kernel extracts	H, SL	1,4,5
Freeze-dried corn foliage	ON	2
Corn flour	H	9,13
Corn meal	AI, H	7,11
Cottonseed flour	H	4,13
Cottonseed protein concentration	SL	10
Extract of cottonseed meal	H	3
Cottonseed hulls	AI	7
Sorghum flour	H	9
Soybean flour	H, SL	4,5,13
Soybean lecithin	SL	10
Soybean meal	SL	10
Apple-pomace	AI	7
Bran	AI	7
Grape	AI	7
Apple grape	AI	7
Citrus pulp	AI	7
Garbanzo bean (chickpea)	H	12
Wheat germ	ON	2
Extracts of wheat germ	ON	2
Extracts of host plants	H	3,4,9
Extracts of cottonseed flour	SL	5
Sucrose	ON, H, SL	2–4,8,10,13,14
Glucose	H,ON	2,13
Fructose	ON	2
Invert sugar	H	3,4,11
Raffinose	ON, H, SL	2,3,5
Galactose	H	3
Lactose	H	13
Maltose	ON, H	2,13
Trehalose	ON	2
Ribose	ON	2
Corn oil	ON, H	2,4,11,13
Cottonseed oil	H, SL	3–6,8,11,13
Soybean oil	H, SL	4,10,13
Yeast extract	H	13
Torula yeast	SL	10
Wheat	H	11
Casein	ON	2
β-Sitosterol	ON	2
Salt mixture	ON	2
Ascorbic acid	ON	2
Vitamin mix	ON	2
Amino acids mix	ON	2

[a] H, *Heliothis* spp; SL, *Spodoptera littoralis* (Egyptian cotton leafworm); ON, *Ostrinia nubilalis* (European corn borer); AI, *Agrotis ipsilon* (black cutworm).

[b] References: 1, Allen and Pate (1966); 2, Bartelt et al. (1990); 3, Bell and Kanavel (1975); 4, Bell and Kanavel (1978); 5, El-Nockrashy et al. (1986); 6, Guerra and Shaver (1969); 7, Gholson and Showers (1979); 8, McLaughlin et al. (1971); 9, Montoya et al. (1966); 10, Navon et al. (1987); 11, Patti and Carner (1974); 12, Potter and Watson (1983b); 13, Smith et al. (1982); 14, Starks et al. (1965).

Table 12.3. *Commercial baits as adjuvants to bio-insecticides for the control of lepidopterous pests*

Brand	Company	Major components
Coax®	CCT Corp.	Vegetable flour/disaccharide
Devour®	Crystal Chemical Inter-America	Vegetable flour/disaccharide
Entice®	Custom Chemicides	Vegetable flour/disaccharide
Pheast®	AgriSense/Biosys	Yeast/vegetable flour/polysaccharides
Gusto®	Atochem North America	Cottonseed flour/disaccharide
Gustol® (discontinued)	Sandoz Corp.	Soybean flour/sucrose
Konsume®	Fermone Corp.	Vegetable flour/disaccharide

The possibility of feeding stimulants for other economically important lepidopterous larva was investigated by El-Nockrashy et al. (1986) on *Spodoptera littoralis,* the cotton leafworm. They evaluated various solvent extracts of cottonseed flour, nontreated soy flour, crude and refined cottonseed oil, ethanol extracts of cottonseed kernels, the sugar raffinose, and a hot water extract of dried corn kernels. Other feeding tests on *S. littoralis* were conducted using various mixtures of cottonseed protein concentrate and cottonseed oil, crude soybean lecithin, soybean meal and soybean oil, Torula yeast, and sucrose. Most preferred by the larvae was an experimental mix, Fest-18, containing cottonseed protein, soybean lecithin, torula yeast, and sucrose (Navon et al., 1987). Extensive behavioral research on feeding stimulants for use in a granular bait was published by Bartelt et al. (1990) on *Ostrinia nubilalis* larvae, the European corn borer. A large number of materials were incorporated in starch granules, including sugars, amino acids, both individually and in a mixture, freeze-dried corn foliage, ingredients of the corn borer rearing diet, and extracts of wheat germ representing lipids and sugars. Flours made from cottonseed, soybeans, corn, or chickpeas had very good feeding stimulatory activity. With the black cutworm, *Agrotis ipsilon,* a variety of low-cost feeding stimulants including apple-pomace bran, bran, cottonseed hulls, citrus pulp, and cornmeal were tested (Gholson and Showers, 1979).

The development of feeding stimulants for insects other than lepidopterous pests has progressed from the experimental stage as well. For mole crickets, *Scapteriscus acletus,* the work has been primarily done by Yu and co-workers (Kepner and Yu, 1987). Their study established that sucrose was the most preferred sugar. Combined with an oil, either crude cottonseed oil or refined soybean oil, the mixture proved to be a much better bait than molasses and amyl acetate, which are currently used in commercial baits for mole crickets. Research on feeding stimulants for coleopteran pests—for example, the cotton boll weevil, *Anthonomus grandis,* and species of diabroticite beetles such as the cucumber beetles and corn root worms—has been in progress for a number of years (McKibben et al., 1990; Metcalf et al., 1987). Since the early 1980s it has been known that the diabroticite beetles can taste a class of triterpenoids, the cucurbitacins, at concentrations as low as a few nanograms per area of approximately 1 cm^2

(Metcalf et al., 1980). The cucurbitacins evoke compulsive feeding and act as arrestants on the beetles. Commercial formulations of these feeding stimulants have been introduced recently, although other formulations are still in the developmental stage (Lance and Sutter, 1990). As mentioned in the Introduction, baits for flies have been developed for fruit flies of agricultural importance. Because experimental traps in the 1930s contained basic materials such as fermentation solution or urea plus sodium hydroxide, research has shown that protein hydrolysate as attractant/food bait was superior (Steiner et al., 1961). Hydrolyzed proteins from different sources and processing methods were field-tested in apple orchards on *Rhagoletis pomonella,* the apple maggot fly, by Neilson (1960). It appeared that enzymatic hydrolysates were preferred by the fly and that the soy hydrolysate was preferred over hydrolysates of yeast, casein, and lactalbumin. Toxic baits for the control of the housefly, *Musca domestica,* in animal rearing facilities have been used together with conventional applications of insecticides by spray or as larvicides in manure. Hopkins Fly Grits (discontinued) was a granular product made from sand coated with sugar and the insecticide bomyl. In traps for cockroaches, the feeding stimulants consist of a source of fat, protein, or carbohydrate or a combination thereof. Barson (1982) used oats or peanut butter in his tests. Rust and Reierson (1981) compared 14 bait mixtures. In this study, the commercial baits contained one of the following materials: corn grit, bran, fine meal, waxed dog food, waxed soybean blocks, and peanut oil. The experimental baits included dry dogfood, peanut butter, boiled raisins, fresh white bread, or chlorofrom extract of white bread. Peanut butter ranked as a moderately effective bait in these tests, and three out of six commercial available cockroach traps tested in another study by Appel (1990) contained this material (It Works®, Holiday®, d-Con d-Stroy®). Other commercial products contain either dog food and a sugar as in Bolt® (S.C. Johnson, Inc.) or corn syrup combined with other sources of sugars (maple syrup, honey, molasses) and oatmeal (Spaulding and Pasarela, 1989).

In describing the numerous experimental materials tested as feeding stimulants for pest insects, the goal was not to be exhaustive but to provide an insight of their origin and the status of commercial products. The success of a feeding stimulant is governed by lowering insecticide use, with savings in insecticide cost and reduced environmental/human exposure, and/or by enhanced insect control resulting in consumer satisfaction. However, last but not least the ease of applying or using the feeding-stimulant–insecticide mixture proves to be important as well. This aspect is not necessarily compatible with the acceptance of the applied material from an insect's point of view.

12.4. Efficacy of Feeding Stimulants in Laboratory Versus Field Situations

The evaluation of feeding stimulants has primarily focused on increasing the efficacy of biopesticides, such as Bt and NPV, and in a few cases that of pyrethroids (Meisner et al., 1984) in laboratory, greenhouse, and field tests. Invariably, the laboratory and greenhouse tests showed the benefits of feeding stimulants in enhancing insecticide efficacy. For example, feeding stimulant mixtures containing soy flour with cottonseed oil and combined with Bt increased the mortality of the cotton leafworm by 10- to 18-fold over Bt alone (El-Nockrashy et al., 1986). Feeding stimulants with NPV or Bt on *H. zea* and *Trichoplusia ni* increased mortality in "on-plant" lab tests significantly (Allen and Pate, 1966; Bell and Kanavel, 1975, 1978; Ignoffo et al., 1976). In contrast, field trials with foliar-applied experimental and commercial bait formulations in combination with biopesticides for control of lepidopterous larvae have not been as consistent. Although some field trials showed increased efficacy with the use of baits (Bell and Romine, 1980; Johnson, 1982; Meisner et al., 1990; Potter and Watson, 1983a; Sneh and Gross, 1983; Sneh et al., 1983; Southern and Jackson, 1984), other trials were inconclusive (Durant, 1984; Luttrell et al., 1983; Lutwama and Matanmi, 1988; Montoya et al., 1966; Schuster, 1979).

The lack of consistency of field trial results was attributed mainly to the impact of environmental conditions, such as degradation by ultraviolet light and rain wash-off, on the biopesticides, as well as to the variability of the feeding stimulants. However, Sneh and Gross (1983) documented protection against degradation of Bt by ultraviolet light provided by the cottonseed flour-based feeding stimulant Coax®, making interpretation of positive results difficult to attribute solely to feeding stimulant activity. The impact of application method on the efficacy of the baited pesticides has been of interest to a few researchers. Lampert and Southern (1987) described several methods of application, and it was determined that the pinch method (i.e., a "home" variety of a granular application) was the most effective in control of *Heliothis* spp. on tobacco (Table 12.4). Sprayable feeding stimulants as baits were less reliable in the pest control.

In beginning to analyze the difficulties of obtaining field efficacy of feeding stimulants, we have to understand how the material is applied to the foliage and consequently how the deposits are presented to the insect pest to feed upon. Spray applications frequently use wetting agents (spreaders) in the spray tank in order to reduce run-off of the sprayed material from the foliage. This may cause the feeding stimulants—the water-soluble as well as water-suspended material—to spread over the leaf surface, reducing their effective concentration. In laboratory studies (Avé, *unpublished data*), sugars present in commercial feeding stimulant formulations or water-soluble materials of their main component, cottonseed or soy flour, were easily leached out, leaving a much less preferred residue for the *H. zea* larvae (Table 12.5). Cottonseed and soy flours and sugars are still the main ingredients of commercial feeding stimulants for lepidopterous larvae; and if these components are not retained in any way with wetting of the leaf on exposure to rain or dew formation, the development of adequate formulations is very much needed.

In addition to sprayable feeding stimulant formulations for agricultural crops, baited granular insecticides are used. Of course, granular applications are limited

Table 12.4. Control of tobacco budworm, Heliothis virescens, *in tobacco with various formulations and application methods of* Bacillus thuringiensis[a]

Treatment	Rate (kg/ha)	Percentage of larvae surviving after treatment (SEM)	
		4 days[b]	11 days[b]
Bait[c]/Dipel 2X WP, pinch	1 g/plant	3.2 (1.6)e	0 (0)e
Bait[c]/Dipel 2X WP, band	0.15	65.6 (8.7)ab	25.9 (1.0)abc
Bait[d]/Dipel 2X WP, band	0.10	33.3 (1.0)cd	14.8 (4.8)bc
Dipel 10G[e], pinch[f]	1 g/plant	3.3 (1.7)e	1.7 (1.7)de
Dipel 10G, band[g]	11.2	25.5 (6.2)	11.2 (5.9)cd
Dipel 2X, spray	0.56	49.0 (9.7)bc	35.7 (3.9)ab
Orthene TIS[h], spray	1.12	23.2 (11.5)	23.2 (11.5)bc
Control	0	80.4 (2.6)a	49.4 (8.1)a

[a]Adapted from Lampert and Southern (1987).

[b]Figures in a column followed by the same letter are not significantly different.

[c]Corn-meal bait with Dipel 2X WP, wettable powder.

[d]Corn-bran bait with Dipel 2X WP, wettable powder.

[e]Dipel 10G granular formulation.

[f]Pinch method consisted of 1 g of bait applied in tobacco bud with teaspoon.

[g]Band: Material applied with granular applicator with 15-cm distributor.

[h]Orthene TIS (acephate) chemical control.

to a suitable crop/pest complex. For European corn borer larvae residing in whorl structures of corn, baited granules can be a very successful control method. Bartelt et al. (1990) have shown in greenhouse trials that equivalent insect control could be obtained with 75% less of the biopesticide Bt if feeding stimulants were added to the granules. The advantage of a granular formulation is that it packages the feeding stimulants so that more of the feeding stimulant is available over time. Granular baits, with either conventional insecticides or bio-insecticides, have been used for grasshopper control (Capinera and Hibbard, 1987; Dunkle and Shasha, 1988; McGuire et al., 1991; Onsager et al., 1981; Quinn et al., 1989).

Current research on treated cornstarch (pregelatinized) by McGuire and Shasha (1990) has focused on sprayable formulations which seem to self-encapsulate and allow much better persistence on foliage. Perhaps this is the future for sprayable feeding stimulants. With interest in ways to increase toxin-dose transfer by optimizing application technology, a growing number of studies (Chapple, 1993) deal with insect movement and feeding on food surfaces treated with chemical as well as biological insecticides. This information is also needed to determine the effects of feeding stimulants and their formulation requirements.

12.5. Conclusion and Outlook

In order to assess the full potential and limitations of feeding stimulants in enhancing efficacy of insecticides, we have to know more about insect feeding

Table 12.5. Dual-choice tests of food preference of Helicoverpa zea neonates with regard to the loss of water-soluble components from soy flour or from the feeding stimulant COAX®

	Larval food preference	
Choice of treatments[a]	Feeding activity[b] [average (SEM)]	Settling[c] [average (SEM)]
1. Leached soy flour	0	3.3 (0.2)
2. Soy flour	9.2 (0.5)	17.2 (2.0)
1. Reconstituted soy flour	5.3 (0.4)	9.5 (0.6)
2. Soy flour	6.7 (0.2)	9.8 (1.0)
1. Leached Coax®	0.8 (0.4)	3.7 (1.3)
2. Coax®	9.5 (0.2)	20.0 (2.2)
1. Reconstituted Coax®	6.2 (0.8)	11.8 (1.3)
2. Coax®	8.2 (0.5)	7 (1.2)

[a]Soy flour and Coax® were suspended in water at agriculture tank-mix rate. After centrifugation, part of the pellet was used as leached material, and the other part was combined with an appropriate amount of supernatant to make the reconstituted material. All test material was incorporated in 3% agar.

[b]Average feeding activity on scale from 0 (no feeding) to 10 (heavy feeding) scored after 18 hr in choice tests.

[c]Average number of larvae found feeding at the end of the testing period.

and intraplant movement of the insect both before and after insecticide applications. Within the diverse microclimate of the crop canopy, insect mobility has been well documented for a few lepidopterous larvae (Hoy et al., 1989, 1990b). However, if "normal" insect–plant interactions—locomotion and feeding—are changed by deposits of insecticide formulations, what are the respective contributions of repellency/deterrency, physiological irritation, and feeding inhibition? Because research on these particular phenomena has been limited and the words repellency, deterrency, and/or feeding inhibition are used when perhaps not appropriate, precise knowledge of these factors is needed to assess the potential of feeding stimulants.

The size and density of the sprayed insecticide deposits are important factors. Bryant and Yendol (1988) determined the efficacy of Bt in relation to the feeding habits of gypsy moth larvae for which small (50–150 μm) droplets of Bt at high density were the most effective for control. Exciting research done by Chapple, Taylor, and Hall at the Ohio State University's research station in Wooster has produced a computer-based pesticide droplet simulation model which addresses Bt uptake by an insect with variables such as rate of feeding and locomotion together with insecticide application parameters including rate of deposit degradation (Chapple, 1993). Two questions remain, are these parameters the same for deposits containing a feeding stimulant adjuvant, and what is the optimal ratio of insecticide and feeding stimulant formulations? Currently, the concentrations of commercial feeding stimulants are recommended over a wide range; for example, manufacturers of several liquid products have recommended the use of 1 pint to 4 quarts of feeding stimulant per acre. In our quest for understanding the insect's response to insecticide deposits with feeding stimulants, we will gain a better understanding of the range of effective concentrations. In addition, much formulation research is needed on maintaining these concentrations in the foliar deposits of insecticides. Thorough investigations on these subjects will assist in designing formulations for feeding stimulants with better efficacy and persistence under field conditions.

Research on feeding stimulants is rooted in studies concerned with learning about the choice of food substrate by insects. The cucurbitacins, which arrest and cause compulsive feeding in diabroticite beetles, are the most powerful feeding stimulants known. We have not been as fortunate with feeding stimulants for other insect species (for reference consult the literature on feeding behavior and sensory discrimination in insects: Blom, 1978; Dethier, 1973; Dethier and Kuch, 1971; Frazier, 1992; Hsiao, 1969; Ma, 1972; Mitchell, 1974; van Drongelen, 1979). Interestingly, the number of individual chemicals that stimulate feeding is not very large; sucrose is the main feeding stimulant for many insect species. Feeding stimulant research has contributed to understanding insect feeding behavior in insect–plant relationships, and it is essential that such a high level of research sophistication is also applied to further the development of feeding stimulant baits for use in insecticidal application. The fact that cottonseed flour, soybean flour, or wheat germ (Bartelt et al., 1990; Gothilf and Beck, 1967) has been found to promote feeding seems to me as intriguing as the insect's choice of a particular host plant.

References

Allen, G. E., and Pate, T. L. (1966). The potential role of a feeding stimulant used in combination with the nuclear polyhedrosis virus of *Heliothis*. *J. Invert. Pathol.* 8, 129–131.

Appel, A. G. (1990). Laboratory and field performance of consumer bait products for German cockroach (Dictyoptera: Blattellidae) control. *J. Econ. Entomol.* 83, 53–59.

Barson, G. (1982). Laboratory evaluation of boric acid plus porridge oats and iodofenphos gel as toxic baits against the German cockroach, *Blattella germanica* (L.) (Dictyoptera: Blattellidae). *Bull. Entomol. Res.* 72, 229–237.

Bartelt, R. J., McGuire, M. R., and Black, D. A. (1990). Feeding stimulants for the European corn borer (Lepidoptera: Pyralidae): additives to a starch-based formulation for Bacillus thuringiensis. *Environ. Entomol.* 19, 182–189.

Bell, M. R., and Kanavel, R. F. (1975). Potential bait formulations to increase effectiveness of nuclear polyhedrosis virus against the pink bollworm. *J. Econ. Entomol.* 68, 389–391.

Bell, M. R., and Kanavel, R. F. (1977). Field tests of a nuclear polyhedrosis virus in a bait formulation for control of pink bollworms and *Heliothis* spp. on cotton in Arizona. *J. Econ. Entomol.* 70, 625–628.

Bell, M. R., and Kanavel, R. F. (1978). Tobacco budworm: development of a spray adjuvant to increase effectiveness of a nuclear polyhedrosis virus. *J. Econ. Entomol.* 71, 350–352.

Bell, M. R., and Romine, C. L. (1980). Tobacco budworm field evaluation of microbial control in cotton using *Bacillus thuringiensis* and nuclear polyhedrosis virus with a feeding adjuvant. *J. Econ. Entomol.* 73, 427–430.

Blaney, W. M., Simmonds, M. S. J., Evans, S. V., and Fellows, L. E. (1984). The role of the secondary plant compound 2,5-dihydroxymethyl-3,4-dihydroxypyrrolidine as a feeding inhibitor for insects. *Entomol. Exp. Appl.* 36, 209–216.

Blaney, W. M., Simmonds, M. S. J., Ley, S. V., and Jones, P. S. (1988). Insect antifeedants: a behavioural and electrophysiological investigation of natural and synthetically derived clerodane diterpenoids. *Entomol. Exp. Appl.* 46, 267–274.

Blom, F. (1978). Sensory activity and food intake: a study of input–output relationships in two phytophagous insects. *Netherlands J. Zool.* 28, 277–340.

Bryant, J. E., and Yendol, W. G. (1988). Evaluation of the influence of droplet size and density of *Bacillus thuringiensis* against gypsy moth larvae (Lepidoptera: Lymantriidae). *J. Econ. Entomol.* 81, 130–134.

Capinera, J. L., and Hibbard, B. E. (1987). Bait formulations of chemical and microbial insecticides for the suppression of crop-feeding grasshoppers. *J. Agric. Entomol.* 4, 337–344.

Chapple, A. C. (1993). Dose transfer of *Bacillus thuringiensis* to the Diamondback moth (*Plutella xylostella*) via cabbage: a synthesis. Ph.D. dissertation, Ohio State University.

Dethier, V. G. (1973). Electrophysiological studies of gustation in lepidopterous larvae. II. Taste spectra in relation to food-plant discrimination. *J. Comp. Physiol.* 82, 103–134.

Dethier, V. G. (1980). Food-aversion learning in two polyphagous caterpillars, *Diacrisia virginica* and *Estigmene congrua*. *Physiol. Entomol.* 5, 321–325.

Dethier, V. G., and Kuch, J. H. (1971). Electrophysiological studies of gustation in lepidopterous larvae. I. Comparative sensitivity to sugars, amino acids, and glycosides. *Z. Vergl. Physiol.* 72, 343–363.

Dethier, V. G., and Yost, M. T. (1979). Oligophagy and absence of food-aversion learning in tobacco hornworms, *Manduca sexta*. *Physiol. Entomol.* 4, 125–130.

Dunkle, R. L., and Shasha, B. S. (1988). Starch-encapsulated *Bacillus thuringiensis:* a potential new method for increasing environmental stability of entomopathogens. *Environ. Entomol.* 17, 120–126.

Durant, J. A. (1984). Cotton insect pests: field evaluation of selected insecticide treatments. *J. Agric. Entomol.* 1, 201–211.

El-Nockrashy, A. S., Salama, H. S., and Taha, F. (1986). Influence of bait formulations on the effectiveness of *Bacillus thuringiensis* against *Spodoptera littoralis* (Boisd) (Lep., Noctuidae). *J. Appl. Entomol.* 101, 381–389.

Farrar, R. R., Jr., and Ridgway, R. L. (1994). Comparative studies of the effects of nutrient-based phagostimulants on six lepidopterous insect pests. *J. Econ. Entomol.* 87, 44–52.

Frazier, J. L. (1986). The perception of plant allelochemicals that inhibit feeding. In: Brattsten, L. B., and Ahmad, S. (eds.), *Molecular Aspects of Insect–Plant Associations.* Plenum Press, New York, pp. 1–42.

Frazier, J. L. (1992). How animals perceive secondary plant compounds. In: Rosenthal, G. A., Berenbaum, M. R. (eds.), *Herbivores: Their Interaction with Secondary Plant Metabolites,* vol. 2. Academic Press, San Diego, pp. 89–134.

Ghassemi, M., Painter, P., Painter, P., Quinlivan, S., and Dellarco, M. (1983). *Bacillus thuringiensis,* nucleopolyhedrosis virus, and pheromones: environmental considerations and uncertainties in large scale insect control. *Environ. Int.* 9, 39–49.

Gholson, L. E., and Showers, W. B. (1979). Feeding behavior of black cutworms on seedling corn and organic baits in the greenhouse. *Environ. Entomol.* 8, 552–557.

Giles, D. P., and Rothwell, D. N. (1983). The sub-lethal activity of amidines on insects and acarids. *Pesticide Sci.* 14, 303–312.

Gist, G. L., and Pless, C. D. (1985). Feeding deterrent effects of synthetic pyrethroids on the fall armyworm, *Spodoptera frugiperda. Florida Entomol.* 68, 456–461.

Gothilf, S., and Beck, S. D. (1967). Larval feeding behaviour of the cabbage looper, *Trichoplusia ni. J. Insect Physiol.* 13, 1039–1053.

Gould, F. (1991). Arthropod behavior and the efficacy of plant protectants. *Annu. Rev. Entomol.* 36, 305–330.

Gould, F., and Anderson, A. (1991). Effects of *Bacillus thuringiensis* and HD-73 delta-endotoxin on growth, behaviour, and fitness of susceptible and toxin-adapted strains of *Heliothis virescens* (Lepidoptera: Noctuidae). *Environ. Entomol.* 20, 30–38.

Gould, F., Anderson, A., Landis, D., and Van Mellaert, H. (1991). Feeding behavior and growth of *Heliothis virescens* larvae on diets containing *Bacillus thuringiensis* formulations or endotoxins. *Entomol. Exp. Appl.* 58, 199–210.

Guerra, A. A., and Shaver, T. N. (1969). Feeding stimulants from plants for larvae of the tobacco budworm and boll worm. *J. Econ. Entomol.* 62, 98–100.

Hall, F. R., Adams, A. J., and Hoy, C. W. (1989). Correlation of precisely defined spray deposit parameters with biological responses of resistant diamond moth (DBM) field populations. *Aspects Appl. Biol.* 21, 125–127.

Haynes, K. F. (1988). Sublethal effects of neurotoxic insecticides on insect behavior. *Annu. Rev. Entomol.* 33, 149–168.

Hoy, C. W., McCulloch, C. E., Shoemaker, C. A., Shelton, A. M. (1989). Transition probabilities for *Trichoplusia ni* (Lepidoptera: Noctuidae) larvae on cabbage as a function of microclimate. *Environ. Entomol.* 18, 187–194.

Hoy, C. H., Adams, A. J., and Hall, F. R. (1990a). Behavioral response of *Plutella xylostella* (Lepidoptera: Plutellidae) populations to permethrin deposits. *J. Econ. Entomol.* 83, 1216–1221.

Hoy, C. H., McCulloch, C. E., Sawyer, A. J., Shelton, A. M., and Shoemaker, C. A.

(1990b). Effect of intraplant insect movement on economic thresholds. *Environ. Entomol.* 19, 1578–1596.

Hsiao, T. H. (1969). Adenine and related substances as potent feeding stimulants for the alfalfa weevil, *Hypera positica*. *J. Insect Physiol.* 15, 1785–1790.

Ignoffo, C. M., Hostetter, D. L., and Smith, D. B. (1976). Gustatory stimulant, sunlight protection, evaporation retardant: three characteristics of a microbial insecticidal stimulant. *J. Econ. Entomol.* 69, 207–210.

Johnson, D. R. (1982). Suppression of *Heliothis* spp. on cotton by using *Bacillus thuringiensis, Baculovirus heliothis* and two feeding adjuvants. *J. Econ. Entomol.* 75, 207–210.

Kepner, R. L., and Yu, S. J. (1987). Development of a toxic bait for control of mole crickets (Orthoptera: Gryllotalpidae). *J. Econ. Entomol.* 80, 659–665.

Lampert, E. P., and Southern, P. S. (1987). Evaluation of pesticide application methods for control of tobacco budworms (Lepidoptera: Noctuidae) on flue-cured tobacco. *J. Econ. Entomol.* 80, 961–967.

Lance, D. R., and Sutter, G. R. (1990). Field-cage and laboratory evaluations of semiochemical-based baits for managing western corn rootworm (Coleoptera Chrysomelidae). *J. Econ. Entomol.* 83, 1085–1090.

Lockwood, J. A., Sparks, T. C., and Story, R. N. (1984). Evolution of insect resistance to insecticides: a re-evaluation of the roles of physiology and behavior. *Bull. Entomol. Soc. Am.* 30, 41–51.

Luttrell, R. G., Yearian, W. C., and Young, S. Y. (1983). Effects of spray adjuvants on *Heliothis zea* (Lepidoptera: Noctuidae) nuclear-polyhedrosis virus efficacy. *J. Econ. Entomol.* 76, 162–167.

Lutwama, J. J., and Matanmi, B. A. (1988). Efficacy of *Bacillus thuringiensis* subsp. *Kurstaki* and *Baculovirus heliothis* foliar applications for suppression of *Helicoverpa armigera* (Hubner) (Noctuidae) and other lepidopterous larvae on tomato in southwestern Nigeria. *Bull. Entomol. Res.* 78, 173–179.

Ma, W. C. (1972). Dynamics of feeding responses in *Pieris brassicae* Linn as a function of chemosensory input: a behavioural, ultrastructural and electrophysiological study. *Meded. Landbouwhogesch. Wageningen* 72–11.

McGuire, M. R., and Shasha, B. S. (1990). Sprayable self-encapsulating starch formulations for *Bacillus thuringiensis*. *J. Econ. Entomol.* 83, 1813–1817.

McGuire, M. R., Street, D. A., and Shasha, B. S. (1991). Evaluation of starch encapsulation for formulation of grasshopper (Orthoptera, Acrididae) entomopoxviruses. *J. Econ. Entomol.* 84, 1652–1656.

McKibben, G. H., Smith, J. W., and McGovern, W. L. (1990). Design of an attract-and-kill device for the boll weevil (Coleoptera, Curculionidae). *J. Entomol. Sci.* 25, 581–586.

McLaughlin, R. E., Andrews, G., and Bell, M. R. (1971). Field tests for control of *Heliothis* spp. with a nuclear-polyhedrosis virus included in a boll weevil bait. *J. Invert. Pathol.* 18, 304–305.

Meisner, J., Ascher, K. R. S., and Eizick, C. (1984). Effect of the commercial phagostimulants Coax and Gustol on the toxicity of cypermethrin and deltamethrin against *Spodoptera littoralis* (Lepidoptera: Noctuidae). *J. Econ. Entomol.* 77, 1123–1126.

Meisner, J., Hadar, D., Wysoki, M., and Harpaz, I. (1990). Phagostimulants enhancing the efficacy of *Bacillus thuringiensis* formulations against the giant looper, *Boarmia (Ascotis) selenaria*, in avocado. *Phytoparasitica* 18, 107–112.

Metcalf, R. L., Metcalf, R. A., Rhodes, A. M. (1980). Cucurbitacins as kairomones for diabroticite beetles. *Proc. Natl. Acad. Sci.* 77, 3769–3772.

Metcalf, R. L., Ferguson, J. E., Lampman, R., and Andersen, J. F. (1987). Dry cucurbitacin-containing baits for controlling Diabroticite beetles (Coleoptera: Chrysomelidae). *J. Econ. Entomol.* 80, 870–875.

Mitchell, B. K. (1974). Behavioural and electrophysiological investigations on the response of larvae of the colorado potato beetle (*Leptinotarsa decemlineata*) to amino acids. *Entomol. Exp. Appl.* 17, 255–264.

Montoya, E. L., Ignoffo, C. M., and McGarr, R. L. (1966). A feeding stimulant to increase effectiveness of, and a field test with, nuclear-polyhedrosis virus of *Heliothis*. *J. Invert. Pathol.* 8, 320–324.

Moriarty, F. (1969). The sublethal effects of synthetic insecticides on insects. *Biol. Rev.* 44, 321–357.

Navon, A., Meisner, J., and Ascher, K. R. S. (1987). Feeding stimulant mixtures for *Spodoptera littoralis* (Lepidoptera: Noctuidae). *J. Econ. Entomol.* 80, 990–993.

Neilson, W. T. A. (1960). Field tests of some hydrolyzed proteins as lures for the apple maggot *Rhagoletis pomonella*. *Can. Entomol.* 92, 464–467.

Onsager, J. A., Rees, N. E., Henry, J. E., Foster, R. N. (1981). Integration of bait formulations of Nosema locusta and carbaryl for control of rangeland grasshoppers. *J. Econ. Entomol.* 74, 183–187.

Patti, J. H., and Carner, G. R. (1974). *Bacillus thuringiensis* investigations for control of *Heliothis* spp. on cotton. *J. Econ. Entomol.* 67, 415–418.

Pluthero, F. G., and Singh, R. S. (1984). Insect behavioural responses to toxins: practical and evolutionary considerations. *Can. Entomol.* 116, 57–68.

Polles, S. G., and Vinson, S. B. (1969). Effect of droplet size on persistence of ULV malathion and comparison of toxicity of ULV and EC malathion to tobacco budworm larvae. *J. Econ. Entomol.* 62, 89–94.

Potter, M. F., and Watson, T. F. (1983a). Timing of nuclear polyhedrosis virus-bait spray combinations for control of egg and larval stages of tobacco budworm (Lepidoptera: Noctuidae). *J. Econ. Entomol.* 76, 446–448.

Potter, M. F., and Watson, T. F. (1983b). Garbanzo bean as a potential feeding stimulant for use with a nuclear polyhedrosis virus of the tobacco budworm (Lepidoptera: Noctuidae). *J. Econ. Entomol.* 76, 449–451.

Quinn, M. A., Kepner, R. L., Walgenbach, D. D., Bohls, R. A., Pooler, P. D., Foster, R. N., Reuter, K. C., and Swain, J. L. (1989). Immediate and 2nd-year effects of insecticide spray and bait treatments on populations of rangeland grasshoppers. *Can. Entomol.* 121, 589–602.

Ramachandran, R., Raffa, K. F., Miller, M. J., Ellis, D. D., and McCown, B. H. (1993). Behavioral responses and sublethal effects of spruce budworm (Lepidoptera: Tortricidae) and fall webworm (Lepidoptera: Arctiidae) larvae to *Bacillus thuringiensis* CryIA(a) toxin in diet. *Environ. Entomol.* 22, 197–211.

Retnakaran, A., Lauzon, H., and Fast, P. (1983). *Bacillus thuringiensis* induced anorexia in the spruce budworm, *Choristoneura fumiferana*. *Entomol. Exp. Appl.* 34, 233–239.

Rombach, M. C., Aguda, R. M., Picard, L., and Roberts, D. W. (1989). Arrested feeding of the Asiatic rice borer (Lepidoptera: Pyralidae) by *Bacillus thuringiensis*. *J. Econ. Entomol.* 82(2), 416–419.

Ruscoe, C. N. E. (1977). The new NRDC pyrethroids as agricultural insecticides. *Pesticide Sci.* 8, 236–242.

Rust, M. K., and Reierson, D. A. (1981). Attraction and performance of insecticidal baits for German cockroach control. *Int. Pest Control* 23, 106–109.

Schuster, D. J. (1979). Adjuvants tank-mixed with *Bacillus thuringiensis* for control of cabbage looper larvae on cabbage. *J. Georgia Entomol. Soc.* 14, 182–186.

Smith, D. B., Hostetter, D. L., Pinnel, R. E., and Ignoffo, C. M. (1982). Laboratory studies of viral adjuvants: formulation development. *J. Econ. Entomol.* 75, 16–20.

Sneh, B., and Gross, S. (1983). Biological control of the Egyptian cotton leafworm *Spodoptera littoralis* (Boisd.) (Lepidoptera, Noctuidae) in cotton and alfalfa fields using a preparation of *Bacillus thuringiensis* ssp *entomocidus* supplemented with adjuvants. *Z. Angew. Entomol.* 95, 418–424.

Sneh, B., Schuster, S., and Gross, S. (1983). Improvement of the insecticidal activity of *Bacillus thuringiensis* var. *entomocidus* on larvae of *Spodoptera littoralis* (Lepidoptera, Noctuidae) by addition of chitinolytic bacteria, a phagostimulant and a UV-protectant. *Z. Angew. Entomol.* 96, 77–83.

Southern, P. S., and Jackson, D. M. (1984). Control of *Heliothis virescens* on flue-cured tobacco using *Bacillus thuringiensis* and a cottonseed flour feeding stimulant. *Tobacco Sci.* 28, 10–13.

Spaulding, L., and Pasarela, N. R. (1989). Non-particulate, non-flowable, non-repellant insecticide-bait composition for the control of cockroaches. US Patent 4845103.

Starks, K. J., McMillian, W. W., Sekul, A. A., and Cox, H. C. (1965). Corn earworm larval feeding response to corn silk and kernel extracts. *Ann. Entomol. Soc. Am.* 58, 74–76.

Steiner, L. F., Rohner, G. G., Ayers, E. L., and Christenson, L. D. (1961). The role of attractants in recent Mediterranean fruit fly eradication program in Florida. *J. Econ. Entomol.* 54, 30–35.

Tan, K-H. (1981). Antifeeding effect of cypermethrin and permethrin at sub-lethal levels against *Pieris brassicae* larvae. *Pesticide Sci.* 12, 619–626.

van Drongelen, W. (1979). Contact chemoreception of host plant specific chemicals in larvae of various *Yponomeuta* species (Lepidoptera). *J. Comp. Physiol.* 134, 265–279.

van Frankenhuyzen, K. (1990). Effect of temperature and exposure time on toxicity of *Bacillus thuringiensis* Berliner spray deposits to spruce budworm, *Choristoneura fumiferana* Clemens (Lepidoptera: tortricidae). *Can. Entomol.* 122, 69–75.

Yendol, W. G., Hamlen, R. A., and Rosario, S. B. (1975). Feeding behavior of gypsy moth larvae on *Bacillus thuringiensis*-treated foliage. *J. Econ. Entomol.* 68, 25–27.

Young, J. R., and McMillian, W. W. (1979). Differential feeding by two strains of fall armyworm larvae on carbaryl treated surfaces. *J. Econ. Entomol.* 72, 202–203.

Zehnder, G. W., and Gelernter, W. D. (1989). Activity of the M-ONE formulation of a new strain of *Bacillus thuringiensis* against the Colorado potato beetle (Coleoptera: Chrysomelidae): relationship between susceptibility and insect life stage. *J. Econ. Entomol.* 82, 756–761.

13

Use of Feeding Inhibitors in Insect Control
J. L. Frazier and S. Chyb

13.1. Introduction

The last 400 million years or so has seen a very complex process of coevolution between plants and phytophagous insects (Ehrlich and Raven, 1964; Futuyma and Keese, 1992). Plants, apart from creating physical and mechanical barriers, rely mostly on chemical protection against insect feeding. Numerous terpenoid, alkaloid, and phenolic feeding inhibitors have evolved in the course of this arms race between plants and insects (Harborne, 1993; Rosenthal and Berenbaum, 1992; Rosenthal and Janze, 1979). The proposed process is that a random mutation within the plant genome led to the synthesis of a new compound which prevented an insect from feeding on the plant. Those insects faced with strong selection pressures exerted by plant allelochemicals evolved a mechanism or mechanisms that allowed them to overcome the deterrent or toxic effects of the new compounds. In this way some defensive compounds were successfully counteracted by one or more insect species which allowed their use as unique feeding stimulants (e.g., cucurbitacins and diabroticite beetles); however, the phagodeterrent activity remained preserved against a great many other insects. The current list of phagodeterrents is very long and their potency varies greatly among insect species, suggesting that there are many highly diverse insect–plant relationships. It appears that phagodeterrents rather than nutrients are a major force in the development of present insect–plant interactions as was proposed more than 30 years ago by Thorsteinson (Thorsteinson, 1960) and has seen renewed emphasis (Berenbaum, 1986).

The practical use of natural product feeding inhibitors in insect control is rapidly becoming a reality. Recent advances in the isolation, structural determination, and synthesis of active partial structures of antifeedants have allowed us to perform some detailed studies of modes of action. There is a growing list of commercial products based on feeding inhibitors (see Table 13.1). There are clear advantages to their applications in field situations. Feeding inhibitors are highly selective and often effective in submillimolar concentrations, suggesting that field use rates of less than 0.5 pound per acre (approximately 600 g·ha^{-1}) will be common. Unlike synthetic pesticides, which are often highly toxic, naturally derived and synthetic feeding inhibitors with varied modes of action may not put as much selection pressure on insects, thus reducing the likelihood of developing resistance. Because direct effects on parasites and predators are limited or nonexistent, feeding inhibitors may be used with additive effects. Feeding inhibitors can thus fit very well into most current integrated pest management (IPM) programs and may become a valuable complement to other approaches. It may soon be possible to clone a wide variety of genes for enzymes synthesizing feeding inhibitors as well as insect-specific toxins into crop plants, and indeed first steps in this direction have been already taken with *Bacillus thuringiensis* endotoxin (Gasser and Fraley, 1989), cowpea protease inhibitors (Gatehouse and Hilder, 1988), and snowdrop lectin (Shi et al., 1994).

Still, many important biological and practical questions remain unanswered— for example, the molecular basis of antifeedant perception by insects, field application routes and use rates, cost-effective design, synthesis, and marketing considerations of the new products (Jain and Tripathi, 1993). It is the purpose of this chapter to outline different modes of action of feeding inhibitors, give examples of recent applications in field trials, and describe some trends in current research. We will also indicate where this area could benefit greatly from a closer cooperation between academic and industrial research scientists. The combined efforts of natural product chemists, insect physiologists, and applied insect ecologists, regardless of institutional affiliation, could result in rapid development of novel and effective approaches to crop protection. In this period of tightening financial resources, a shared agenda may offer the best solutions for all parties.

13.2. Mechanisms of Feeding Inhibition

Feeding in phytophagous insects is a repetitive process with several distinct behavioral and physiological components (Fig. 13.1; compare Chapter 6). The act of food consumption is preceded by locomotion combined with orientation. Perception of host plants is accomplished through visual and olfactory stimulation which leads to cessation of locomotory activity and initiation of feeding, usually in the form of a few test bites or probes. The presence of feeding stimulants in the plants provides sensory input for the maintenance of feeding, while detection of phagodeterrents may cause shortened bouts or even termination of feeding. The entomological literature abounds in terminology describing classes of chemical compounds involved in feeding decision-making processes, and these are reviewed in Chapter 4. Here we would like to focus on further distinction among compounds inhibiting insect feeding.

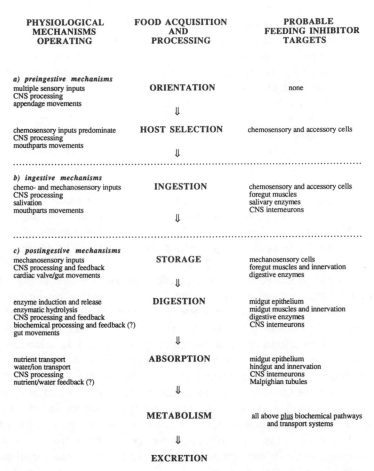

PHYSIOLOGICAL MECHANISMS OPERATING	FOOD ACQUISITION AND PROCESSING	PROBABLE FEEDING INHIBITOR TARGETS
a) preingestive mechanisms multiple sensory inputs CNS processing appendage movements	ORIENTATION ⇓	none
chemosensory inputs predominate CNS processing mouthparts movements	HOST SELECTION ⇓	chemosensory and accessory cells
b) ingestive mechanisms chemo- and mechanosensory inputs CNS processing salivation mouthparts movements	INGESTION ⇓	chemosensory and accessory cells foregut muscles salivary enzymes CNS interneurons
c) postingestive mechansisms mechanosensory inputs CNS processing and feedback cardiac valve/gut movements	STORAGE ⇓	mechanosensory cells foregut muscles and innervation digestive enzymes
enzyme induction and release enzymatic hydrolysis CNS processing and feedback biochemical processing and feedback (?) gut movements	DIGESTION ⇓	midgut epithelium midgut muscles and innervation digestive enzymes CNS interneurons
nutrient transport water/ion transport CNS processing nutrient/water feedback (?)	ABSORPTION ⇓	midgut epithelium hindgut and innervation CNS interneurons Malpighian tubules
	METABOLISM ⇓	all above plus biochemical pathways and transport systems
	EXCRETION	

Figure 13.1. Stages of food acquisition and processing in insects. Food ingestion by insects is preceded by orientation and host selection, and it is followed by storage in the initial portion of the alimentary canal, digestion and absorption in the insect's midgut, entry of nutrients into the metabolic cycles, and excretion of waste and undigested materials. Likely feeding inhibitor targets change with the physiological mechanisms operating at each level. While preingestive inhibition is rapid and acute, postingestive inhibition takes a longer time and produces more chronic effects.

Plant chemicals that modify insect behavior are called *allelochemicals*. They are classified according to whether the signals conveyed by them are advantageous to the source plant (*allomones*), the target insect (*kairomones*), or both (*synomones*). The compounds that prevent, block, or otherwise interfere with food selection and consumption belong to the first group and will be referred to as *feeding inhibitors*.

Preingestive inhibition would be associated with the physiological mechanisms involved in orientation, searching, and host selection. The action of preingestive feeding inhibitors is rapid, occurring within the first few seconds or minutes of contact with a potential food source. Target sites for preingestive feeding inhibitors are localized within the contact chemosensilla (Fig. 13.2), resulting in messages to the central nervous system that signal a lack of acceptability (see Chapter 4). *Antifeedants* or *feeding deterrents* (*phagodeterrents*) constitute a class of preingestive feeding inhibitors that act through gustatory receptors and evoke rejection of the plant material by changing the sensory code from "acceptable" to "unacceptable."

Ingestive inhibition entails block of synthesis, release or action of the salivary enzymes, and innervation of musculature responsible for movement of head appendages or transport of food in esophagus and foregut. This process has an intermediate time course between that of pre- and postingestive inhibition.

Postingestive inhibition involves physiological processes operating after food has already been ingested by the insect. Effects, slower but more persistent than in cases of preingestive and ingestive inhibitors, are achieved through the action of three functional classes of compounds. *Digestive inhibitors* act at gut epithelial cells and block synthesis, release, and action of digestive enzymes. *Feedback inhibitors* interfere with the neural and hormonal processes through which the operation of the alimentary canal is controlled by the central nervous system. Finally, *processing inhibitors* target the activity of the interneurons in the processing circuitry within the central nervous system.

It is extremely important to distinguish pre- from postingestive effects of feeding inhibitors. The most powerful approach to study preingestive effects is offered by electrophysiological recordings which allows monitoring the interactions of putative feeding inhibitors with the gustatory cell(s) in real time (Frazier and Hanson, 1986). Postingestive effects are usually estimated through feeding bioassays; their design should carefully consider the functional levels and numerous targets at which feeding inhibition may take place (Berenbaum, 1986; Lewis and Van Emden, 1986).

13.2.1. Preingestive Inhibitors

Food acceptance or rejection by insects is mediated by sensory structures localized mostly on mouthparts, in the food canal, and on the antennae (Zacharuk, 1980; Frazier, 1985). These structures usually take the form of elongated cuticular appendages housing two or more gustatory neurons. The detailed description of events underlying food selection in insects can be found in Chapter 5. Here we illustrate the potential molecular target sites for feeding inhibitors acting at the level of the contact chemosensillum (Fig. 13.2). Much of this description is based on our understanding of details in chemosensory systems of other animals, with some direct evidence from insects (Frazier, 1992). Host and nonhost plant compounds enter the sensillum through a terminal pore and diffuse in the sensillar

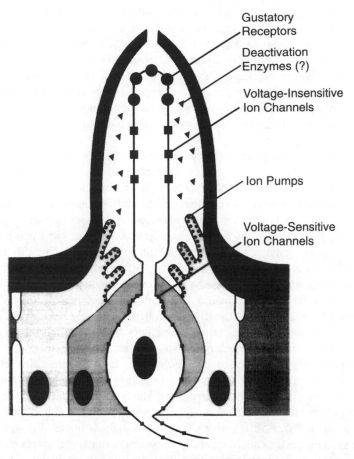

Figure 13.2. Potential molecular targets for feeding inhibitors in insect contact chemo-sensillum. Five classes of likely targets are shown: gustatory receptors and associated voltage-insensitive ion channels in the distal dendritic segment, voltage-sensitive channels localized in the proximal dendritic segment, the cell body and along the axon, ion pumps in the accessory cell's plasma membrane lining the sensillar sinus, and putative deactivation enzymes in the sensillar fluid bathing the dendrite. For simplicity, only one chemosensory neuron is shown here.

fluid before they reach receptor molecules embedded in the plasma membrane of the distal dendritic segments. Upon binding to these proteins, a process of gustatory signal transduction is initiated. Change in the conformation of receptor proteins may either open an ion channel that is part of this molecule (ligand-gated ion channels), or, through a second messenger system which amplifies the signal, open a set of distant, voltage-insensitive ion channels. Either way, this leads to an ionic current that may depolarize or hyperpolarize the gustatory neuron's plasma membrane. These receptor potentials are subsequently summated in time and space, and the resulting potential reaches the spike initiation zone in the proximal dendritic segment. A pattern of impulses coding for the stimulus is formed and propagated to the processing centers in the insect's central nervous system by axonal voltage-sensitive ion channels. Action of taste ligands is most probably terminated by deactivation enzymes present in the sensillar lymph, similar to mechanisms occurring in insect olfactory sensilla (Prestwich and Blomquist, 1987; Stengl et al., 1992). Proper operation of the gustatory neurons depends on changes in membrane potential, resulting from ion gradients, which are established and maintained by ion pumps in accessory cells. There are three important points we would like to make here: (1) Insect taste reception is exclusively a peripheral process; (2) there is a very limited number of cellular elements; and (3) gustatory sensilla present a whole array of molecular targets for feeding inhibitors (Fig. 13.2). Each of the receptor molecules, voltage-sensitive and -insensitive ion channels, ion pumps, and deactivation enzymes described above may be a potential target for novel ways to control insects.

Antifeedants, the most common preingestive feeding inhibitors, act via molecular ligand interactions with gustatory receptors (Mullin et al., 1994). Some of these compounds may cause response inhibition of gustatory neurons tuned to feeding stimulants by competitive or noncompetitive receptor block. Other antifeedants are detected by dedicated gustatory neurons (deterrent cells) which they directly stimulate (Schoonhoven, 1982; also see Chapter 4). Yet another mechanism of deterrency is to evoke bursting or temporally modified response from gustatory neurons, the molecular basis of which is largely unknown (Mitchell, 1987). Most feeding deterrents are effective against a limited number of usually closely related insect species. One notable exception is azadirachtin, a tetranortriterpenoid isolated from the seeds of the Indian neem tree *Azadirachta indica* A. Juss. This compound has been shown to prevent feeding by a very large number of insect species and, not surprisingly, has been adapted for commercial use (Table 13.1).

A number of known neurotoxins act selectively on different types and subtypes of ligand receptors, ion channels, and ion pumps (Hille, 1992; Nomura et al., 1993). To date, only few of these compounds have been used to investigate molecular mechanisms of insect gustatory transduction (Chyb and Frazier, *unpublished data*) or have been used as models for potential products for insect control (Hayashi and Gard, 1994; Nathanson, 1984).

13.2.2. Ingestive Inhibitors

Food ingestion by insects follows host acceptance and is accompanied by salivation and movement of mouthparts and foregut. Both enzyme release and muscular contraction underlying the passage of food through esophagus and foregut are under neural control. For example, antifeeding azadirachtin has been found to reduce the frequency of muscular contractions and cause a decrease in gut motility of nymphs of *Locusta migratoria* (Mordue et al., 1985).

Initial stages of enzymatic processing which occur during ingestion may have a significant effect on food acceptability to insects. For example, salivary enzymes may transform some of the chemosensory inert molecules present in the food source into "secondary" stimulants or deterrents. To our knowledge, no study has addressed this problem. In contrast, we have substantial evidence of changes in chemical composition of plant leaves following insect attack (Baldwin, 1994; Tallamy and Raupp, 1991). Presently, the targets of these induced compounds are mostly unknown, but many of these chemicals might have potential for use in insect control.

13.2.3. Postingestive Inhibitors

Postingestive inhibition relies primarily on blocking the physiological mechanisms involved in food storage, digestion, and absorption. Following ingestion, food is stored in the insect's alimentary canal and subsequently subjected to the hydrolytic action of digestive enzymes, proteases, carbohydrases, lipases, and so on.

Nutrients, such as sugars or proteins, may cause stimulation or inhibition of insect digestive enzymes. Gossypol and related terpenoids, as well as saponins and tannins, are among numerous plant allelochemicals directly affecting the activity of insect midgut proteases (Ishaaya, 1986). One example is a protease inhibitor isolated from cowpea. This was genetically transferred into a tobacco plant and found to effectively block feeding by a number of insect pests, such as *Diabrotica undecimpunctata* (Gatehouse and Hilder, 1988). Several plant allelochemicals—many phenolics, for example—interact with nutrients in the insect's gut lumen and decrease their digestion. Interestingly, stability and activity of these allelochemicals depends on physicochemical conditions in the gut (Appel, 1994).

Several proteins produced by a bacterium, *Bacillus thuringiensis* (Bt toxins), interfere with proper functioning of insect midgut cells, causing disintegration of the midgut epithelium and eventually insect death. Although the details of the mechanisms through which they act are not yet completely understood, the generally accepted view is that ionic, osmotic, and pH gradients between midgut epithelial cells and the gut lumen are destroyed when Bt toxins insert into the apical plasma membranes and form ion-unselective pores (Lambert and Peferoen, 1992).

It may be also possible to control insects via inhibitors of enzymes crucial to their metabolism—for example, trehalase (Jain and Tripathi, 1993). Trehalose is an insect-specific, physiologically inert, disaccharide circulating in the hemolymph. The inhibition of trehalose hydrolysis could result in eliminating the insect's access to its own internal energy stores.

An indirect block of feeding may be accomplished by preventing or delaying metamorphosis. One of the effects of azadirachtin is growth disruption. This may be manifested in insect species whose feeding is not inhibited by this compound as well as in those that are strongly deterred by its presence in the diet (Mordue and Blackwell, 1993). Conversely, speeding up pupation and/or adult emergence would shorten the duration of the larval stage, a period usually associated with intensive food consumption.

Another approach is to interfere with the neural feedback between midgut and the peripheral nervous system (feedback inhibitors). Following a meal which distends the foregut, the insect's endocrine system releases a factor that causes an overall reduction in the sensitivity of the peripheral nervous system and particularly its contact chemosensory component. Bernays et al. (1972) reported changes in the tip region of locust's contact chemosensilla; diameter of the terminal pore becomes smaller following a meal and does not increase until at least 2 hr after the meal. This subject is discussed further in Chapter 4.

Alternatively, feeding inhibition at the postingestive level could be accomplished by targeting the sensory processing centers localized in the central nervous system. Messages from the contact chemoreceptors are processed, integrated with the input from other sensory modalities, and subsequently used by the insect to make feeding decisions. Even though the message from contact chemoreceptors may reach the central nervous system in its original form, it would be possible to deform the way in which it is later used in the decision-making process. The most likely mechanism would involve block of synaptic transmission from gustatory cells or at the level of higher-order neuronal elements.

13.3. Advantages and Limitations of Feeding Inhibitors in Insect Control

Feeding inhibitors offer several advantages over traditional agrochemicals, making them attractive for use within IPM strategies. The pest specificity of antifeedant compounds may be modifiable; that is, the range of species targeted may be defined either by the chemical structure of the antifeedant or by the composition of the mixture of antifeedants if different feeding inhibitors are effective against species within the range. For example, polygodial is a selective aphid antifeedant (Pickett et al., 1987). Because many feeding inhibitors are transient chemicals acting rapidly on the affected insects, their residual activity and environmental impact may be very limited, although much remains to be done with formulations or modes of delivery (Griffiths, 1988, 1990; Norris, 1986). Feeding inhibition may be achieved by combining a feeding stimulant into a bait mixed with pesticides and delivered such that it outcompetes the crop plant for attracting and stimulating the insects for feeding (Fleischer and Kirk, 1994). Natural parasites and predators may remain, for the most part, unaffected by feeding inhibitors directed at their hosts, although tritrophic effects are well-documented on a dose-dependent manner (Beckage et al., 1988). The target sites of feeding inhibitors are different from those of classic insecticides, and therefore products based on them will be effective on pesticide-resistant populations and will also allow multiple tactics to slow resistance development (Holloway and McCaffery, 1988). Although feeding inhibitors need not be toxic to be effective, the probability of resistance development in targeted insects may be lower than with traditional agrochemicals, especially because disruption of feeding behavior is followed by a wide variety of mortality factors (Griffiths, 1990). Genetic engineering may also allow us to incorporate genes for synthesis of deterrent compounds into economically important crops. Only the beginning of this potential has been realized (Gasser and Fraley, 1989; Gatehouse and Hilder, 1988).

The use of feeding inhibitors is, however, not without some limitations. Overcoming these weaknesses may be the key to effective use of this emerging technology. For example, some antifeedants may act on molecular targets common to insects and mammals, including humans (Mullin et al., 1994). These targets are not identical, reflecting evolutionary distance between these animal groups. The design of antifeedants can incorporate these subtle structural differences that produce both pest specificity and human safety. A similar rationale has been used to find the toxins that would selectively act on pest target sites only, and has been the basis of biorational approaches in the past (Hayashi and Gard, 1994). In addition, targets in insect taste sensilla are located peripherally; comparable targets in mammals are mostly in the central nervous system. As a result, fewer competing processes are involved in the insect system. Thus simplified or unique pharmacodynamics may be necessary to reach the target site in insects and permit a simpler design process.

Feeding deterrents are usually nonvolatile and unstable molecules, and hence their action may be transient both environmentally and within the pest (Griffiths et al., 1988). This offers wide latitude in the choice of delivery routes, as well as combinations with other compounds to develop pest selectivity while ensuring minimal environmental impacts. As with all the other compounds acting through taste receptors, they will cause habituation, a decrease in response over time, which, depending upon the species and potency, may or may not be a limiting factor under field conditions (Raffa and Frazier, 1988). Even azadirachtin with its multiple modes of action is not an exception with respect to its chemosensory actions (Simmonds and Blaney, 1984). Topical applications of antifeedants may therefore have very limited effectiveness (Berenbaum, 1986; Feeny, 1992), which can be an advantage in their specificity of action. We currently understand little about the details of antifeedant interactions with plant surface molecules, with enzymes present in insect saliva, or their postingestional effects. This brings a rather large arena of added advantages or risks in their product development, depending upon your viewpoint. Some new approaches to understanding the mode of action indicate that a greater knowledge of the chemical and molecular basis of action can expand our control options (Appel, 1994). Antifeeding compounds may affect some of the beneficial insects (or vertebrates) depending upon the compound or dose, again requiring further development of its actual field use delivery system (Mordue and Blackwell, 1993).

Feeding inhibitors affecting more than one type of target (e.g., different types of ion channels) may be more effective across species but would also bring a greater risk of involving some vertebrate target site actions with them. Still, octopamine site-directed compounds have recently been patented with insect control as an intended payoff (see Table 13.2). Any compounds active at synaptic sites may block the impulse transmission in several points along the way, and, given the indications that many neuropeptides are involved in neural control circuits, many options remain to be developed.

Several feeding inhibitors are both deterrent and toxic to a given insect pest. Combining these two features in one molecule reduces the likelihood of developing resistance because two or more sites of action are involved (Jermy, 1990). Bernays and Chapman (1977) argued that deterrency is not enough to prevent insect feeding reliably under field conditions, and antifeedants would need to be toxic to be practical. While many other factors come into play in the final efficacy under field conditions, many feeding deterrents have been shown to effectively inhibit food consumption despite the lack of toxicity. Insects deprived of their chemosensory apparatus consume these compounds readily without significant negative effects (e.g., de Boer and Hanson, 1987). Certainly, reduced consumption can result in reduced reproductive fitness or increased movement by larval stages, thereby exposing them to a variety of mortality factors. Clearly, more field-level efficacy studies with feeding inhibitors of clearly defined modes of action are required before any limitations to their effectiveness can be well-documented. Formulations, timing, mode of delivery, and combinations are all important variables that are yet untested.

13.4. Patented Feeding Inhibitors

It is within the patent arena that a much closer cooperation among academic and industrial scientist could achieve greater rewards for both. The development of chemical patents is as much an art as a science, and companies differ widely in their philosophies and scope of patenting activity. Universities likewise differ greatly in their approaches. Because patents must be granted in all countries of intended use, few universities are in a position to do this justice, while most agricultural companies would insist on some protection that is relevant to their international marketplace. Industrial patenting activities are far better developed than at most universities, and the slight delays in publication of a few months required to secure patents for feeding inhibitors are well worth the investment on the part of an academic scientist. Having been on both sides of this equation, one of us (JLF) found far more resistance and hesitation in developing agreements that would allow new discoveries to be patented on the university side than on the industrial side. A spirit of cooperation often exhibited by the academic scientist was too easily overcome by a restrictive university lawyer or policy. Some modifications here would pay great dividends for both potential partners.

Chemical patents require a new chemical structure and a use for it that is not obvious to someone skilled in the art. It is here that the experience of the industrial partners can be invaluable in developing a solid position with a thorough background check of the international patenting activity in the area. This is by far the patent of choice to be developed; and once the structure of a new natural product is published, even without a biological activity known, it can no longer be patented. Patenting a new compound for feeding inhibitor use as a "method of use" patent can be so easily overcome by a competitor that it is considered of little value by most companies. The ideal solution is for both partners to work together with a mutual understanding of the option to patent being given to an industrial partner with an agreeable time for securing this as the only needed delay in publication. This can be done prior to detailed investigations to the advantage of both partners.

Table 13.1 contains several commercial products based on neem oil and its well-documented active ingredient azadirachtin. Because this compound is known to have multiple modes of action, and because neem oil contains several related molecules, the consequent mechanisms that come into play under field conditions may be several (Ley et al., 1993; Mordue and Blackwell, 1993). It is, however, certain that at least the feeding inhibitor effect is a part of the commercially relevant action of these products. While a partial listing of patented antifeedants in Table 13.2 gives some indication of the activity to date, the table is not meant to be exhaustive, but rather indicates the various countries and kinds of molecules so far patented for feeding inhibitory action within a recent time period. It is

Table 13.1. Some commercial products for insect control based on insect feeding deterrence

Product	Active material	Target pest	Manufacturer
Margosan-O	Neem extract	Ornamental pests	W. R. Grace, Cambridge, MA
Azatin EC	Azadirachtin	White flies, leafminers	Agridyne Technologies, Salt Lake City, UT
Align	Azadirachtin	Vegetable, fruit, nut pests	Agridyne Technologies, Salt Lake City, UT
BioNeem	Neem extract	Ornamental pests	Ringer Corp., Minneapolis, MN
Neemazal	Neem extract	Many pests	Trifolio M, Lahnau, Germany
Neemguard	Neem extract	Many pests	Gharda Chemicals, Bombay, India

Table 13.2.   Some patented antifeedants for insect control

Compound(s)	Target insect	Patentee and date
Warburganal	African armyworm	Research Corp., UK, 1981
Warburganal intermediate	African armyworm	Rikagaku Kenkyusho (Japanese patent 57-200786), 1982
Erythro-9,10-dihydroxyoctadecen-1-ol acetate	Boll weevil	USDA patent, 1983
Azadirachtin partial structure	Corn earworm	FMC Corp., Philadelphia, PA (US patent)
Neem extract	Leafminers	Vikwood, Ltd. (US patent), 1985
Tetrahydrofuraniial acetogenins	Mosquitoes, aphids, mites, nematodes	Mikolajczak et al. (USDA patent), 1986
Mixed esters of succinic acid	Termites	Kahovova et al. (Czech patent 250,588), 1988
Azido-substituted octopamine agonists	Tobacco hornworm and other invertebrates	J. A. Nathanson (US patent 180,758), 1988
Organic nitriles	Fall armyworm and European corn borer	Powell et al. (US patent 321,829), 1990
Elaeodendron buchananii isolated compound I	African armyworm	Tsjino et al., Jpn. Kokai Tokkyo Koho (Japanese patent 02,233,686), 1990
2',3',20,21,22,23-Hexahydrosalanin and derivatives	Colorado potato beetle	James A. Klocke and Ronald B. Yamasaki (US patent 4,960,791), 1988
Acetyl, valeryl, and lauryl sucrose esters	Diamondback moth	Jouy et al., Fr. Demande (French patent 2,663,815) 1992
trans-2-*N,N*-dimethylaminopinan-3-one oxime and salts	Barley leaf beetles	A. M. Chibiraev et al. (German patent 4,121,335), 1990
trans-2-Dimethylaminopinocamphone oxime-HCl	Colorado beetle larvae	Nedelkina et al. USSR patent (SU 1,768,106), 1992
Chrysanthemic acid amides	Granary pests	Sobotka et al. (Polish patent 154,799), 1992
Chloroethyl phenylamine-aryl-1,4-dialk(en)-piperazines	Tobacco hornworm and invertebrates	James Nathanson, PCT Int. Appl. WO 93 00,811 1993

our view that this very fertile area is just beginning to be developed and offers much promise for novel additions to IPM systems.

13.5. Conclusions and Future Outlook

Several classes of plant-derived compounds possess the ability to inhibit insect feeding either at the preingestional (i.e., contact chemosensory), ingestional (i.e., salivary enzyme), or postingestional (i.e., midgut physiology) level. Hundreds of secondary plant metabolites have been and will be screened against economically important insect pests, although the viable arena need not be confined to plants (Jain and Tripathi, 1993; Lindow et al., 1989; Nair, 1994). Certain structural motifs are likely to be found that are of significance for feeding inhibition, but to date only isolated variations are known, the database is simply too limited to identify or predict such motifs (Lajide et al., 1993; Luthria et al., 1993). Perhaps a linking of insect chemosensory electrophysiology with molecular connectivity indices will provide the ultimate approach to QSAR for determining such molecular motifs (Lam and Frazier, 1990; Luco et al., 1994). Likewise, some other chemical features may entail toxic capabilities or environmental persistence, yet these too are presently unknown within the feeding inhibitor chemistry. A "mix and match" approach may be feasible with feeding inhibitors of variable specificity in order to create customized formulations targeted against specific insect pests, thus taking advantage of plants' own natural defense mechanisms with some knowledge of their biological and environmental consequences. The application of plant compound formulations need not be limited to insects for utility, as exemplified by a neem-based product for the golden snail (Maini and Morallo-Rejesus, 1993), or limited to the plant parts on which insects normally feed, as exemplified by the birch bark derivatives of betulin (Lugemwa et al., 1990).

It is possible that engineering genes encoding elements of pathways for the synthesis of feeding inhibitory compounds may be cloned into economically important crops, thereby rendering them resistant to insect damage (Gatehouse and Hilder, 1988), or this approach may be coupled with other sources of unique compounds or agents (Gasser and Fraley, 1989; Lindow et al., 1989). With this approach, the active chemical is located within the plant part to be protected at the appropriate time. This strategy is not without certain ecological risks or associated plant energy costs that need to be carefully considered (Griffiths, 1990). Transgenic crops may hybridize with wild relatives and thus release transgenes into the nontargeted natural plant communities (Abbott, 1994).

This area of applied entomology may benefit greatly from progress achieved in other areas of biological sciences. For example, a number of different controlled release strategies have been developed by modern pharmacology: delayed release, targeted release, and extended release (Wilkins, 1990). This could prolong the chemical and biological activity of antifeedants and other feeding inhibitors which are transient in their potency, and thus their release has to be timed to be effective (Griffiths, 1990). For example, Greene and Meyers (1990) described a temperature-controlled release system depending upon increases in soil temperature during the growth season: an agrochemical is chemically coupled to a temperature-sensitive polymer and the pesticide is released when soil temperature exceeds the melting point of the polymer. This release coincides in time with the growth of the plant and thus ensures an effective delivery.

Feeding inhibitors may be found in plant materials regarded as industrial waste. For example, Bentley et al. (1988) have identified several antifeedants for Colorado potato beetle (*Leptinotarsa decemlineata*) in by-products of orange juice processing. Certainly, there may be other instances where valuable compounds could be extracted from sources regarded as waste material. Few plant extracts are readily available for crop protection, but they may contain bioactive compounds from which more efficient and environmentally safe insecticides may be developed (Escoubas et al., 1994).

Use of feeding inhibitors in pest management enables us to exploit natural defenses evolved by plants and other organisms in novel ways and thereby reduce the risks associated with synthetic pesticides. Although much of this approach needs to be refined before commercialization of products can be achieved, significant effort is underway and the promise of a new dimension in the tool box of integrated pest management tactics is soon to be realized. A closer cooperation among academic and industrial scientists could significantly enhance this effort.

References

Abbott, R. J. (1994). Ecological risks of transgenic plants. *Trends Ecol. Evol.* 9, 280–283.

Appel, H. M. (1994). The chewing herbivore gut lumen physicochemical conditions and their impact on plant nutrients, allelochemicals, and insect pathogens. In: Bernays, E. A. (ed.), *Insect–Plant Interactions,* vol. 5. CRC Press, Boca Raton, FL, pp. 209–222.

Baldwin, I. T. (1994). Chemical changes rapidly induced by folivory. In: Bernays, E. A. (ed.), *Insect–Plant Interactions,* vol. 5. CRC Press, Boca Raton, FL, pp. 2–23.

Beckage, N. E., Metcalf, J. S., Nielson, B. D., and Nesbit, D. J. (1988). Disruptive effects of azadirachtin on development of *Cotesia congregata* in host tobacco hornworm larvae. *Arch. Insect Biochem. Physiol.* 9, 47–65.

Bentley, M. D., Rajab, M. S., Alford, A. R., Mendel, M. J., and Hassanali, A. (1988). Structure–activity studies of modified citrus limonoids as antifeedants for Colorado potato beetle larvae, *Leptinotarsa decemlineata* (Say). *Entomol. Exp. Appl.* 49, 189–193.

Berenbaum, M. (1986). Postingestive effects of phytochemicals on insects: on Paracelsus and Plant products. In: Miller, J. R., and Miller, T. A. (eds.), *Insect/Plant Interactions.* Springer-Verlag, New York, pp. 121–154.

Bernays, E. A., and Chapman, R. F. (1977). Deterrent chemicals as a basis of oligophagy in *Locusta migratoria* (L.). *Ecol. Entomol.* 2, 1–18.

Bernays, E. A., Blaney, W. M., and Chapman, R. F. (1972). Changes in chemoreceptor sensilla on the maxillary palps of *Locusta migratoria* in relation to feeding. *J. Exp. Biol.* 57, 745–753.

de Boer, G., and Hanson, F. E. (1987). Differentiation of roles of chemosensory organs in food discrimination among host and non-host plants by the larvae of the tobacco hornworm, *Manduca sexta. Physiol. Entomol.* 12, 387–398.

Ehrlich, P. R., and Raven, P. H. (1964). Butterflies and plants: a study in co-evolution. *Evolution* 18, 586–608.

Escoubas, P., Lajide, L., and Mizutani, J. (1994). Insecticidal and antifeedant activities of plant compounds. Potential leads to novel pesticides. In: Hedin, P. A., Menn, J. J., and Hollingsworth, R. M. (eds.), *Natural and Engineered Pest Management Agents. Am. Chem. Soc. Symp.* 551, 162–171.

Feeny, P. (1992). The evolution of chemical ecology: contributions from the study of herbivorous insects. In: Rosenthal, G. A., and Berenbaum, M. R. (eds.), *Herbivores: Their Interactions with Secondary Plant Metabolites,* vol. 2. Academic Press, New York, pp. 1–44.

Fleischer, S. J., and Kirk, D. (1994). Kairomonal baits: effect on acquisition of a feeding indicator by diabroticite vectors in cucurbits. *Environ. Entomol.* 23, 1138–1149.

Frazier, J. L. (1985). Nervous system: sensory system. In: Blum, M. S. (ed.), *Fundamentals of Insect Physiology.* John Wiley & Sons, New York, pp. 287–356.

Frazier, J. L. (1992). How animals perceive secondary plant compounds. In: Rosenthal, G. A., and Berenbaum, M. R. (eds.), *Herbivores: Their Interactions with Secondary Plant Metabolites,* vol. 2. Academic Press, New York, pp. 89–134.

Frazier, J. L., and Hanson, F. E. (1986). Electrophysiological recording and analysis of insect chemosensory responses. IN: Miller, J. R., and Miller, T. A. (eds.), *Insect/ Plant Interactions.* Springer-Verlag, New York, pp. 285–330.

Futuyma, D. J., and Keese, M. C. (1992). Evolution and coevolution of plants and phytophagous arthropods. In: Rosenthal, G. A., and Berenbaum, M. R. (eds.), *Herbivores: Their Interactions with Secondary Plant Metabolites,* vol. 2. Academic Press, New York, pp. 440–477.

Gasser, C. S., and Fraley, R. T. (1989). Genetically engineered plants for crop improvement. *Science* 244, 1293–1307.

Gatehouse, A. M. R., and Hilder, V. A. (1988). Introduction of genes conferring insect resistance. In: *Proceedings of the Brighton Crop Protection Conference,* Farnham, England, pp. 1245–1254.

Greene, L., and Meyers, P. (1990). Temperature controlled pesticide release systems. In: *Proceedings of the Brighton Crop Protection Conference,* Farnham, England, pp. 593–598.

Griffiths, D. C. (1990). Opportunities for control of insects in arable crops using semiochemicals and other unconventional methods. In: *Proceedings of the Brighton Crop Protection Conference,* Farnham, England, pp. 487–496.

Griffiths, D. C., Hassanali, A., Merritt, L. A., Mudd, A., Pickett, J. A., Shah, S. J., Smart, L. E., Wadhams, L. J., and Woodcock, C. M. (1988). Highly effective antifeedants against coleopteran pests. In: *Proceedings of the Brighton Crop Protection Conference,* Farnham, England, pp. 1041–1046.

Griffiths, W. (1988). Crop protection realism: The role of agrochemicals. In: *Proceedings of the Brighton Crop Protection Conference,* Farnham, England, pp. 111–120.

Harborne, J. B. (1993). *Introduction to Ecological Biochemistry.* Academic Press, London.

Hayashi, J. H., and Gard, I. (1994). Insect-selective toxins as biological control agents. In: Borkovec, A. B., and Loeb, M. J. (eds.), *Insect Neurochemistry and Neurophysiology*. CRC Press, Boca Raton, FL, pp. 73–87.

Hille, B. (1992). *Ionic Channels of Excitable Membranes*. Sinauer, Sunderland.

Holloway, G. J., and McCaffery, A. R. (1988). Reactive and preventative strategies for the management of insecticide resistance. In: *Proceedings of the Brighton Crop Protection Conference*, Farnham, England, pp. 465–469.

Ishaaya, I. (1986). Nutritional and allelochemic insect–plant interactions relating to digestion and food intake: some examples. In: Miller, J. R., and Miller, T. A. (eds.), *Insect/ Plant Interactions*. Springer-Verlag, New York, pp. 191–223.

Jain, D. C., and Tripathi, A. K. (1993). Potential of natural products as insect antifeedants. *Phytother. Res.* 7, 327–334.

Jermy, T. (1990). Prospects of antifeedant approach to pest control—a critical review. *J. Chem. Ecol.* 16, 3151–3166.

Lajide, L., Escoubas, P., and Mizutani, J. (1993). Antifeedant activity of metabolites of *Aristolochia albida* against the tobacco cutworm, *Spodoptera litura*. *J. Agric. Food Chem.* 45, 669–673.

Lam, P. Y-S., and Frazier, J. L. (1990). Rational approach to glucose taste chemoreceptor inhibitors as novel insect antifeedants. In: Baker, D. R., Fenyes, J. G., and Moberg, W. K. (eds.), *Synthesis and Chemistry of Agrochemicals II: Am. Chem. Soc. Symp.* 443, 400–412.

Lambert, B., and Peferoen, M. (1992). Insecticidal promise of *Bacillus thuringiensis*. *Bioscience* 42, 112–122.

Ley, S. V., Denholm, A. A., and Wood, A. (1993). The chemistry of azadirachtin. *Nat. Prod. Rep.* 109–157.

Lewis, A. C., and Van Emden, H. F. (1986). Assays for insect feeding. In: Miller, J. R., and Miller, T. A. (eds.), *Insect/Plant Interactions*. Springer-Verlag, New York, pp. 95–119.

Lindow, S. E., Panopoulos, N. J., and McFarland, B. L. (1989). Genetic engineering of bacteria from managed and natural habitats. *Science* 244, 1300–1307.

Luco, J. M., Sosa, M. E., Cesco, J. C., Tonn, C. E., and Giordano, O. S. (1994). Molecular connectivity and hydrophobicity in the study of antifeedant activity of clerodane diterpenoids. *Pesticide Sci.* 41, 1–6.

Lugemwa, F. N., Huang, F.-Y., Bentley, M. D., Mendel, M. J., and Alford, A. R. (1990). A *Heliothis zea* antifeedant from the abundant birchbark triterpene betulin. *J. Agric. Food Chem.* 38, 493–496.

Lutria, D. L., Ramakrishan, V., and Banerji, A. (1993). Insect antifeedant activity of furochromones: structure–activity relationships. *J. Nat. Prod.* 56, 671–675.

Maini, P. N., and Morallo-Rejesus, B. (1993). Antifeedant activity of the crude and formulated product from *Azadirachta indica* to golden snail (*Pomacea* spp.). *Philippine J. Sci.* 122, 117–128.

Mitchell, B. K. (1987). Interactions of alkaloids with galeal chemosensory cells of Colorado potato beetle. *J. Chem. Ecol.* 13, 2009–2022.

Mordue (Luntz), A. J., and Blackwell, A. (1993). Azadirachtin: an update. *J. Insect Physiol.* 39, 903–924.

Mordue (Luntz), A. J., Cottee, P. K., and Evans, K. A. (1985). Azadirachtin: its effect on gut motility, growth and moulting in *Locusta*. *Physiol. Entomol.* 10, 431–437.

Mullin, C. A., Chyb, S., Eichenseer, H., Hollister, B., and Frazier, J. L. (1994). Neuroreceptor mechanisms in insect gustation: a pharmacological approach. *J. Insect Physiol.* 40, 913–931.

Nair, M. G. (1994). Natural products as sources of potential agrochemicals. In: Hedin, J. A., Menn, J. J., and Hollingsworth, R. M. (eds.), *Natural and Engineered Pest Management Agents. Am. Chem. Soc. Symp.* 551, 145–161.

Nathanson, J. A. (1984). Caffeine and related methylxanthines: possible naturally occurring pesticides. *Science* 226, 184–187.

Nomura, Y., Kitamura, Y., Tohda, M., and Tokumitsu, Y. (1993). Drugs acting on the intracellular signalling system. *Medicinal Res. Rev.* 13, 1–60.

Norris, D. M. (1986). Anti-feeding compounds. In: Bowers, W. S., Ebing, W., Fukuto, T. R., Martin, R. W., and Yamamoto, I. (eds.), *Chemistry of Plant Protection: Sterol Biosynthesis, Inhibitors and Anti-feeding Compounds.* Springer-Verlag, Berlin, pp. 97–146.

Pickett, J. A., Dawson, G. W., Griffiths, D. C., Hassanali, A., Merritt, L. A., Mudd, A., Smith, M. C., Wadhams, L. J., Woodcock, C. M. and Zhang, Z. (1987). Development of plant antifeedant for crop protection. In: Greenhalgh, R., and Roberts, T. R. (eds.), *Pesticide Science and Biotechnology.* Blackwell, Oxford, pp. 125–128.

Prestwich, G. D., and Blomquist, G. J. (eds.) (1987). *Pheromone Biochemistry.* Academic Press, Orlando, FL.

Raffa, K. F., and Frazier, J. L. (1988). A generalised model for quantifying behavioral desensitization to antifeedants. *Entomol. Exp. Appl.* 46, 93–100.

Rosenthal, G. A., and Janzen, D. H. (eds.) (1979). *Herbivores: Their Interaction with Secondary Plant Metabolites,* first edition. Academic Press, New York.

Rosenthal, G. A., and Berenbaum, M. R. (eds.) (1992). *Herbivores: Their Interactions with Secondary Plant Metabolites,* second edition, vol. 2. Academic Press, New York.

Schoonhoven, L. M. (1982). Biological aspects of antifeedants. *Entomol. Exp. Appl.* 31, 57–69.

Shi, Y., Wang, M. B., Powell, K. S., Van Damme, E., Hilder, V. A., Gatehouse, A. M. R., Boulder, D., and Gatehouse, J. A. (1994). Use of the rice sucrose synthase-1 promoter to direct phloem-specific expression of β-glucuronidase and snowdrop lectin genes in transgenic tobacco plants. *J. Exp. Botany* 45, 623–631.

Simmonds, M. S. J., and Blaney, W. M. (1984). Some effects of azadirachtin on lepidopterous larvae. In: Schmutterer, H., and Ascher, K. R. S. (eds.), *Proceedings of the 2nd International Neem Conference.* GTZ Eschbom, Germany, pp. 163–180.

Stengl, M., Hatt, H., and Breer, H. (1992). Peripheral processes in insect olfaction. *Annu. Rev. Physiol.* 54, 665–681.

Tallamy, D. W., and Raupp, M. J. (eds.) (1991). *Phytochemical Induction by Herbivores*. John Wiley and Sons, New York.

Thorsteinson, A. J. (1960). Host selection in phytophagous insects. *Annu. Rev. Entomol.* 5, 193–218.

Wilkins, R. M. (1990). Controlled release technology: safety and environmental benefits. *Proceedings of the Brighton Crop Protection Conference,* Farnham, England, pp. 1043–1052.

Zacharuk, R. Y. (1980). Ultrastructure and function of insect chemosensilla. *Annu. Rev. Entomol.* 25, 27–47.

Taxonomic Index*

---

*References to illustrations and tables are bold unless there are two or more successive pages.

Subject Index*

abdominal distension, 162, 167, 178–179, 215, 218, 225, 234, 332
abdominal nerve, 216, 332
ablation, 112, 373
absorption, 64, 272, 310, 370
acceptability hierarchy, 290
acceptor (receptor) sites, 285, 368–369
accessory
  cell, of chemoreceptor, 369
  gland, of male, 320
acetic acid, 170
acetone, 170
acetylcholine, 82
across-fiber patterning, 124–125
adaptation, sensory, 147, 164, 217, 229, 279–280
adenosine diphosphate (ADP), 80–81, 83, 174–175, 177
adenosine nucleotide, 113–114, 126, 174–176
adenosine triphosphate (ATP), 83, 114, 118, 175–177
adenosine triphosphate diphosphohydrolase, see apyrase
adipocyte satiety factor, 142
adrenalin, 142
L-alanine, 113, 116, 118, 170
alkaloid, 118–119, 123
allatectomy, 320–322, 328–329, 332
allelochemical, 190–191, 205, 364, 366, 370
allomone, 366
allyl isothiocyanate, 114
allyl sulfide, 114

amino acid
  in food, 257, 259, 261, 269–271
  in hemolymph, 260–261, 280, see also hemolymph
  in nectar, 227
  in phloem, 64, 204–205
  in saliva, 79
  in stigmatic exudate, 213
  intake rate, 64
  sensitivity to, 112–114, 116, 118–119, 123, 126, 227
γ-aminobutyric acid, 113
aminopeptidase, 78–79, 272
ammonia, 173
amphetamine, 236
amyl acetate, 352
amylase, 78–79
anautogeny, 162, 318, 322, 324–326, 329
anemotaxis, 162, 171
antenna, 109, 111–112, 282, 287, 296
  sensilla on, 102, **104,** 159, 170–172, 228–229, 232, 282, 367
antennal lobe, 109–110, 121, 282, 296
anticoagulant, 84, 86
antifeedant, see deterrent
antimicrobial, 88
antiplatelet, 83–85
antithrombin, 84
apolysis, 314–315
aposematic color, 298
apyrase, 83–87, 90, 174
arachidonic acid, 80
area-concentrated search, 146
